AF358476

Oxidative Stress and Antioxidant Strategies: Relationships and Cellular Pathways for Human Health

Oxidative Stress and Antioxidant Strategies: Relationships and Cellular Pathways for Human Health

Guest Editors

Alessia Remigante
Rossana Morabito

Basel • Beijing • Wuhan • Barcelona • Belgrade • Novi Sad • Cluj • Manchester

Guest Editors

Alessia Remigante
Department of Biomedical,
Dental and Morphological
and Functional Imaging
University of Messina
Messina
Italy

Rossana Morabito
Department of Chemical,
Biological, Pharmaceutical
and Environmental Sciences
University of Messina
Messina
Italy

Editorial Office
MDPI AG
Grosspeteranlage 5
4052 Basel, Switzerland

This is a reprint of the Special Issue, published open access by the journal *Cells* (ISSN 2073-4409), freely accessible at: www.mdpi.com/journal/cells/special_issues/W0GWG0MK88.

For citation purposes, cite each article independently as indicated on the article page online and using the guide below:

Lastname, A.A.; Lastname, B.B. Article Title. *Journal Name* **Year**, *Volume Number*, Page Range.

ISBN 978-3-7258-3596-6 (Hbk)
ISBN 978-3-7258-3595-9 (PDF)
https://doi.org/10.3390/books978-3-7258-3595-9

Contents

About the Editors

Alessia Remigante

Alessia Remigante is a Physiology Professor at the University of Messina, Italy. My research topic is focused on how oxidant molecules, transported by the blood stream, exert their action on the plasma membrane with possible effects on transport systems and, in turn, on erythrocyte homeostasis. For this reason, increasing attention has been given to ion transport system function and underlying pathways, with the aim of considering them as parameters to verify at which level oxidants act and, possibly, to define novel targets for new drug development.

Rossana Morabito

Rossana Morabito is a Physiology Professor at the University of Messina, Italy. My research topic is focused on how the oxidant molecules, transported by the blood stream, exert their action on the plasma membrane with possible effects on transport systems and, in turn, on erythrocyte homeostasis. For this reason, increasing attention has been given to ion transport system function and underlying pathways, with the aim of considering them as parameters to verify at which level oxidants act and, possibly, to define novel targets for new drug development.

cells

MDPI

Editorial

Oxidative Stress and Antioxidant Strategies: Relationships and Cellular Pathways for Human Health

Alessia Remigante *[ID] **and Rossana Morabito** [ID]

Department of Chemical, Biological, Pharmaceutical and Environmental Sciences, University of Messina, 98166 Messina, Italy; rmorabito@unime.it
* Correspondence: aremigante@unime.it

Citation: Remigante, A.; Morabito, R. Oxidative Stress and Antioxidant Strategies: Relationships and Cellular Pathways for Human Health. *Cells* **2024**, *13*, 1871. https://doi.org/10.3390/cells13221871

Received: 25 October 2024
Accepted: 6 November 2024
Published: 11 November 2024

Chronic diseases and aging have increased significantly in recent decades. These pathological states are produced by different causes, and a common factor in most of them is oxidative stress. Oxidative stress is defined as an imbalance between the oxidative state, mainly due to the formation of reactive species (RS), and antioxidant defense mechanisms. However, when excess oxidants are produced, or when the antioxidant defenses that regulate them are ineffective, this balance can be disturbed, resulting in oxidative conditions. Oxidative products are highly reactive and can directly or indirectly modulate the functions of many enzymes and transcription factors through a complex signaling cascade. This phenomenon increases with age and affects the normal functioning of numerous cells and tissues. Due to the broad and profound biological effects of RS, numerous experimental and clinical studies have focused their attention on the involvement of oxidative stress as a key regulator in the chronic disease state and aging.

This Special Issue will investigate the molecular mechanisms underlying oxidative stress, as well as pathophysiological consequences in cell and tissue function, in order to open new avenues in therapy and drug design (natural or synthetic). Here, we offer an overview of the content of this Special Issue, which contains two reviews and seven original articles:

1. Chlorine (Cl_2) exposure poses a significant risk to ocular health, with the cornea being particularly susceptible to its corrosive effects. Antioxidants, known for their ability to neutralize reactive oxygen species (ROS) and alleviate oxidative stress, were explored as potential therapeutic agents to counteract chlorine-induced damage. In vitro experiments using human corneal epithelial cells showed decreased cell viability due to chlorine-induced ROS production, which was reversed by antioxidant incubation. The mitochondrial membrane potential decreased due to both low and high doses of Cl_2 exposure; however, it was recovered through antioxidants. The wound scratch assay showed that antioxidants mitigated impaired wound healing after Cl_2 exposure. In vivo and ex vivo, after Cl_2 exposure, increased corneal fluorescein staining indicates damaged corneal epithelial and stromal layers of mice corneas. Likewise, Cl_2 exposure in human ex vivo corneas led to corneal injury characterized by epithelial fluorescein staining and epithelial erosion. However, antioxidants protected against Cl_2-induced damage. These results highlight the effects of Cl_2 on corneal cells using in vitro, ex vivo, and in vivo models while also underscoring the potential of antioxidants, such as vitamin A, vitamin C, resveratrol, and melatonin, as protective agents against acute chlorine toxicity-induced corneal injury. Further investigation is needed to confirm the antioxidants' capacity to alleviate oxidative stress and enhance the corneal healing process [1].

2. Moderate levels of reactive oxygen species (ROS), such as hydrogen peroxide (H_2O_2), fuel tumor metastasis and invasion in a variety of cancer types. Conversely, excessive ROS levels can impair tumor growth and metastasis by triggering cancer cell death. In order to cope with the oxidative stress imposed by the tumor microenvironment, malignant cells exploit a sophisticated network of antioxidant defense mechanisms. Targeting the

antioxidant capacity of cancer cells and enhancing their sensitivity to ROS-dependent cell death represent promising strategies for alternative anticancer treatments. Transient Receptor Potential Ankyrin 1 (TRPA1) is a redox-sensitive non-selective cation channel that mediates extracellular Ca^{2+} entry upon an increase in intracellular ROS levels. The ensuing increase in intracellular Ca^{2+} concentration can in turn engage a non-canonical antioxidant defense program or induce mitochondrial Ca^{2+} dysfunction and apoptotic cell death, depending on the cancer type. Herein, the authors describe the opposing effects of ROS-dependent TRPA1 activation on cancer cell fate and propose the pharmacological manipulation of TRPA1 as an alternative therapeutic strategy to enhance cancer cell sensitivity to oxidative stress [2].

3. Nitric oxide (NO) represents a crucial mediator for regulating cerebral blood flow (CBF) in human brain both under basal conditions and in response to somatosensory stimulation. An increase in intracellular Ca^{2+} concentrations ($[Ca^{2+}]i$) stimulates the endothelial NO synthase to produce NO in human cerebrovascular endothelial cells. Therefore, targeting the endothelial ion channel machinery could represent a promising strategy to rescue endothelial NO signaling in traumatic brain injury and neurodegenerative disorders. Allyl isothiocyanate (AITC), a major active constituent of cruciferous vegetables, was found to increase CBF in non-human preclinical models, but it is still unknown whether it stimulates NO release in human brain capillary endothelial cells. In the present investigation, the authors showed that AITC evoked a Ca^{2+}-dependent NO release in the human cerebrovascular endothelial cell line, hCMEC/D3. The Ca^{2+} response to AITC was shaped by both intra- and extracellular Ca^{2+} sources, although it was insensitive to the pharmacological blockade of transient receptor potential ankyrin 1, which is believed to be among the main molecular targets of AITC. In accord, AITC failed to induce transmembrane currents or elicit membrane hyperpolarization, although NS309, a selective opener of small- and intermediate-conductance Ca^{2+}-activated K+ channels, induced significant membrane hyperpolarization. The AITC-evoked Ca^{2+} signal was triggered by the production of cytosolic, but not mitochondrial, reactive oxygen species (ROS) and supported by store-operated Ca^{2+} entry (SOCE). Conversely, the Ca^{2+} response to AITC did not require Ca^{2+} mobilization from the endoplasmic reticulum, lysosomes, or mitochondria. However, pharmacological manipulation revealed that AITC-dependent ROS generation inhibited plasma membrane Ca^{2+}-ATPase (PMCA) activity, thereby attenuating Ca^{2+} removal across the plasma membrane and resulting in a sustained increase in $[Ca^{2+}]i$. Moreover, the AITC-evoked NO release was driven by ROS generation and required ROS-dependent inhibition of PMCA activity. These data suggest that AITC could be exploited to restore NO signaling and restore CBF in brain disorders that feature neurovascular dysfunction [3].

4. Cardiac lipo-toxicity is an important contributor to cardiovascular complications during obesity. Given the fundamental role of the endoplasmic reticulum (ER)-resident Selenoprotein T (SELENOT) for cardiomyocyte differentiation and protection and for the regulation of glucose metabolism, the authors took advantage of a small peptide (PSELT) derived from the SELENOT redox-active motif to uncover the mechanisms through which PSELT could protect cardiomyocytes against lipo-toxicity. To this end, the authors modeled cardiac lipo-toxicity by exposing H9c2 cardiomyocytes to palmitate (PA). The results showed that PSELT counteracted PA-induced cell death, lactate dehydrogenase release, and the accumulation of intracellular lipid droplets, while an inert form of the peptide (I-PSELT) lacking selenocysteine was not active against PA-induced cardiomyocyte death. Mechanistically, PSELT counteracted PA-induced cytosolic and mitochondrial oxidative stress and rescued SELENOT expression that had been downregulated by PA through FAT/CD36 (cluster of differentiation 36/fatty acid translocase), the main transporter of fatty acids in the heart. Immunofluorescence analysis indicated that PSELT also relieved the PA-dependent increase in CD36 expression, while in SELENOT-deficient cardiomyocytes, PA exacerbated cell death, which was not mitigated by exogenous PSELT. On the other hand, PSELT improved mitochondrial respiration during PA treatment and regulated mitochondrial biogenesis and dynamics, preventing the PA-provoked decrease in PGC1-α and

increases in DRP-1 and OPA-1. These findings were corroborated by transmission electron microscopy (TEM), revealing that PSELT improved the cardiomyocyte and mitochondrial ultra-structures and restored the ER network. Spectroscopic characterization indicated that PSELT significantly attenuated infrared spectral-related macromolecular changes (i.e., content of lipids, proteins, nucleic acids, and carbohydrates) and also prevented the decrease in membrane fluidity induced by PA. The findings further delineate the biological significance of SELENOT in cardiomyocytes and indicate the potential of its mimetic PSELT as a protective agent for counteracting cardiac lipo-toxicity [4].

5. Mercury is a toxic heavy metal widely dispersed in the natural environment. Mercury exposure induces an increase in oxidative stress in red blood cells (RBCs) through the production of reactive species and alteration of the endogenous antioxidant defense system. Recently, among various natural antioxidants, the polyphenols from extra-virgin olive oil (EVOO), an important element of the Mediterranean diet, have generated growing interest. Here, the authors examined the potential protective effects of hydroxytyrosol (HT) and/or homovanillyl alcohol (HVA) on an oxidative stress model represented by human RBCs treated with $HgCl_2$ (10 μM, 4 h of incubation). Morphological changes, as well as markers of oxidative stress, including thiobarbituric acid reactive substance (TBARS) levels, the oxidation of protein sulfhydryl (-SH) groups, methemoglobin formation (% MetHb), apoptotic cells, a reduced glutathione/oxidized glutathione ratio, Band 3 protein (B3p) content, and anion exchange capability through B3p, were analyzed in RBCs treated with $HgCl_2$ with or without 10 μM of HT and/or HVA pre-treatment for 15 min. These data show that 10 μM HT and/or HVA pre-incubation impaired both acanthocyte formation, due to 10 μM $HgCl_2$, and mercury-induced oxidative stress injury, as well as restored the endogenous antioxidant system. Interestingly, $HgCl_2$ treatment was associated with a decrease in the rate constant for SO_4^{2-} uptake through B3p, as well as MetHb formation. Both alterations were attenuated by pre-treatment with HT and/or HVA. These findings provide mechanistic insights into benefits derived from the use of naturally occurring polyphenols against oxidative stress induced by $HgCl_2$ on RBCs. Thus, dietary supplementation with polyphenols might be useful in populations exposed to $HgCl_2$ poisoning [5].

6. Aging is a process characterized by a general decline in physiological functions. The high bioavailability of reactive oxygen species (ROS) plays an important role in the aging rate. Due to the close relationship between aging and oxidative stress (OS), functional foods rich in flavonoids are excellent candidates to counteract age-related changes. This study aimed to verify the protective role of Açaì extract in a d-Galactose (d-Gal)-induced model of aging in human erythrocytes. Markers of OS, including ROS production, thiobarbituric acid reactive substance (TBARS) levels, the oxidation of protein sulfhydryl groups, and the anion exchange capability through Band 3 protein (B3p) and glycated haemoglobin (A1c), were analyzed in erythrocytes treated with d-Gal for 24 h, with or without pre-incubation for 1 h with 0.5–10 μg/mL Açaì extract. The results show that the extract avoided the formation of acanthocytes and leptocytes observed after exposure to 50–100 mM d-Gal, respectively, prevented d-Gal-induced OS damage, and restored alterations in the distribution of B3p and CD47 proteins. Interestingly, d-Gal exposure was associated with an acceleration of the rate constant of SO_4^{2-} uptake through B3p, as well as A1c formation. Both alterations have been attenuated by pre-treatment with the Açaì extract. These findings contribute to clarifying the aging mechanisms in human erythrocytes and propose functional foods rich in flavonoids as natural antioxidants for the treatment and prevention of OS-related disease conditions [6].

7. The second-most common cause of dementia is vascular dementia (VaD). The majority of VaD patients experience cognitive impairment, which is brought on by oxidative stress and changes in autophagic function, which ultimately result in neuronal impairment and death. In this study, the authors examined a novel method for reversing VaD-induced changes brought on by açai berry supplementation in a VaD mouse model. The purpose of this study was to examine the impact of açai berries on the molecular mechanisms underlying VaD in a mouse model of the disease that was created by repeated

ischemia–reperfusion (IR) of the whole bilateral carotid artery. Here, the authors found that açai berry was able to reduce VaD-induced behavioral alteration, as well as hippocampal death, in CA1 and CA3 regions. These effects are probably due to the modulation of nuclear factor erythroid 2-related factor 2 (Nrf-2) and Beclin-1, suggesting possible crosstalk between these molecular pathways. In conclusion, the protective effects of açai berry could be a good supplementation in the future for the management of vascular dementia [7].

8. The Nrf2 gene encodes a transcription factor best known for regulating the expression of antioxidant and detoxification genes. A long list of small molecules has been reported to induce Nrf2 protein via Keap1 oxidation or alkylation. Many of these Nrf2 inducers exhibit off-target or toxic effects due to their nature as electrophiles. In searching for non-toxic Nrf2 inducers, the authors found that a culture medium change to fresh DMEM is capable of inducing Nrf2 protein in HeLa, HEK293, AC16, and MCF7 cells. Testing the components of DMEM led to the discovery of L-Cystine as an effective Nrf2 inducer. L-Cystine induces a dose-dependent increase in Nrf2 protein, ranging from 0.1 to 1.6 mM. RNA-seq analyses and RT-PCR revealed the induction of multiple Nrf2 downstream genes, including NQO1, HMOX1, GCLC, GCLM, SRXN1, TXNRD1, AKR1C, and OSGIN1, by 0.8 mM of L-Cystine. The induction of Nrf2 protein was dependent on L-Cystine entering cells via the cystine/glutamate antiporter and the presence of Keap1. The half-life of Nrf2 protein increased from 19.4 min to 30.9 min with 0.8 mM L-Cystine treatment. L-Cystine was capable of eliciting cyto-protection by reducing ROS generation and protecting against oxidant- or doxorubicin-induced apoptosis. As an amino acid derivative, L-Cystine is considered a non-toxic Nrf2 inducer that exhibits the potential for protection against oxidative stress and tissue injury [8].

9. Either extracts, cell-free suspensions, or bacterial suspensions are used to study bacterial lipid peroxidation processes. Along with gas chromatography–mass spectrometry, liquid chromatography–mass spectrometry, and several other strategies, the thiobarbituric acid test is used for the determination of malondialdehyde (MDA) as the basis for the commercial test kits and the colorimetric detection of lipid peroxidation. The aim of the study was to evaluate lipid peroxidation processes levels in the suspensions, extracts, and culture supernatants of Escherichia coli and Salmonella Derby strains. The dependence of the formation of thiobarbituric acid-reactive substance levels in the cell extracts, suspensions, and cell-free supernatants on bacterial species, as well as their concentrations and growth phases, were revealed. The effect of bacterial concentrations on MDA formation was also found to be more pronounced in bacterial suspensions than in extracts, probably due to the dynamics of MDA release into the intercellular space. This study highlights the possible importance of MDA determination in both cell-free suspensions and extracts, as well as in bacterial suspensions, to elucidate the role of lipid peroxidation processes in bacterial physiology, bacteria–host interactions, and host physiology [9].

Author Contributions: A.R. and R.M. writing—original draft preparation. All authors have read and agreed to the published version of the manuscript.

Funding: This research received no external funding.

Conflicts of Interest: The authors declare that this study was conducted in the absence of any commercial or financial relationships that could be construed as potential conflicts of interest.

References

1. An, S.; Anwar, K.; Ashraf, M.; Han, K.Y.; Djalilian, A.R. Chlorine-Induced Toxicity on Murine Cornea: Exploring the Potential Therapeutic Role of Antioxidants. *Cells* **2024**, *13*, 458. [CrossRef] [PubMed]
2. Moccia, F.; Montagna, D. Transient Receptor Potential Ankyrin 1 (TRPA1) Channel as a Sensor of Oxidative Stress in Cancer Cells. *Cells* **2023**, *12*, 1261. [CrossRef] [PubMed]
3. Berra-Romani, R.; Brunetti, V.; Pellavio, G.; Soda, T.; Laforenza, U.; Scarpellino, G.; Moccia, F. Allyl Isothiocianate Induces Ca^{2+} Signals and Nitric Oxide Release by Inducing Reactive Oxygen Species Production in the Human Cerebrovascular Endothelial Cell Line hCMEC/D3. *Cells* **2023**, *12*, 1732. [CrossRef] [PubMed]

4. Rocca, C.; De Bartolo, A.; Guzzi, R.; Crocco, M.C.; Rago, V.; Romeo, N.; Perrotta, I.; De Francesco, E.M.; Muoio, M.G.; Granieri, M.C.; et al. Palmitate-Induced Cardiac Lipotoxicity Is Relieved by the Redox-Active Motif of SELENOT through Improving Mitochondrial Function and Regulating Metabolic State. *Cells* **2023**, *12*, 1042. [CrossRef] [PubMed]
5. Perrone, P.; Spinelli, S.; Mantegna, G.; Notariale, R.; Straface, E.; Caruso, D.; Falliti, G.; Marino, A.; Manna, C.; Remigante, A.; et al. Mercury Chloride Affects Band 3 Protein-Mediated Anionic Transport in Red Blood Cells: Role of Oxidative Stress and Protective Effect of Olive Oil Polyphenols. *Cells* **2023**, *12*, 424. [CrossRef] [PubMed]
6. Remigante, A.; Spinelli, S.; Straface, E.; Gambardella, L.; Caruso, D.; Falliti, G.; Dossena, S.; Marino, A.; Morabito, R. Acai (*Euterpe oleracea*) Extract Protects Human Erythrocytes from Age-Related Oxidative Stress. *Cells* **2022**, *11*, 2391. [CrossRef] [PubMed]
7. Impellizzeri, D.; D'Amico, R.; Fusco, R.; Genovese, T.; Peritore, A.F.; Gugliandolo, E.; Crupi, R.; Interdonato, L.; Di Paola, D.; Di Paola, R.; et al. Acai Berry Mitigates Vascular Dementia-Induced Neuropathological Alterations Modulating Nrf-2/Beclin1 Pathways. *Cells* **2022**, *11*, 2616. [CrossRef] [PubMed]
8. Dai, W.; Chen, Q.M. Fresh Medium or L-Cystine as an Effective Nrf2 Inducer for Cytoprotection in Cell Culture. *Cells* **2023**, *12*, 291. [CrossRef] [PubMed]
9. Tsaturyan, V.; Poghosyan, A.; Toczylowski, M.; Pepoyan, A. Evaluation of Malondialdehyde Levels, Oxidative Stress and Host-Bacteria Interactions: Escherichia coli and Salmonella Derby. *Cells* **2022**, *11*, 2989. [CrossRef] [PubMed]

Article

Unveiling Smyd-2's Role in Cytoplasmic Nrf-2 Sequestration and Ferroptosis Induction in Hippocampal Neurons After Cerebral Ischemia/Reperfusion

Daohang Liu [1] and Yizhun Zhu [1,2,*]

1 School of Pharmacy, Shanghai Key Laboratory of Bioactive Small Molecules, Fudan University, Shanghai 201203, China; dhliu@cdutcm.edu.cn
2 School of Pharmacy, Macau University of Science and Technology, Macau 999078, China
* Correspondence: yzzhu@must.edu.mo

Abstract: SET and MYND Domain-Containing 2 (Smyd-2), a specific protein lysine methyltransferase (PKMT), influences both histones and non-histones. Its role in cerebral ischemia/reperfusion (CIR), particularly in ferroptosis—a regulated form of cell death driven by lipid peroxidation—remains poorly understood. This study identifies the expression of Smyd-2 in the brain and investigates its relationship with neuronal programmed cell death (PCD). We specifically investigated how Smyd-2 regulates ferroptosis in CIR through its interaction with the Nuclear Factor Erythroid-2-related Factor-2 (Nrf-2)/Kelch-like ECH-associated protein (Keap-1) pathway. Smyd-2 knockout protects HT-22 cells from Erastin-induced ferroptosis but not TNF-α + Smac-mimetic-induced apoptosis/necroptosis. This neuroprotective effect of Smyd-2 knockout in HT-22 cells after Oxygen–Glucose Deprivation/Reperfusion (OGD/R) was reversed by Erastin. Smyd-2 knockout in HT-22 cells shows neuroprotection primarily via the Nuclear Factor Erythroid-2-related Factor-2 (Nrf-2)/Kelch-like ECH-associated protein (Keap-1) pathway, despite the concurrent upregulation of Smyd-2 and Nrf-2 observed in both the middle cerebral artery occlusion (MCAO) and OGD/R models. Interestingly, vivo experiments demonstrated that Smyd-2 knockout significantly reduced ferroptosis and lipid peroxidation in hippocampal neurons following CIR. Moreover, the Nrf-2 inhibitor ML-385 abolished the neuroprotective effects of Smyd-2 knockout, confirming the pivotal role of Nrf-2 in ferroptosis regulation. Cycloheximide (CHX) fails to reduce Nrf-2 expression in Smyd-2 knockout HT-22 cells. Smyd-2 knockout suppresses Nrf-2 lysine methylation, thereby promoting the Nrf-2/Keap-1 pathway without affecting the PKC-δ/Nrf-2 pathway. Conversely, Smyd-2 overexpression disrupts Nrf-2 nuclear translocation, exacerbating ferroptosis and oxidative stress, highlighting its dual regulatory role. This study underscores Smyd-2's potential for ischemic stroke treatment by disrupting the Smyd-2/Nrf-2-driven antioxidant capacity, leading to hippocampal neuronal ferroptosis. By clarifying the intricate interplay between ferroptosis and oxidative stress via the Nrf-2/Keap-1 pathway, our findings provide new insights into the molecular mechanisms of CIR and identify Smyd-2 as a promising therapeutic target.

Keywords: Smyd-2; Nrf-2; ischemic stroke; ferroptosis; lipid peroxidation; hinge and latch

check for updates

Citation: Liu, D.; Zhu, Y. Unveiling Smyd-2's Role in Cytoplasmic Nrf-2 Sequestration and Ferroptosis Induction in Hippocampal Neurons After Cerebral Ischemia/Reperfusion. *Cells* **2024**, *13*, 1969. https://doi.org/10.3390/cells13231969

Academic Editors: Alessia Remigante and Rossana Morabito

Received: 25 October 2024
Revised: 19 November 2024
Accepted: 20 November 2024
Published: 28 November 2024

1. Introduction

Strokes have become the second leading cause of death and disability worldwide, following only heart disease, with high mortality and morbidity rates [1–4]. Approximately 80% of all strokes are ischemic, primarily caused by the focal occlusion or stenosis of intracranial/extracranial arteries [5,6]. Minimally invasive endovascular therapy or thrombolytic drugs to a certain extent can restore blood supply in the brain. However, this blood flow simultaneously brings a series of endogenous and exogenous free radicals resulting from reduction–oxidation (redox) imbalance, eventually leading to brain edema and hemorrhagic transformation [7–9]. These neuropathological lesions are known as cerebral ischemia/reperfusion (CIR) injuries. Thus, the intervention of reactive oxygen species (ROS) can effectively salvage the function of the ischemic penumbra.

Ferroptosis, a novel type of PCD, occurs in addition to neuronal apoptosis and necrosis after CIR [10,11]. The pathological hallmark of ferroptosis is the peroxidation of phospholipids containing long-chain unsaturated fatty acids on the cytoplasmic and organelle membranes. This process is triggered by the depletion of intracellular glutathione (GSH) and the collapse of the glutathione peroxidase-4 (GPX-4) system in the presence of ferrous ions [12–14]. Nrf-2 negatively regulates neuronal ferroptosis by regulating GSH systems, NADPH regeneration systems, dominant iron storage ferritin, and iron transporter ferroportin [15]. As an essential member of the cap-'n'-collar (CNC) transcription factor family that belongs to leucine zipper proteins (bZIPs) in mammals, Nrf-2 is expressed chiefly in apparatus with high metabolic activity and vital detoxification functions, especially in the brain [16]. Under physiological conditions, the lysine-rich Neh-2 domain of Nrf-2 interacts with the ETGE (high-affinity) and DLG (low-affinity) motifs of Keap-1 to form inactive dimers in the cytoplasm [17–20]. Keap-1 bundles the ring box protein-1 (RBP-1) and Cullin-3-Rbx-1-E3 ligase complex through its BTB domain to continuously ubiquitinate and degrade Nrf-2, maintaining Nrf-2 activity at a low level [21–23].

Interestingly, the constitutive stabilization of Nrf-2 by Keap-1 is interpreted as the "Hinge-and-Latch" recognition model [17,24]. However, during stroke progression, deleterious factors such as oxidative stress and electrophilic agents can modify critical sulfhydryl groups in Keap-1. Simultaneously, phosphorylated Protein Kinase C-δ (PKC-δ) phosphorylates Nrf-2 at serine residue 40 (Ser-40), inducing a conformational change in the Nrf-2-Keap-1 dimer. This alteration rapidly disrupts the ubiquitin-dependent degradation of Nrf-2, allowing it to dissociate from Keap-1 [25,26]. Once released, Nrf-2 translocates into the nucleus, where it preferentially forms heterodimers with small Maf (sMaf) proteins. These heterodimers bind to antioxidant response elements (AREs) through the bZip domain, initiating the transcription of target genes involved in the antioxidative stress response. This mechanism protects neurons from the gradual and cumulative ROS damage that occurs throughout the stroke cycle [27–29]. Moreover, Nrf-2 undergoes various posttranslational modifications (PTMs), such as phosphorylation and acetylation, which are critical for its functional regulation. Phosphorylation, in particular, plays a pivotal role in Nrf-2's release from Keap-1 and subsequent nuclear translocation. Notably, the potential interactions among PTMs—such as adjacent phosphorylated serine/threonine residues and methylated lysine/arginine residues—appear to be mutually exclusive. These posttranslational modifications in the Neh-2 domain of Nrf-2 significantly influence its nuclear translocation and functional activity, offering promising avenues for the development of novel therapeutic strategies to mitigate CIR-induced neuronal impairment.

In the central nervous system (CNS), epigenetic modification mechanisms were first revealed in the processes of neuronal development and maturation in the frontal lobe, such as the dynamic regulation of DNA methylation in neurogenesis and the regulation of non-histone protein CpH methylation in neuronal maturation [30,31]. SET and MYND Domain-Containing 2 (Smyd-2) is a unique protein lysine methyltransferase (PKMT) from the Myeloid-Nervy-DEAF1 domain protein family. Smyd-2 is primarily engaged in methylating histone H3 lysine K4 (H3K4) and histone H3 lysine K36 (H3K36) and covalently modifies non-histone proteins [32,33]. While the non-histone methylation mechanisms of Smyd-2 have been extensively studied in contexts such as heart development, skeletal muscle formation, cardiovascular diseases, and tumorigenesis, its role in CNS pathology remains largely unexplored. Notably, Smyd-2 has been implicated in regulating the growth and migration of gonadotropin-releasing hormone (GnRH) neurons in the vomeronasal organ (VNO) of mice [34,35]. Our previous research demonstrated that Smyd-2 might contribute to blood–brain barrier (BBB) disruption during CIR injury [36]. This finding aligns with the broader involvement of other Myeloid-Nervy-DEAF1 domain family members in CNS pathology. For instance, Smyd-1 has been reported to regulate skNAC transcription factors, thereby influencing inflammatory activation in the mouse cortical striatum. Similarly,

RNA-binding proteins such as TDP-43, FUS, and TLS can downregulate Smyd-3 precursor mRNA, leading to motor neuron death, neuroinflammatory cascades, and PCD.

Non-histone methylation at the same or adjacent lysine residues introduces complexity into the regulation of pathological processes in vulnerable neurons across various CNS diseases. However, current studies remain too preliminary to definitively delineate the precise role of Smyd-2 in the brain. It is unclear whether Smyd-2, like its homologous family members, participates in CIR progression through endogenous PCD regulatory mechanisms. The alterations in non-histone lysine methylation sites and the extent of methylation saturation by Smyd-2 under CIR stimuli also require further validation. Understanding how these dynamic methylation events affect the function of specific non-histone proteins in CIR is a key area for future research. Nuclear-translocated proteins, such as Nrf-2, contain several positively charged lysine residues in their Neh-2 domain that may interact or compete with Smyd-2. Methylation at these lysine residues could potentially influence Nrf-2's subcellular localization and stability by modulating its phosphorylation-dependent activation or ubiquitin-mediated degradation. Thus, we aim to offer a more detailed investigation of the role and mechanism of Smyd-2 in CIR, highlighting the relationship between Smyd-2 and ferroptosis and unraveling the molecular basis of Nrf-2 ubiquitylation, phosphorylation, and nuclear translocation.

To address the above gaps, our study systematically investigates the role of Smyd-2 in ferroptosis regulation and its interaction with the Nrf-2/Keap-1 signaling pathway during CIR. Using both in vivo (MCAO model) and in vitro (OGD/R model) systems, we explored Smyd-2's expression patterns and its impact on hippocampal neuron survival. Central to our investigation is the interplay between the Smyd-2-mediated lysine methylation of Nrf-2 and its phosphorylation and nuclear translocation, key processes for anti-lipid peroxidation and redox balance. Through gene targeting approaches and pharmacological interventions, we analyzed the modulation of ferroptosis by Smyd-2 and its influence on neuronal viability, unveiling a novel epigenetic mechanism that exacerbates ferroptosis under CIR conditions. Our findings highlight Smyd-2's potential as a therapeutic target for ischemic stroke, advancing our understanding of ferroptosis and oxidative stress in neuronal injury.

2. Materials and Methods

2.1. Chemical Reagents

Radio immunoprecipitation assay (RIPA), phenylmethanesulfonyl fluoride (PMSF), phosphatase inhibitor, loading buffer, tween-20, glycine, BCA protein assay kit, diethylpyrocarbonate (DEPC), sodium dodecyl sulfate (SDS), and goat serum were purchased from Beyotime Biotech, Shanghai, China. Bovine serum albumin (BSA), goat serum, paraformaldehyde (4%), phosphate-buffered saline (PBS), sodium chloride, isopropanol, β-mercaptoethanol (BME), and Anti-Fluorescence Quenching DAPI were purchased from Servicebio Technology Co., Ltd., Wuhan, China. DCFH-DA fluorescent probe and nuclear/cytoplasmic protein extraction kits were purchased from Yeasen Biotech, Shanghai, China. Enhanced chemiluminescence (ECL) and BODIPY 581/591 C11 fluorescent probe were purchased from Millipore, Bedford, MA, USA. Trizol, phenol-chloroform, 2, 3, 5-triphenyl tetrazole chloride (TTC), and Triton-100 were purchased from Sigma-Aldrich, USA. ML-385, cycloheximide (CHX), and DDO-7263 were purchased from MedChemExpress (MCE) Co., Ltd., USA. Fluoro-Jade B® (FJB) fluorescence probe was bought from QiMing Biotechnology Co., Ltd., Shanghai, China. Antibodies (Rabbit) against Smyd-2, FTH-1, COX-2, GPX-4, GAPDH, HO-1, Keap-1, Nrf-2, histone H-3, and methylated lysine were obtained from Proteintech company, Wuhan, China. Antibodies (Rabbit) containing 15-LOX, NQO-1, PGC-1α, PKC-δ, phosphorylated Nrf-2 (p-Nrf-2/Ser-40), phosphorylated PKC δ (p-PKC-δ/Ser359), and the secondary antibodies were provided by Abcam, Cambridge, UK. Antibodies (Mouse) against Smyd-2 were obtained from Santa Cruz Biotechnology, MA, USA. EndoFectin™ Max was provided by iGene Biotechnology Co., Ltd., Guangzhou,

China. Annexin V–FITC/PI kits were purchased from Applygen Technologies Inc., Beijing, China. Erastin was obtained from Selleck Chemicals, Houston, TX, USA.

2.2. In Vivo Experiments and Grouping

Male C57BL/6J mice aged 6–8 weeks with a weight range of 20 ± 5 g were purchased from Beijing Vital River Laboratory Animal Technology Co., Ltd. The mice were housed in a specific-pathogen-free (SPF) facility at the Animal Experiment Center of the School of Pharmacy, Fudan University. They were kept in individual cages with ad libitum food and water. The temperature and humidity were strictly controlled at 23 ± 2 °C and 50–60%, respectively, with a 12 h light/dark cycle to maintain the circadian rhythm of the mice. The mice were acclimatized in the SPF facility for one week before the start of the experiments. All laboratory animal procedures were conducted in accordance with the Laboratory Animal Operation Procedures and the Laboratory Animal Management Regulations of the Laboratory Animal Center, School of Pharmacy, Fudan University. The mice were fasted and deprived of water for 12 h before surgery. The mice were randomly divided into sham, CIR, and respective medication groups, with 5 to 6 mice in each experimental group.

2.3. Establishment of Middle Cerebral Artery Occlusion (MCAO) Model in Mice

The middle cerebral artery occlusion (MCAO) model is widely employed to study the pathophysiology of CIR due to its ability to replicate the key pathological features of ischemic stroke and subsequent reperfusion injury [37]. Wild-type male C57BL/6 mice (aged 8–10 weeks) were fasted for 12 h with free access to water before MCAO. Mice were anesthetized with 0.3% pentobarbital sodium (0.1–0.2 mL/10 g body weight). A surgical incision was made along the midline of the neck to expose the bilateral carotid arteries. The sternocleidomastoid muscle was located and separated from the muscle and fascia, fully exposing the common carotid artery (CCA), external carotid artery (ECA), and internal carotid artery (ICA). The CCA was ligated at the proximal end with a suture, and the ICA was clamped with a vascular clip. The ECA was opened using ophthalmic scissors near its bifurcation with the CCA. A silicone thread (Jialing Biotech, Guangzhou, China) was inserted into the middle cerebral artery (MCA) at a horizontal angle of 45 degrees. The silicone thread was advanced approximately 2 cm into the MCA, effectively blocking blood flow for 1 h. The silicone thread and the surgical suture clamping the CCA were removed to restore blood flow to the ipsilateral hemisphere of the mice. The mice's body temperature was kept constant during the surgery. Tissue samples were collected according to the specified reperfusion times following the MCAO.

2.4. Lateral Ventricle Localization Injection of Inhibitors and Adeno-Associated Virus in Mice

The mice were anesthetized and fixed by clamping their upper incisors onto a crossbar and stabilizing the nose bar by turning a knob while gently placing the ear bars in the mouse's ear canals and adjusting them to keep the skull level. Adhesions to the meninges were removed with hydrogen peroxide to identify the Bregma point. One of the injection sites was located 2.5 mm to the right of the Bregma point. The neuronal promoters and inhibitors for the expression of the mouse Smyd-2 were packaged into recombinant adeno-associated virus (serotype 9) vectors (Genomeditech, Shanghai, China). The mice were transduced with adeno-associated virus–Smyd-2 mixed with saline at a rate of 2 μL/min, with 2 μL per mouse, 10 days before MCAO modeling. Sham mice received an equal amount of control viral vector for the same duration. ML-385 was injected into the lateral ventricle at a dose of 40 pmol/4 μL 24 h before MCAO modeling. After injection, the syringe needle was left in place at the injection site for 3 min before being slowly removed to promote the absorption of the adeno-associated virus and inhibitors.

2.5. Evaluation of Neurological Function

The Adhesive Tape Removal Test was used to assess sensorimotor deficits in mice [38]. The experimental mice were taken out, and a 4 mm diameter medical adhesive tape was randomly attached to both forelimbs. In normal mice, the tape removal process is typically completed within approximately 20 s. The tape removal time was measured at specific reperfusion time points following MCAO. Times exceeding 180 s were considered invalid and excluded from analysis.

The Corner Test assessed unilateral sensory and motor deficits in mice [39]. The experimental mice were placed between two square plates ($30 \times 30 \times 1$ cm^3) connected at a 30° angle with an opening at the other end for the mouse to enter smoothly. The spatial obstruction prompts the mice's forward and upward reflex action, followed by a turn to face the open end. The probability of turning left or right in normal mice is almost equal, while MCAO/R mice tend to roll over to the contralateral side. The testing was conducted at specific time points according to the reperfusion time point after MCAO. Attempts where the mouse failed to turn upright were not counted toward the experimental trials.

2.6. Evaluation of Cerebral Infarct Size

The 2,3,5-Triphenyl Tetrazolium Chloride (TTC) staining method was used to determine the infarct volume in mice following MCAO [40]. The brain tissue was quickly removed and frozen at -80 °C for 5 min after reperfusion. The tissue was then solidified and sliced along the coronal plane on ice, with a slice thickness of 1 mm. The brain tissue slices were incubated in preheated TTC solution at 37 °C to ensure adequate staining. Subsequently, they were fixed in 4% paraformaldehyde solution at 4 °C for 24 h and then photographed. The infarct area (%) was calculated as the infarct volume (white infarct lesion) divided by the total brain volume, multiplied by 100% through Image Pro Plus (IPP) 6.0 software.

2.7. Isolation of Nuclear Protein and Cytoplasmic Protein

A mixture of cell lysis buffers A and B from a cell nuclear/cytoplasmic protein extraction kit was prepared in a 20:1 ratio, and PMSF was added to the mixture to a final concentration of 1 mM. The hippocampus was separated and weighed, and a lysis buffer was added to the tissue in a ratio of 200 μL per 60 mg of hippocampal tissue. Magnetic beads were applied to homogenize the mixture with a homogenizer for 30 s at 4 °C, repeated four times. The mixture was then incubated on ice for 15 min and centrifuged at 2000 rpm for 5 min at 4 °C. The precipitate was mixed with cell lysis buffer A containing PMSF at a ratio of 1:20 for 15 min. Cell lysis buffer B was added to the mixture at a ratio of 10 μL per 100 μL of the precipitate. The mixture was then centrifuged at 12,000 rpm for 15 min at 4 °C. The supernatant was collected and subsequently incubated with cell nuclear protein extraction buffer C mixed with PMSF. The mixture was centrifuged at 12,000 rpm for 15 min at 4 °C, and the supernatant was collected in 1.5 mL centrifuge tubes.

2.8. Western Blot

Separation and concentration gels were prepared at appropriate concentrations based on the molecular weight of the target protein. Protein loading was conducted according to experimental groups. Gels and electrophoresis apparatus were assembled for electrophoresis at 80 V. The voltage was then increased to 120 V, and electrophoresis was continued for an additional hour with a constant voltage once the clear separation of the protein bands was observed. The proteins were then transferred to a polyvinylidene fluoride (PVDF) membrane (Millipore, USA) based on the target protein's molecular weight, with a constant current of 250 mA being applied. After completing the protein transfer, the membranes were washed four times with TBST at room temperature for 5 min each and blocked with 5% BSA at room temperature for 2 h. The membranes were incubated with the primary antibody, diluted with BSA, and left on a shaker (Eppendorf, Framingham, MA, USA) in a refrigerator at 4 °C overnight, followed by four washes with TBST. Next, the membranes were incubated with the secondary antibody specific to the primary antibody species for 1 h and washed four times with TBST. ECL detection was used to visualize the protein bands, and ImageJ was used to analyze the grayscale values of the bands.

2.9. Immunoprecipitation (IP)

The beads with methylated lysine and Nrf-2 antibodies were bonded and incubated overnight at 4 °C. The antibody-coupled magnetic beads were then placed in the tissue lysate to allow the antibodies to bind to the proteins of interest overnight at 4 °C. A magnetic rack (Thermo, USA) was used to separate the magnetic bead–antigen–antibody complexes from the protein samples. Subsequently, the collected complexes were washed three times with pre-cooled RIPA buffer and were boiled with a 5× loading buffer protein sample for 5 min at 100 °C to purify the targeted proteins. The supernatant was extracted for electrophoresis. Protein expression was analyzed by Western blot.

2.10. Assessment of Cerebral Histopathology

Mice were securely fixed on a mouse board and then underwent thoracotomy. The cannula was slowly inserted into the anterior wall near the apex until it reached the left ventricle. The right atrium was then incised, and an appropriate amount of PBS mixed with heparin was infused into the heart. Next, 4% paraformaldehyde was perfused to fix the brain in situ. The fixed brain was then extracted and placed into a 4% paraformaldehyde solution for another 12 h of fixation, followed by dehydration in a gradient glucose solution, paraffin embedding, and sectioning using a Leica1900 slicer (Leica Microsystems Nussloch GmbH, Nussloch, Germany).

Fluoro-jade B (FJB) stain: Brain tissue slices were first deparaffinized. Background fluorescence was blocked, and contrast was optimized before the slides containing brain tissue slices were incubated in a working solution of FJB. After that, the slides of brain tissues were placed in a Coplin jar to incubate in darkness and low light for 10 min. The degeneration of neurons was observed and analyzed using Nikon immunofluorescence microscopy.

LFB (Luxol fast blue) stain: Brain tissue sections were placed in Luxol fast blue solution and left overnight at 56 °C after deparaffinization and the removal of lipids. The morphology of neuronal myelin was observed under a microscope. After staining, the sections were washed in 95% ethanol to remove excess stain and differentiate the tissue, followed by counterstaining with a solution of cresyl violet to visualize the neuronal nuclei. The myelin appeared stained blue, while other structures, such as the nuclei and cytoplasm, were stained pink by the cresyl violet under the microscope. Images were captured and assessed with a Leica microscopic imaging system.

Perls stain: Perls stain was used to visualize iron accumulation in brain tissue samples. Deparaffinized brain tissue sections were incubated in a mixture of potassium ferrocyanide and hydrochloric acid for 10 min. The sections were washed three times with distilled water for 5 min each time. After that, the sections were stained with neutral red for 1 min and finally dehydrated in absolute alcohol and observed under a microscope. In the presence of ferric iron (Fe^{3+}), the potassium ferrocyanide solution reacts to form a blue precipitate (Prussian blue). The staining intensity indicates the concentration of iron in the tissue. The counterstaining with hematoxylin helps to distinguish different cell types and tissue structures.

2.11. In Vitro Experiments and Grouping

HT-22 cells were cultured in a cell incubator at 37 °C, 5% CO_2, and 100% humidity. The culture medium consisted of DMEM (Dulbecco's Modified Eagle's Medium) supplemented with 10% FBS (fetal bovine serum), 100 U/mL penicillin, and 100 µg/mL streptomycin. The cells were randomly divided into control, OGD/R, and respective medication groups.

2.12. Establishment of Oxygen–Glucose Deprivation/Reperfusion Model in HT-22 Cells

After HT-22 cells were seeded onto culture plates, they were washed with PBS three times and then incubated in glucose-free DMEM. The culture dish was then transferred to an anaerobic chamber to simulate hypoxia for 2 h. Following hypoxia, the glucose-free DMEM was replaced with fresh DMEM, and the plates were placed back into the incubator for reperfusion. The duration of reperfusion was adjusted according to specific experimental requirements [41].

2.13. RNA Interference

HT-22 cells in good growth status and at an appropriate density were selected and cultivated for transfection. Cell counting was performed to ensure that the number of cells during transfection was sufficient and that the cell confluence was close to 75% before transfection. EndoFectin™-Max transfection reagent was diluted in DMEM and mixed with an appropriate amount of siRNA at room temperature for 30 min to prepare the transfection complex. The complex was added to the DMEM culture medium without antibodies and then transfected into HT-22 cells for 2 h. Subsequently, a serum-containing culture medium was added to occupy about 50% of the original culture medium volume. Transfection was carried out for 24 h and processed according to experimental requirements.

2.14. Measurement of Neuronal Viability

In total, 1000–1500 HT-22 cells were seeded per well in a 96-well plate. The CCK8 assay was performed according to the manufacturer's instructions after OGD/R. Briefly, 20 µL of CCK-8 solution was added to each 96-well plate well and incubated at 37 °C for 2 h. The optical density (OD) was measured at 450 nm using a Thermo Scientific Microplate Reader (Thermo Scientific, Waltham, MA, USA), and the percentage of neuronal viability was calculated by comparing the background-corrected OD of the treatment group to the control group.

2.15. Assessment of Lactate Dehydrogenase (LDH) Release

Following the cell seeding procedure described above, HT-22 cells were pretreated with indicated stimulations. The LDH assay was performed according to the manufacturer's instructions to detect the content of LDH in the supernatant of HT-22 cells. The OD was measured at 490–500 nm using a Thermo Scientific Microplate Reader (Thermo Scientific, Waltham, MA, USA). LDH release rate (%) = (sample OD − low control OD)/(high control OD − low control OD) × 100%.

2.16. Enzyme Assay Kits

The contents of Superoxide Dismutase (SOD), Malondialdehyde (MDA), and GSH in HT-22 cells were determined using enzyme assay kits. The supernatant of each HT-22 cell group was collected by centrifugation at $1000 \times g$ for 15 min after being subjected to modeling or other treatment conditions. The supernatant was then used to measure the OD value at 450 nm to calculate the SOD, MDA, and GSH content in HT-22 cells. The specific kits used were the Total SOD Detection Kit (product ID: S0101S, Beyotime, China), the MDA Detection Kit (product ID: S0131S, Beyotime, China), and the GSH Detection Kit (product ID: G4305-48T, Servicebio, China).

2.17. Determination of the Cellular Lipid Peroxidation

HT-22 cells were seeded in a 6-well plate and treated according to the experimental requirements. The level of reactive oxygen species (ROS) in HT-22 cells was measured using the BODIPY-C11 581/591 probe [42]. The BODIPY 581/591 C11 fluorescent probe was dissolved in a serum-free culture medium to make a stock solution with a 10 mmol/L concentration. After adding 1.5 mL of culture medium containing 1 µmol/L BODIPY 581/591 C11 fluorescent probe, HT-22 cells were incubated in a light-avoiding 37 °C incubator for 30 min. The lipid peroxidation intensity was observed under a fluorescence microscope. Red fluorescence indicates that HT-22 cells are in a reduced state, while green fluorescence indicates that cells are in a highly oxidative state due to the ROS-induced oxidation of the unsaturated butadiene in the BODIPY 581/591 C11 fluorescent probe.

2.18. Determination of the Mitochondrial Lipid Peroxidation

HT-22 cells were seeded in a 6-well plate and treated according to the experimental requirements. Mitochondrial ROS in HT-22 cells were detected using the DCFH-DA probe [43]. The DCFH-DA probe was initially diluted to a concentration of 10 μmol/L using serum-free culture medium, and 1.5 mL was added to each well. This process was carried out in the absence of light. Fluorescence was observed using Nikon confocal microscopy, and the intensity of green fluorescence was proportional to the expression level of DCF, the oxidation product of DCFH-DA. Higher levels of DCF resulted in stronger green fluorescence, indicating a greater accumulation of mitochondrial ROS.

2.19. Identification of Programmed Cell Death in HT-22 Cells

The occurrence of programmed cell death in HT-22 cells was examined using the Annexin V–FITC assay kit. HT-22 cells were resuspended in 200 μL Binding Buffer. In total, 4 μL of 0.5 mg/mL PI and 2 μL of Annexin V–FITC solution were added to the incubator for 15 min. The confocal fluorescence microscope was employed to count the number of Annexin V-positive cells, PI-positive cells, and Annexin V/PI double-stained cells. Annexin V can label cells undergoing early apoptosis and cell death, while PI labels necrotic cells [44].

2.20. Mitochondrial Morphology

The HT-22 cells were plated onto a 24-well plate. The supernatant was discarded, and 500 μL of 2.5% glutaraldehyde fixative was added to the wells to fix HT-22 cells at 4 °C for 12 h. The mitochondrial morphology was observed and compared under the Hitachi HT780 transmission electron microscope (Japan).

2.21. Statistical Analysis

Statistical analysis was performed using Prism 8.3. The experimental results are presented as mean ± standard deviation (SD). One-way ANOVA was used to compare the differences between multiple groups. $p < 0.05$ was considered statistically significant.

3. Results

3.1. OGD/R and CIR Induce an Increase in Smyd-2 Expression in Hippocampal Neurons

We backed up these findings by establishing an OGD/R model for CIR insult. HT-22 cells were subjected to OGD/R as the mice were subjected to CIR. The in vivo experiment results showed that the Smyd-2 protein and mRNA expression reached the highest level and then decreased in a time-dependent manner when mice underwent ischemia for 1 h and reperfusion for 24 h, but there was no significant difference in Smyd-2 mRNA expression after reperfusion for 24 h to 72 h (Figure 1A). Compared with the control group, the Smyd-2 mRNA level and protein expression level of Ht-22 cells gradually increased at the 6th hour after OGD and reached the peak level at the 24th hour (Figure 1B). It is worth noting that the mRNA expression level of Smyd-2 showed no statistical significance from the 24th hour after OGD, which is consistent with the in vivo experiments (Figure 1C). Therefore, we selected Ht-22 cells to undergo OGD for 2 h and then reperfusion for 24 h as the follow-up experimental conditions. The above results show that the protein and mRNA expression of cellular Smyd-2 peaked 24 h after reperfusion.

3.2. The Inhibition of Smyd-2 Expression Delays the Progression of CIR Impairment

To clarify the mechanism and role of the increased expression of Smyd-2 in the course of CIR, we transfected Smyd-2-overexpressing adeno-associated virus, Smyd-2-silenced adeno-associated virus, and a vector through the intra-cerebroventricular (ICV) injection of mice. The PCR results showed that compared with the sham group, the mRNA expression of Smyd-2 was significantly increased in the ICV-Smyd-2 group, while it was almost not expressed in the ICV-Smyd-2 (KO) group (Figure 2A). These results showed that the ICV injection with adeno-associated virus successfully interfered with the overexpression and silencing of Smyd-2 in the hippocampus of the mouse brain, and this condition and dose was applied to subsequent experiments. The experimental results showed that Smyd-2 knockout could reduce the cerebral infarction volume, improve the neurologic deficit scores, alleviate neuronal demyelination, and reduce the number of degenerative neurons in the hippocampus after CIR (Figure 2B,C,E). However, Smyd-2 overexpression significantly increased the volume of cerebral infarction, significantly decreased the neurologic deficit scores, aggravated neuronal demyelination and degeneration, and increased vacuole-like structures between neurons induced by CIR (Figure 2B,C,E). In vitro, Si-RNA-mediated Smyd-2 knockout (Si-Smyd-2) notably remitted the OGD/R-induced decrease in the viability of HT-22 cells and increased the LDH release (Figure 2G). SOD measures the antioxidant capacity and the neutralization of superoxide radicals, while MDA is typically used to detect lipid peroxidation. The cytosolic SOD levels significantly decreased, while the stable cytosolic metabolite MDA levels markedly increased in OGD/R-induced HT-22 cells (Figure 2G). OGD/R-induced HT-22 cells transfected with adenoviral Smyd-2 expression vectors (AD-Smyd-2) showed double the SOD content and half the MDA content compared to the activity measured in the non-transfection and mock-vehicle groups (Figure 2G). The above data suggests that Smyd-2 is involved in the pathophysiologic procedures of CIR/OGDR. Smyd-2 knockout in hippocampal neurons has neuroprotective potential against CIR/OGDR insult, and Smyd-2 overexpression, in turn, could exacerbate CIR/OGDR impairment.

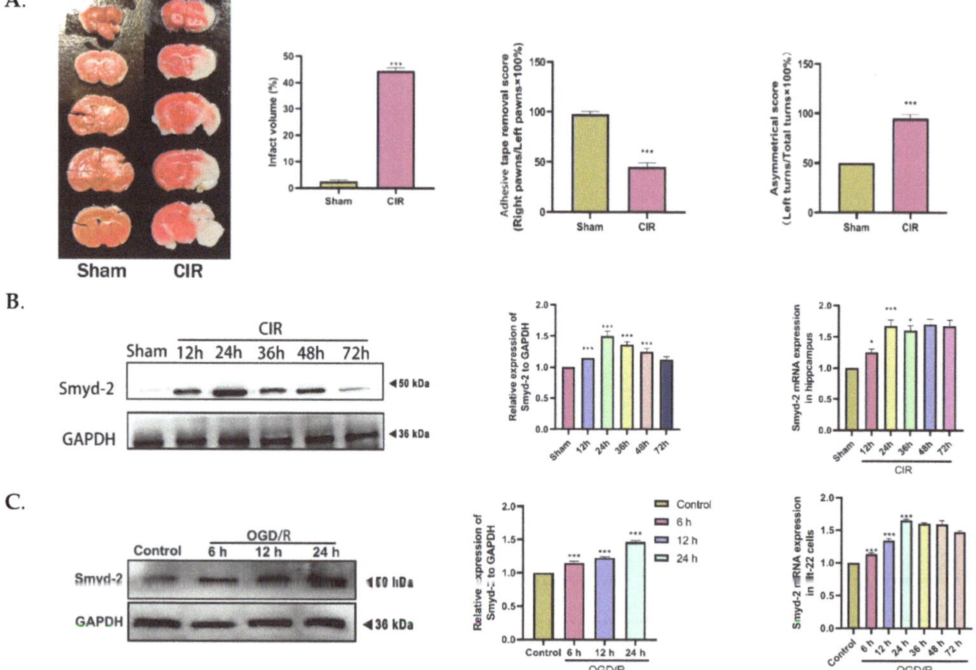

Figure 1. MCAO and OGD/R conduce to Smyd-2 activation in the hippocampus and HT-22 cells. (**A**). Representative images of TTC staining, quantitative analysis of infarct volume, asymmetrical test scores, and the adhesive removal scores in MCAO mice (*n* = 5). (**B**). Smyd-2 expression in mouse brain after CIR (*n* = 5). (**C**). Smyd-2 expression in Ht-22 cells challenged with OGD/R (*n* = 5). * *p* < 0.05 vs. sham group, *** *p* < 0.001 vs. sham group; *** *p* < 0.0001 vs. control group.

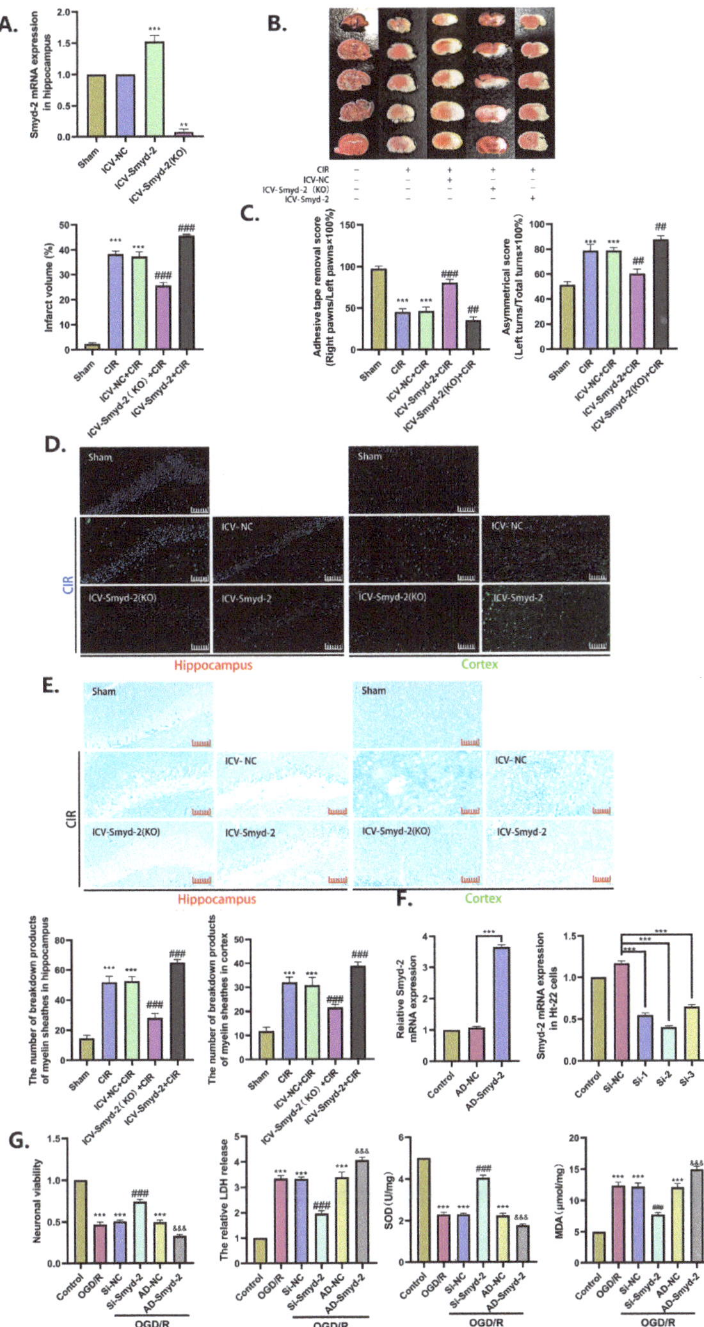

Figure 2. The inhibition of Smyd-2 expression delays the progression of CIR impairment. (**A–C**) Effects of Smyd-2 on cerebral infarction volume and neurobehavioral function in MCAO mice (*n* = 5). (**D**) Representative images of FJB staining slices of the hippocampus and cortex; scale bars = 40 μm (*n* = 5). (**E**) Representative images of LFB staining slices of the hippocampus and cortex and quantitative analysis of the breakdown products of myelin sheathes in the hippocampus

and cortex; scale bars = 40 μm (*n* = 5). (**F**) The transfection effects of Smyd-2-overexpressing adenovirus and siRNA in Ht-22 cells (*n* = 5). (**G**) The neuronal viability, LDH release, SOD level, and MDA level of Ht-22 cells challenged with Smyd-2 siRNA and Smyd-2-overexpressing adenovirus after OGD/R (*n* = 5). ** *p* < 0.01 vs. sham group, *** *p* < 0.001 vs. sham group; ## *p* < 0.01, ### *p* < 0.001 vs. CIR group and CIR + IVC-NC group; *** *p* < 0.0001 vs. control group; ### *p* < 0.0001 vs. OGD/R group and OGD/R + Si-NC group; &&& *p* < 0.0001 vs. OGD/R group and OGD/R + AD-NC group.

3.3. Smyd-2 Knockout Counteracts the Effect of OGD/R on Lipid Peroxidation in Ht-22 Cells

It has been determined that free radicals in HT-22 cells generated by OGD/R oxidized the unsaturated butadiene backbone of the BODIPY-581/591-C11 probe, resulting in a shift in the fluorescence emission from 590 nm to 510 nm, and oxidized DCFH-DA into 2′-7′-dichlorofluorescein (DCF) with a green fluorescent label. Both of these observations suggest that OGD/R can lead to high levels of intracellular ROS and lipid peroxidation products. Si-Smyd-2 in HT-22 cells increased the resistance to lipid peroxidation and reduced the creation of intracellular free radicals when exposed to OGD/R; correspondingly, HT-22 cells transfected with Smyd-2-overexpressing adenovirus appeared to have enhanced lipid peroxidation and generated prodigious quantities of highly reactive free radicals after OGD/R (Figure 3A,B). GSH serves as the detoxifying substrate for GPX-4, which converts cytotoxic phospholipid hydroperoxides into non-toxic alcohols. GPX-4 is a crucial reductase in the pathogenesis of ferroptosis. The results demonstrated that the OGD/R treatment substantially reduced the content of GSH by 50% compared to control HT-22 cells, but OGD/R-induced HT-22 cells transfected with Smyd-2 Si-RNA regained a GSH content of up to 75% (Figure 3A,B). Moreover, double fluorescence staining with Smyd-2 and GPX-4 indicated a significant negative correlation between Smyd-2 expression and GPX-4 activation (Figure 3C). The data above show that the Smyd-2 expression was positively correlated with the ferroptosis index in HT-22 cells after OGD/R.

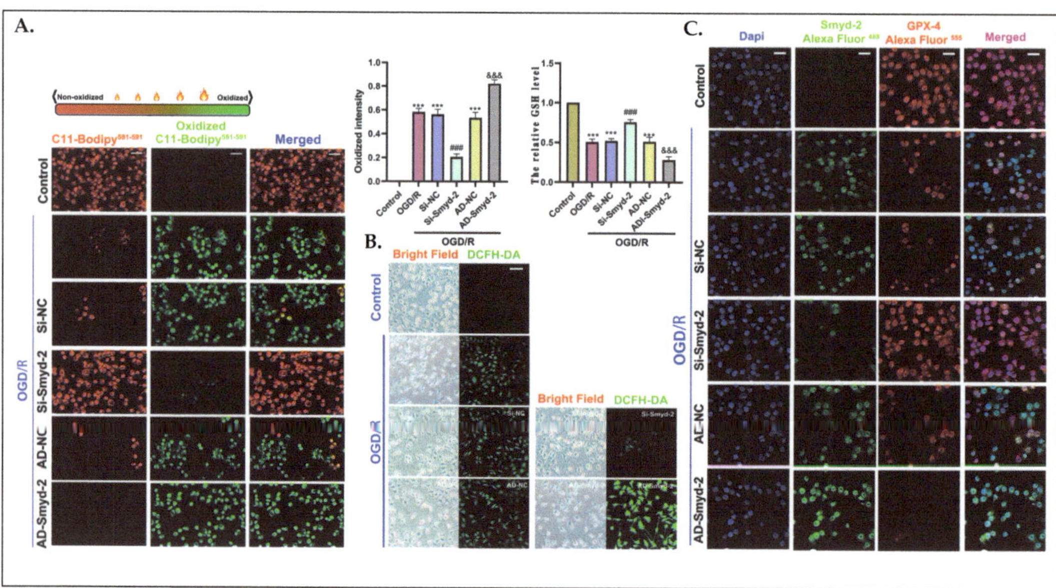

Figure 3. Smyd-2 knockout counteracts the effect of OGD/R on lipid peroxidation in Ht-22 cells. (**A**) BODIPY-581/591-C11 staining was applied to analyze and quantify the effect of Smyd-2 siRNA and adenovirus-mediated Smyd-2 on HT-22 cell ferroptosis challenged with OGD/R. The representative images were obtained with an optical microscope at 400× magnification; scale bars = 50 μm. The GSH

level of HT-22 cells challenged with AD-Smyd-2 and Si-Smyd-2 after OGD/R. (**B**) DCFH-DA staining was applied to analyze and quantify the effect of Smyd-2 siRNA and adenovirus-mediated Smyd-2 on HT-22 cell ferroptosis challenged with OGD/R. The representative images were obtained with a confocal microscope at $200\times$ magnification; scale bars = 100 μm (n = 6). (**C**) Immunofluorescence method was applied to investigate Smyd-2 and GPX-4 protein expression and localization in HT-22 cells and their relation to neuronal ferroptosis after OGD/R. The representative images were obtained with a confocal microscope at $400\times$ magnification; scale bars = 50 μm. *** p < 0.0001 vs. control group; ### p < 0.0001 vs. OGD/R group and OGD/R + Si-NC group; &&& p < 0.0001 vs. OGD/R group and OGD/R + AD-NC group.

3.4. Smyd-2 Regulates the Abnormality in Neuronal Ferroptosis Caused by CIR

Perls blue staining revealed the massive accumulation of ferrocyanide in the CA1 region of the hippocampus in CIR-challenged mice, whereas ICV-Smyd-2 (KO) minimized this accumulation (Figure 4A). Western blot was used to measure ferroptosis-related proteins. The results showed that CIR could increase the expression of Smyd-2, ASCL-4, Nrf-2 (total protein, TP), p-Nrf-2 (phosphorylated Nrf-2), nucleus Nrf-2, 15-LOX, COX-2, HO-1, and NQO-1 and downregulate the expression of SLC7A11, FTH-1, GPX-4, and Keap-1 in mouse hippocampus. In contrast, ICV-Smyd-2 (KO) downregulated the expression of Smyd-2, ASCL-4, 15-LOX, and COX-2 while upregulating the expression of Nrf-2 (total protein, TP), p-Nrf-2 (phosphorylated Nrf-2), nucleus Nrf-2, HO-1, and NQO-1 (Figure 4B–D). Notably, there was no significant difference in the expression of Keap-1 compared to the CIR group.

These results imply that CIR causes ferroptosis in hippocampus neurons, which could be diminished by Smyd-2 knockout, restoring the balance of oxidative stress.

3.5. Smyd-2 Knockout Delays the Progression of Erastin-Induced Neuronal Ferroptosis

To further illustrate the role of Smyd-2 in neuronal PCDs, Erastin was applied to establish the ferroptosis model by incubating HT-22 cells for 12 h. Additionally, tumor necrosis factor (TNF-α), which can control the activation of the transcription factor NF-κB, was used in conjunction with a Smac mimetic that mimics the function of the pro-apoptotic protein Smac/Diablo. HT-22 cells were incubated with these agents for 15 h to induce apoptosis and necrosis. The optimal concentrations of Erastin, TNF-α, and Smac mimetic to inhibit the growth of HT-22 cells were 5 μMol/L, 100 ng/mL, and 120 nMol/L, respectively, as determined by CCK-8 and LDH colorimetric assays (Figure 5A). Accordingly, the viability and SOD content of HT-22 cells significantly decreased, while the LDH release and MDA content increased in both TNF-α + Smac mimetic and Erastin treatments. Interestingly, knocking down the Smyd-2 expression with siRNA succeeded in driving neuronal viability, increasing SOD content, decreasing MDA content, and minimizing LDH release in the Erastin preconditioning group but failed to have any effect on the neuronal viability, LDH release, and MDA content except for slightly improving the SOD content in the TNF-α and Smac mimetic preconditioning group (Figure 5B,C). Next, propidium iodide (PI) was used to specifically label various types of necrotic cells, as it cannot enter apoptotic cells. Conversely, Annexin V binds to phosphatidylserine (PS), which is translocated from the inner leaflet of the plasma membrane to the cell surface in apoptotic cells, in a Ca^{2+}-dependent manner. Based on the established Erastin-induced neuronal ferroptosis and TNF-α + Smac-mimetic-induced neuronal apoptosis/necroptosis, Annexin V/PI immunofluorescence double staining was applied to estimate the PCDs mediated by Smyd-2. Of note, it was validated that both Erastin and TNF-α + Smac mimetic could label HT-22 with Annexin V/PI positivity, with the latter showing a more vigorous intensity of Annexin V-positive fluorescence (Figure 5D). Smyd-2 knockout conspicuously decreased the PI-positive cells without interfering with the Annexin V-positive cells and Annexin V/PI double-stained cells in Erastin-induced HT-22 cells; however, such a decline in PI-positive cells was almost invisible in TNF-α + Smac-mimetic-induced HT-22 cells (Figure 5D). Consistently, Smyd-2 was not significantly involved in excessive apoptosis and necroptosis induced by TNF-α + Smac mimetic but was identified as a critical factor in Erastin-induced neuronal ferroptosis.

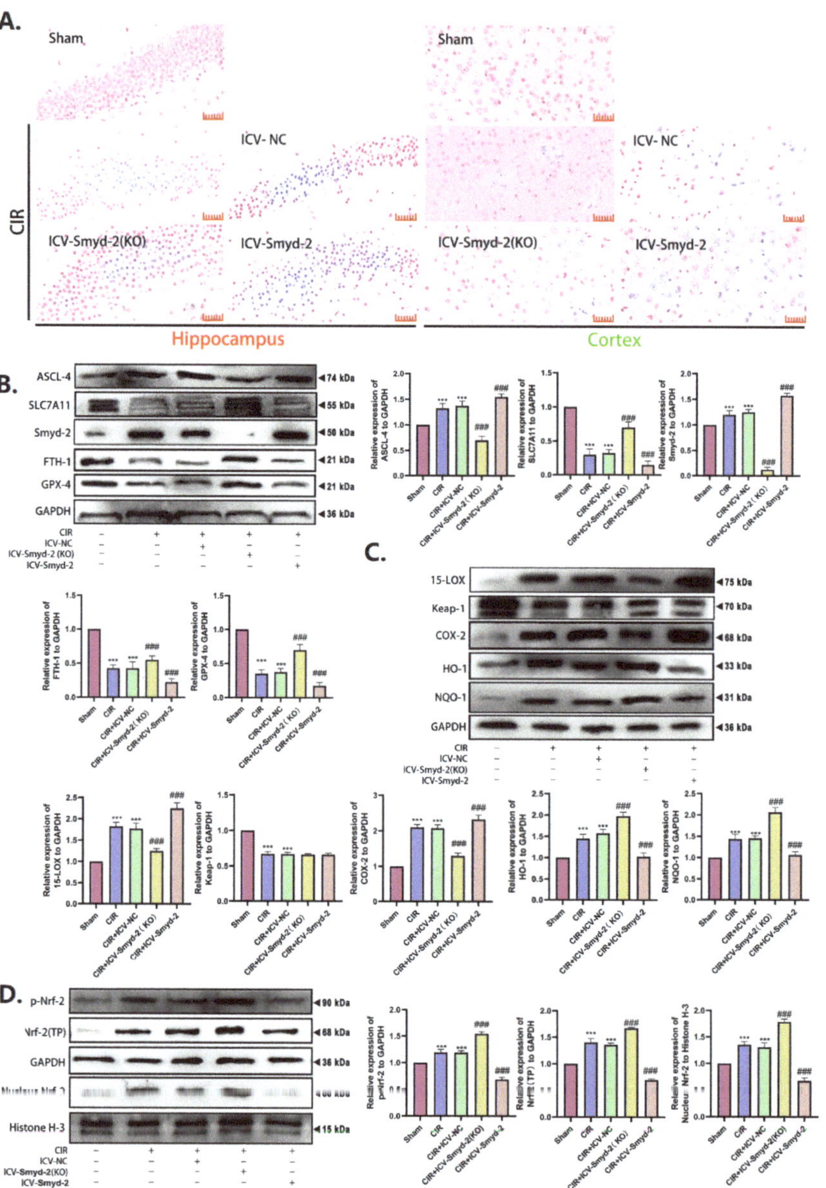

Figure 4. Smyd-2 regulates the abnormality in neuronal ferroptosis caused by CIR. (**A**) Representative images of Perls staining slices of the hippocampus and cortex. Scale bars = 40 μm (*n* = 5). (**B–D**) Representative images of Western blot and quantitative analysis of the expression of Smyd-2, GPX-4, FTH-1, SLC7A11, ACSL-4, 15-LOX, COX-2, NQO-1, Keap-1, HO-1, nucleus Nrf-2, *p*-Nrf-2, and Nrf-2 (TP) in the hippocampus. *** $p < 0.001$ vs. sham group; ### $p < 0.001$ vs. CIR group and CIR + ICV-NC group.

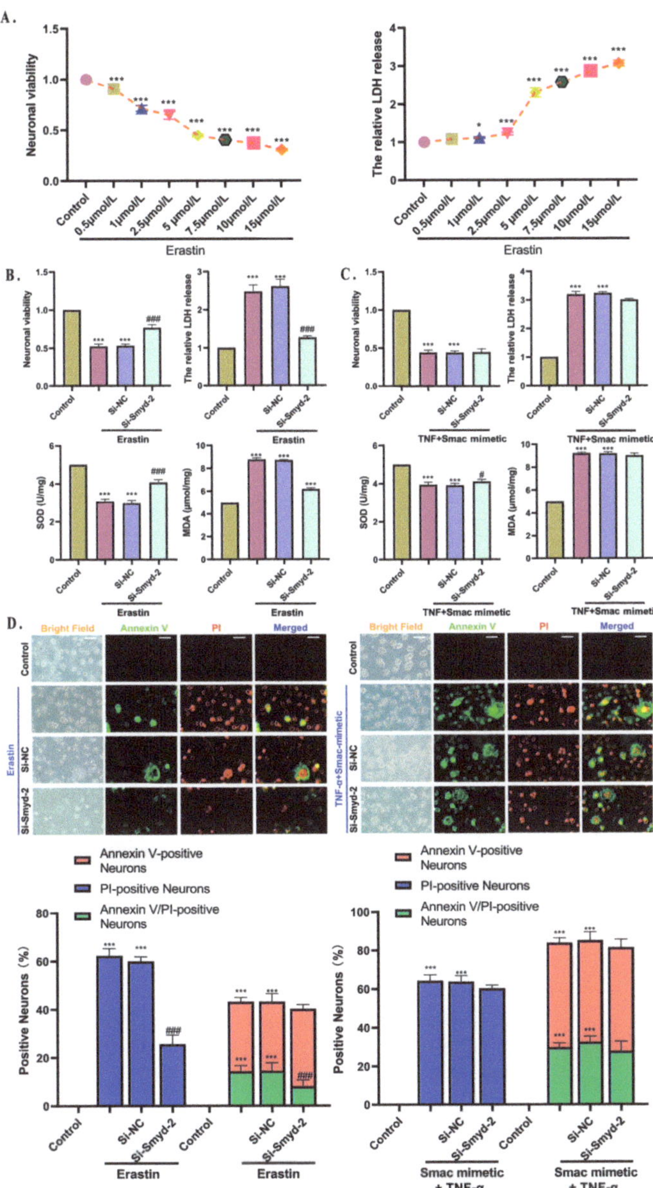

Figure 5. The effect of Smyd-2 overexpression on programmed cell death of HT-22 cells induced by Erastin and Smac mimetic + TNF-α. (**A**) The neuronal viability and LDH release of HT-22 cells challenged with Erastin. (**B,C**) The neuronal viability and LDH release of HT-22 cells challenged with Erastin. (c,d,e,f,g,h,i,j) The neuronal viability, LDH release, SOD level, and MDA level of Si-Smyd-2 HT-22 cells challenged with Erastin and Smac mimetic ($n = 6$). (**D**) Annexin V/PI staining was applied to observe and analyze the effect of Erastin and Smac mimetic on different types of programmed cell death in HT-22 cells induced by OGD/R. The representative images were obtained with an optical microscope at 400× magnification; scale bars = 50 μm ($n = 6$). *** $p < 0.0001$ vs. control group, * $p < 0.05$ vs. control group; ### $p < 0.0001$ vs. OGD/R group and OGD/R + Si-NC group, # $p < 0.05$ vs. OGD/R + Si-NC group.

3.6. The Neuroprotective Effect of Smyd-2 Knockout in OGD/R Can Be Abolished by the Ferroptosis Agonist Erastin

To further investigate the mechanisms of Smyd-2 in OGD/R-induced hippocampal neuronal ferroptosis, the ferroptosis agonist Erastin (5 μMol/L) was applied to incubate HT-22 cells for 12 h, beginning at the 12th hour after OGD. The experimental results demonstrated that Erastin reversed the antioxidant effects conferred by Si-Smyd-2, as indicated by the recovery of MDA content and LDH release and a reduction in SOD content, GSH, and neuronal viability in HT-22 cells after OGD/R (Figure 6A,B). We then verified the mitochondrial morphology through TEM: the analysis demonstrated that Si-Smyd-2 restored the OGD/R-damaged mitochondrial structure in HT-22 cells, which was primarily manifested as reduced irregular swelling, improved mitochondrial membrane potential, and a more distinct crista structure. However, the maintenance of regular mitochondrial morphology by Si-Smyd-2 was significantly reversed by the application of Erastin (Figure 6C). Finally, Western blot analysis was performed to explore the specific mechanisms by which Erastin counteracts the anti-ferroptosis effects associated with Smyd-2 knockout. The results showed that Erastin impaired the lipid membrane repair system restored by Si-Smyd-2 in HT-22 cells after OGD/R. This impairment was reflected in the downregulation of glutathione reduction system proteins GPX-4 and SLC7A11 and the decreased expression of anti-lipid-peroxidation-related proteins FTH-1, Nrf-2 (total protein), p-Nrf-2, and nucleus Nrf-2, as well as the upregulation of the lipid metabolic enzyme ACSL-4, which converts polyunsaturated fatty acids (PUFAs) into the ferroptosis substrate PUFA–phospholipids (PUFA-PLs) (Figure 6D). Nrf-2 transcriptionally regulates FTH-1 and SLC7A11; it seems that Nrf-2 is the most likely downstream regulatory candidate for Smyd-2. Exceeding our expectations, the expression of Smyd-2 and Keap-1 was unaffected by the Erastin treatment in the Si-Smyd-2 HT-22 cells after OGD/R. These findings indicate that Erastin effectively counteracted the anti-ferroptosis capacity conferred by Smyd-2 knockout in OGD/R-challenged HT-22 cells.

3.7. Smyd-2 Knockout Attenuates Neuronal Ferroptosis Induced by CIR/OGDR and Is Mainly Dependent on Nrf-2-Mediated Lipid Metabolic Detoxification

We confirmed that Smyd-2 knockout in hippocampal neurons showed robust resistance to CIR/OGDR-induced ferroptosis which was probably engaged in the activation of Nrf-2, but the patterns and mechanisms of how Smyd-2 interacts with Nrf-2 need further validation. ML-385 is a novel and specific inhibitor of Nrf-2, which directly binds to the Neh-1 domain of Nrf-2 to reduce its transcriptional activity. Similar to Erastin's pro-ferroptosis effects, 2 μmol/L ML-385 scaled up the LDH release and MDA content, leading to reduced viability, GSH content, and SOD levels in Si-Smyd-2 HT-22 cells after OGD/R (Figure 7A). Additionally, ML-385 reduced the amount of ferroptosis in HT-22 cells while ramping up lipid metabolites and ROS accumulation in the mitochondria and cytoplasm of Si-Smyd-2 HT-22 cells after OGD/R (Figure 7B,C,E). It was confirmed that the Nrf-2 inhibitor effectively abolished the antioxidant effect conferred by Smyd-2 knockout. Next, the ML-385 intervention altered the protein expression patterns in Si-Smyd-2 HT-22 cells after OGD/R, leading to a decreased expression of GPX-4, FTH-1, SLC7A11, p-Nrf-2, Nrf-2 (TP), and nucleus Nrf-2, without affecting the expression of Smyd-2. Meanwhile, it is remarkable that the ML-385 treatment decreased the expression of Keap-1, while Erastin did not have much effect on Keap-1 downregulation (Figure 7D).

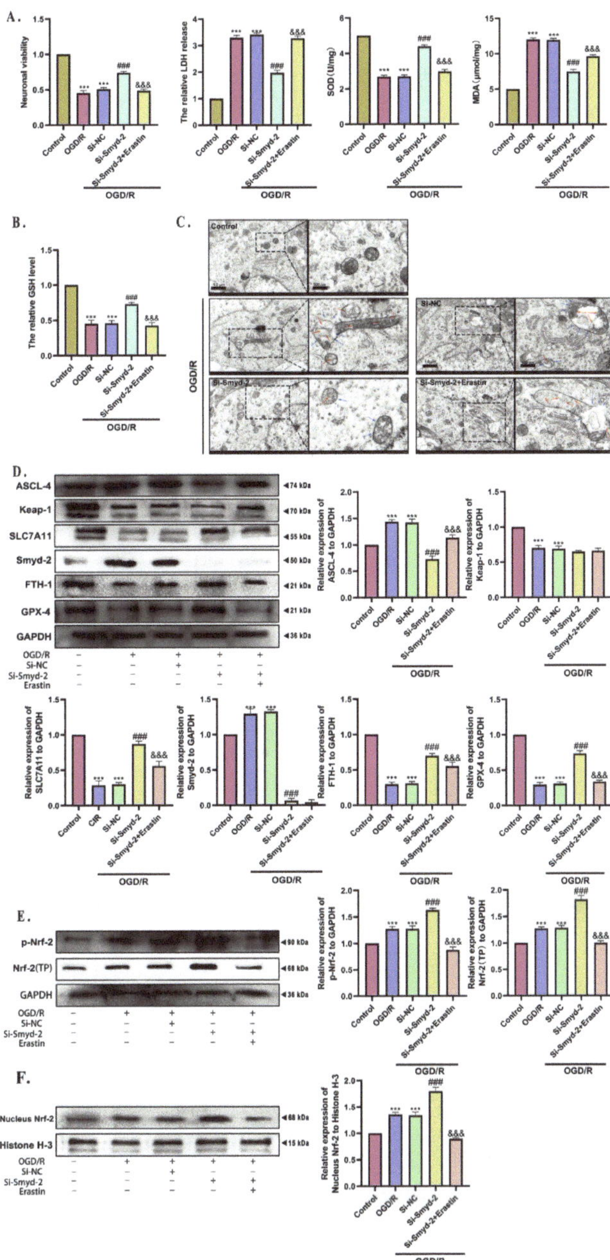

Figure 6. The effect of Si-Smyd-2 combined with Erastin on the ferroptosis of HT-22 cells induced by OGD/R. (**A**) The neuronal viability, LDH release, SOD, and MDA of HT-22 cells challenged with Si-Smyd-2 and Erastin after OGD/R. (**B**) GSH kit was applied to analyze and quantify the effect of the Si-RNA-mediated Smyd-2 and Erastin on HT-22 cell ferroptosis challenged with OGD/R ($n = 5$). (**C**) Pathophysiological and physiological morphologies of mitochondria in each HT-22 cell group

were observed by transmission electron microscopy. The red arrows mark the increased electron density of the matrix and fractured and vague cristae. The blue arrows mark vacuoles in mitochondria. The enlarged region bounded by a rectangular dotted box conduces to obtaining a more detailed view of the mitochondria for each experimental condition. The representative images were obtained with an optical microscope at 8k× magnification; scale bars = 1 μm. The representative enlarged images were obtained with an optical microscope at 20k× magnification; scale bars = 500 nm ($n = 5$). (**D–F**) Representative Western blots and quantitative evaluation of Smyd-2, SLC7A11, ACSL-4, FTH-1, GPX-4, Nrf-2, Keap-1, p-Nrf-2, and nucleus Nrf-2 expression levels in each HT-22 cell group. Data normalized to the loading control GAPDH are expressed as % of control ($n = 5$). *** $p < 0.0001$ vs. control group; ### $p < 0.0001$ vs. OGD/R group and OGD/R + Si-NC group; &&& $p < 0.0001$ vs. OGD/R + Si-Smyd-2 group.

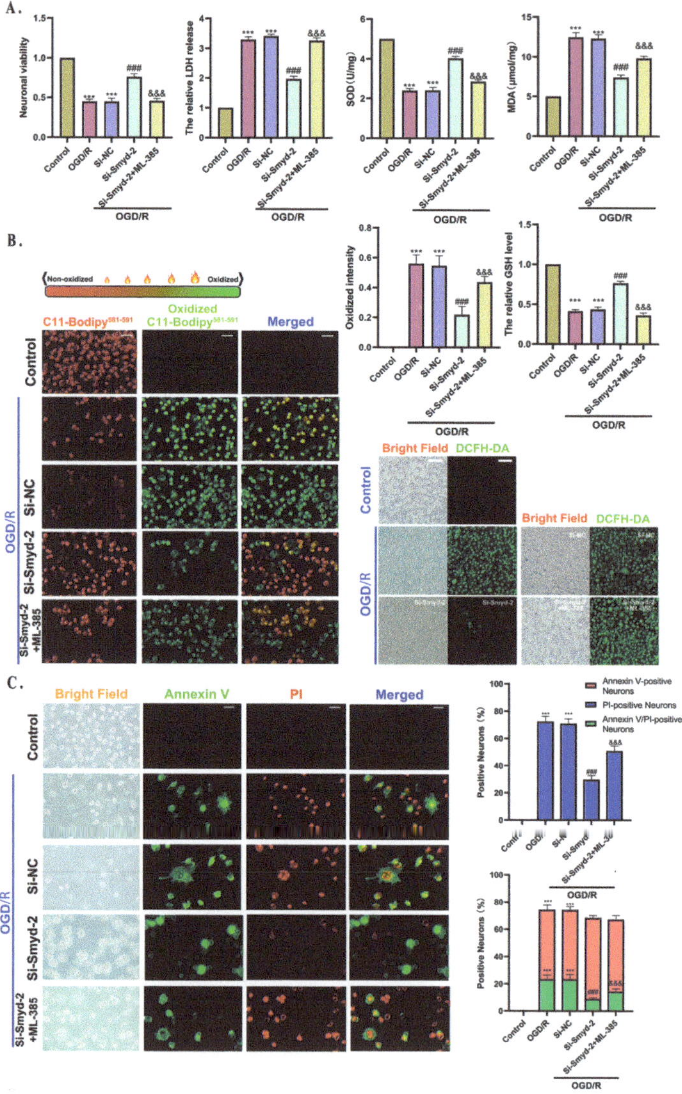

Figure 7. *Cont.*

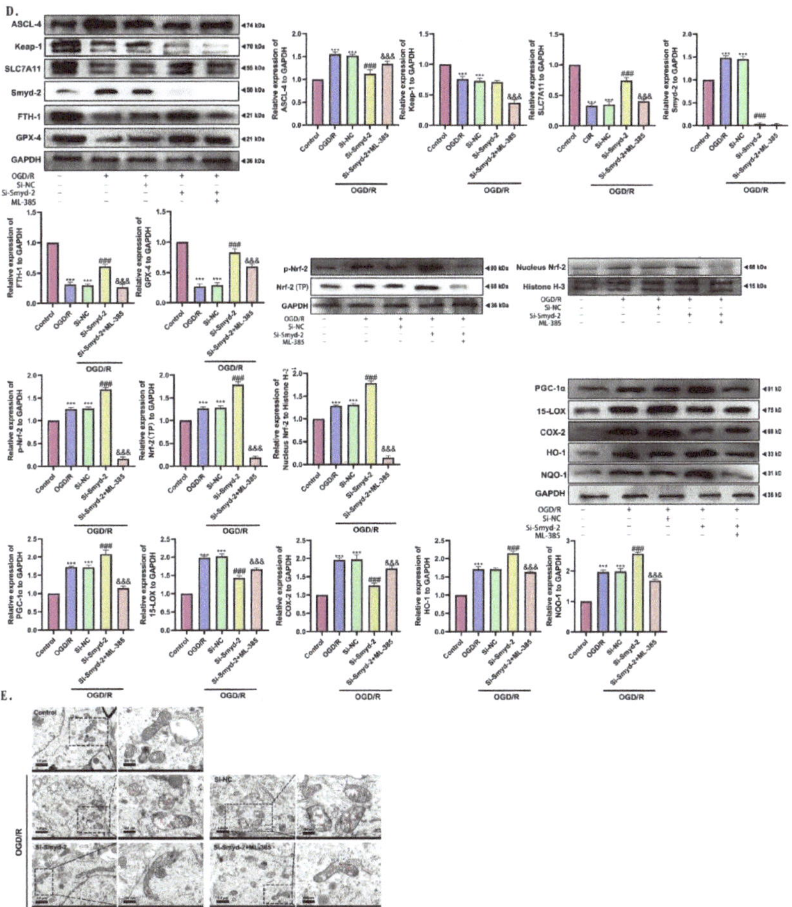

Figure 7. Effects of Smyd-2 (KO) combined with Nrf-2 inhibitor ML-385 on ferroptosis and lipid peroxidation in MCAO mice and OGD/R-induced HT-22 cells. (**A**) The neuronal viability, LDH release, SOD, and MDA of HT-22 cells were challenged with Si-Smyd-2 and ML-385 after OGD/R (n = 5). (**B**) BODIPY-581/591-C11 staining, GSH kit, and DCFH-DA staining were used to analyze and quantify the effect of Si-RNA-mediated Smyd-2 and ML-385 on lipid peroxidation of HT-22 cells after OGD/R. The representative images of BODIPY-581/591-C11 staining were obtained with an optical microscope at 400× magnification; scale bars = 50 μm (n = 5). The representative images of DCFH-DA staining were obtained with an optical microscope at 200× magnification; scale bars = 100 μm (n = 5). (**C**) Annexin V/PI double fluorescence staining was used to study the effect of Si-RNA-mediated Smyd-2 and ML-385 on a different form of programmed cell death in HT-22 cells induced by OGD/R. The representative images were obtained with a confocal microscope at 400× magnification; scale bars = 50 μm (n = 5). (**D**) The effect of Si-RNA-mediated Smyd-2 combined with ML-385 on ferroptosis-related proteins in HT-22 cells induced by OGD/R. Representative Western blots and quantitative evaluation of ACSL-4, Keap-1, SLC7A11, Smyd-2, FTH-1, GPX-4, p-Nrf-2, Nrf-2 (TP), nucleus Nrf-2, PGC-1α, COX-2, 15-LOX, NQO-1, and HO-1 expression levels in each HT-22 cell group. Data normalized to the loading control GAPDH are expressed as % of control (n = 5). (**E**) Pathophysiological and physiological morphologies of mitochondria in each HT-22 cell group were observed by TEM. The red arrows mark the increased electron density of the matrix and

fractured and vague cristae. The blue arrows mark vacuoles in mitochondria. The zoom region bounded by a rectangular dotted box allows a more detailed view of mitochondria for each experimental condition. The representative images were obtained with an optical microscope at 8k× magnification; scale bars = 1 μm (n = 5). The representative enlarged images were obtained with an optical microscope at 20k× magnification; scale bars = 500 nm. *** $p < 0.001$ vs. control group; ### $p < 0.001$ vs. OGD/R group and Si-NC + OGD/R group; &&& $p < 0.001$ vs. Si-Smyd-2 + OGD/R group.

3.8. Smyd-2 Methylates Nrf-2 (Lys-508) to Inhibit OGD/R-Induced Nrf-2 (Ser-40) Phosphorylation and Nuclear Translocation

The role of Smyd-2 in regulating the synthesis or degradation of Nrf-2 remains unclear. Firstly, we focused on whether Nrf-2 methylated by Smyd-2 induced Nrf-2 dissociation from the Nrf-2-Keap-1 system or promoted de novo Nrf-2 synthesis. Therefore, HT-22 cells treated with cycloheximide (CHX) (15 μmol/L) were used to inhibit new Nrf-2 synthesis. CHX directly binds to the site on the 80S subunit of the neuronal ribosome to interrupt the translocation of transfer RNA (tRNA), thereby inhibiting the synthesis of most proteins. Notably, Si-RNA-mediated Smyd-2 knockout increased the expression of Nrf-2, p-Nrf-2, and nucleus Nrf-2 after OGD/R, even in the presence of CHX (Figure 8A). This suggests that Smyd-2 has a minimal effect on Nrf-2 synthesis but may influence Nrf-2 degradation pathways.

Subsequently, we inferred that some Smyd-2-associated Nrf-2 degradation mechanisms drive the expression of Nrf-2, p-Nrf-2, and nucleus Nrf-2 induced by OGD/R. It is known that Protein Kinase C-δ (PKC-δ) phosphorylates Nrf-2 at serine residue 40 (Ser-40), promoting its nuclear translocation and preventing Keap-1-Cul-3 ubiquitin E-3 ligase from polyubiquitinating Nrf-2 for degradation. To further investigate this, we used DDO-7263, a potent Nrf-2-ARE activator, which blocks the assembly of the 26S proteasome, inhibiting the degradation of ubiquitinated Nrf-2 and promoting its compensatory upregulation. Laser confocal microscopy was employed to observe the nuclear translocation of Nrf-2 in OGD/R-induced HT-22 cells pretreated with 35 μMol/L DDO-7263. The fluorescence intensity of Smyd-2 and Nrf-2 remained low and localized within the boundaries under normal conditions. The OGD/R challenge boosted the fluorescence intensity of Smyd-2 and Nrf-2, which also manifested Smyd-2 anchored at the neuron cytoplasm and nuclei while activated Nrf-2 mainly functioned in the nuclei (Figure 8B). Smyd-2 overexpression blocked the OGD/R-induced nuclear translocation of Nrf-2, resulting in decreased nuclear expression (Figure 8B). However, DDO-7263 reduced the Nrf-2 nuclear translocation blocking effect of Smyd-2 (Figure 8B). A further investigation was conducted to rule out the possible mechanisms between Smyd-2 and PKC-δ within OGD/R. Western blot analysis showed that the adenovirus-mediated overexpression of Smyd-2 had no effect on the expression of PKC-δ and p-PKC-δ. Moreover, DDO-7263 upregulated the expression of p-Nrf-2, Nrf-2, and nucleus Nrf-2 in Smyd-2-overexpressing HT-22 cells after OGD/R, whereas the expression of Keap-1 and Smyd-2 remained unchanged (Figure 8C). Unexpectedly, DDO-7263 slightly decreased the levels of PKC-δ and p-PKC-δ in the Smyd-2-overexpressing HT-22 cell group induced by OGD/R, suggesting the presence of negative feedback regulation between Nrf-2 and PKC-δ (Figure 8C).

Western blot analysis was performed using an Nrf-2 antibody after immunoprecipitation with anti-methylated lysine in Si-RNA-mediated Smyd-2 knockout HT-22 cells. The results showed that the lysine methylation levels in the Nrf-2 Neh-1 domain increased after CIR/OGD but decreased in Si-RNA-mediated Smyd-2 knockout HT-22 cells after OGD/R (Figure 8D). Meanwhile, HDOCK software (V1.1) was used to set the protein as rigid, the docking contact site as the full surface, and the number of generated conformations to 100. The scoring function was used to select the conformations with the most negative energy, and PyMOL 2.1 software was used for visualization to simulate the Smyd-2-Nrf-2 docking process. The predicted docking site for Smyd-2 on Nrf-2 was the Lys-508 residue in the Neh-1 domain. Importantly, Smyd-2 methylated the Lys-508 residue of Nrf-2 to increase the steric hindrance and create a "cage effect" that inhibits the phosphorylation of Ser-40 on Nrf-2 and also "hijacked" Nrf-2 in the neuronal cytosol with the Keap-1 homodimer facilitating the ubiquitination and proteasomal degradation of Nrf-2, which may be one of the cytoplasmic Nrf-2 stabilizing mechanisms (Figure 8E).

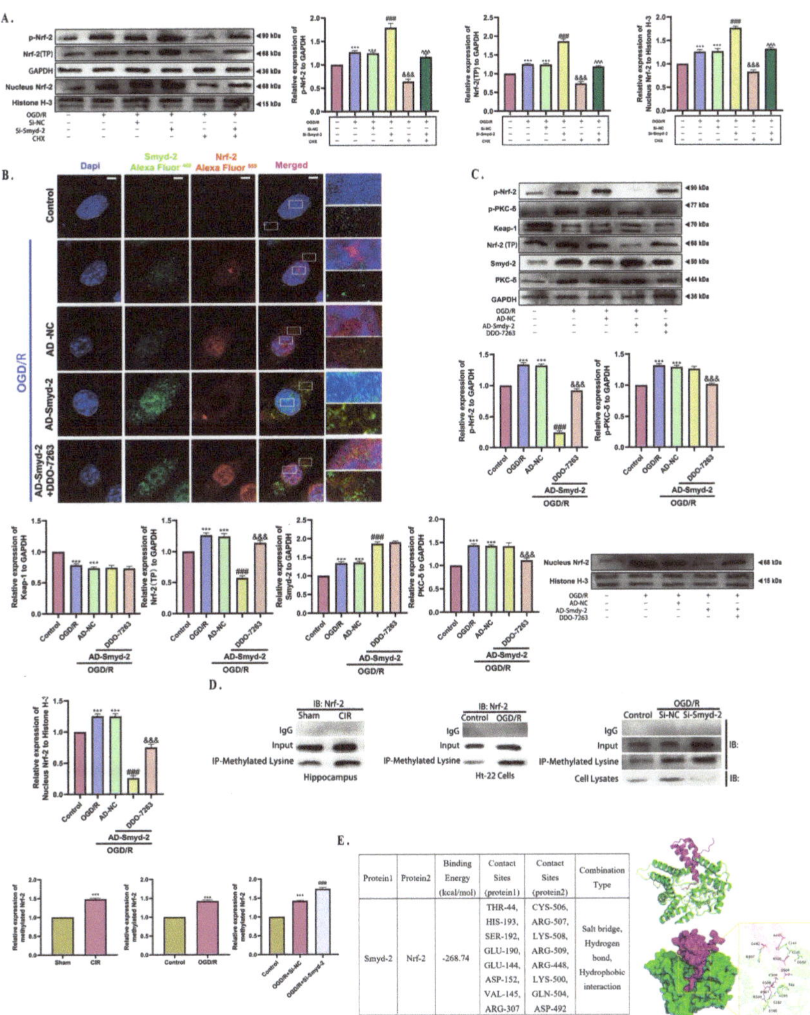

Figure 8. Smyd-2 methylates Nrf-2 (Lys-508) to inhibit OGD/R-induced Nrf-2 (Ser-40) phosphorylation and nuclear translocation. (**A**) Representative Western blot and quantitative evaluation of p-Nrf-2, Nrf-2 (TP), and nucleus Nrf-2 expression levels in each HT-22 cell group. Data normalized to the loading control GAPDH are expressed as % of control. Data normalized to the loading control GAPDH and histone H-3 are expressed as % of control (*n* = 5). (**B**) Confocal images of the localization with immunofluorescence-stained Smyd-2 (green) and immunofluorescence-stained Nrf-2 (red) in various HT-22 cell groups after OGD/R. The zoom region bounded by rectangular boxes represents Smyd-2-Nrf-2 binding in the cytoplasm of HT-22 cells, and the Nrf-2 was transported to the cell nucleus. Scale bars = 10 μm (*n* = 5). (**C**) Representative Western blots and quantitative evaluation of Nrf-2 (TP), p-Nrf-2, nucleus Nrf-2, PKC-δ, p-PKC-δ, Smyd-2, and Keap-1 expression levels in each HT-22 cell group. Data normalized to the loading control histone H-3 are expressed as % of GAPDH

($n = 5$). (**D**) Quantitative analysis of the expression of the methylation level of Nrf-2 ($n = 5$). (**E**) The possible docking sites of two target proteins, Smyd-2/Nrf-2. The binding mode of the complex Nrf-2 with Smyd-2. *** $p < 0.0001$ vs. control group; ### $p < 0.0001$ vs. OGD/R + Si-NC group; &&& $p < 0.0001$ vs. OGD/R group; ^^^ $p < 0.0001$ vs. OGD/R + CHX group. *** $p < 0.0001$ vs. control group; ### $p < 0.0001$ vs. OGD/R group and OGD/R + AD-Smyd-2 group; &&& $p < 0.0001$ vs. OGD/R + AD-Smyd-2 group.

The data above indicate that OGD/R promotes the phosphorylation and nuclear translocation of Nrf-2, while Smyd-2 methylates the Lys-508 residue of Nrf-2 to sequester it in the cytoplasm, thereby exacerbating the ferroptosis regulation during OGD/R.

4. Discussion

The pathological and physiological mechanisms underlying ischemic-stroke-induced hippocampal injury have been extensively explored and studied. Lipid peroxidation injury in hippocampal neurons has been identified as a key biomarker for ischemic stroke, accompanied by increased ROS accumulation and reactive nitrogen species, leading to PCD [45,46]. Another risk factor contributing to stroke progression is the disruption of ferrous iron (Fe^{2+}) homeostasis in hippocampal neurons. Fe^{2+} can catalyze the removal of hydrogen atoms from the long-chain double bonds of polyunsaturated fatty acids (PUFAs) by lipid peroxides (LPOs) through a Fenton reaction [47,48]. Hippocampal neurons are rich in polyunsaturated fatty acids that are particularly vulnerable to these LPOs, and neutralizing toxic LPOs consumes vast amounts of phase II enzymes, leaving the hippocampal neurons with dire antioxidant factor shortages [49–51]. CIR injury wreaks havoc with Fe^{2+} metabolic homeostasis, promotes lipid peroxidation, and ultimately leads to the ferroptosis of hippocampal neurons.

Our previous study showed that the Smyd-2 expression is significantly increased in the cerebral vascular endothelial cells of CIR model mice, suggesting that Smyd-2 may be involved in the destruction of the blood–brain barrier (BBB) caused by CIR. To perfect the role of Smyd-2 in the vast region of the hippocampus that coordinates with the neocortex, we further explored the expression changes and possible mechanisms of Smyd-2 in the hippocampus and cortex by constructing the same mouse model of CIR.

The ischemic phase of cerebral ischemia/reperfusion (CIR) establishes the foundation for neuronal pathology during reperfusion. Upon reperfusion, the restoration of oxygen and glucose supplies generates free radicals, causing significant lipid peroxidation in neuronal membranes [52]. Additionally, CIR induces intracellular calcium overload and releases excitatory glutamate into the synaptic cleft, leading to neurotoxic accumulation in postsynaptic neurons and disruptions in neuronal microcirculation [53,54]. Due to the hippocampal neuron HT-22 cell line's sensitivity to excitatory glutamate, we selected OGD/R as the preferred in vitro model to simulate CIR [55,56]. Histone methyltransferases, such as mixed-lineage leukemia-2 (MLL2), play critical roles in hippocampal neuroplasticity and neuronal homeostasis. The conditional deletion of MLL2 in mouse hippocampal pyramidal neurons leads to hippocampal-dependent memory loss and severe synaptic plasticity defects, highlighting the close link between these enzymes and cognitive processes [57]. Additionally, Chisholm et al. reported age-related differences in non-histone methylation in an MCAO rat model: histone H3K4 methylation was more abundant than H3K9 in adult astrocytes, while this pattern was reversed in aged rats. Notably, older rats had worse post-MCAO outcomes, potentially due to age-related modifications in astrocyte methylation [58].

In our study, we observed Smyd-2 expression primarily in hippocampal neurons following CIR, with limited expression in the cortex. Notably, Smyd-2 mRNA and protein levels increased in the hippocampus in a time-dependent manner, peaking 24 h after reperfusion, aligning with previous findings that Smyd-2 expression increases in neurovascular endothelial cells in the BBB during OGD/R [36]. However, the expression of Smyd-2 did not remain consistent across models. Unlike in the in vivo results, Smyd-2 mRNA in HT-22 cells stabilized from 24 to 72 h after OGD/R, whereas its hippocampal expression in mice declined after 24 h of reperfusion. This difference may result from the temporal gap between transcription and translation or differences in mRNA degradation timelines in mouse neurons versus HT-22 cells. Our findings suggest that Smyd-2 expression varies across brain regions and experimental conditions in response to CIR. Similarly, its expression patterns differ under various pathological conditions: Smyd-2 is downregulated in neonatal rat myocardial cell injury models induced by cobalt chloride, affecting p53 methylation and stability and promoting myocardial apoptosis [59]. Likewise, in cardiac myocytes (H9c2), Smyd-2 downregulation disrupts myofibril integrity under ROS stimulation [60]. Thus, the upregulation of Smyd-2 in CIR may offer valuable insights into the molecular mechanisms underlying ischemic stroke pathogenesis.

To investigate Smyd-2's role in CIR, we administered lateral intra-cerebroventricular injections of Smyd-2 adeno-associated virus in mice. In the Smyd-2 knockout group, we observed significant reductions in the cerebral infarction volume and histopathological damage, the absence of hippocampal ferrous ion deposition, and a notable improvement in neurobehavioral scores following CIR. Conversely, in the Smyd-2 overexpression group, the cerebral infarction volume increased markedly, the neurobehavioral scores declined, and there was extensive neuronal damage and diffuse ferrous ion deposition in the hippocampus and cortex. These results suggest that Smyd-2 exacerbates CIR-induced neuronal impairment by promoting ferrous ion accumulation in the hippocampus.

In vitro experiments provided further evidence of Smyd-2's role. Smyd-2 knockout via adenovirus-delivered siRNA significantly enhanced HT-22 cell viability, reduced mortality, and improved superoxide radical scavenging after OGD/R. In contrast, Smyd-2 overexpression in OGD/R-induced HT-22 cells increased cell mortality, elevated lipid peroxidation, and disrupted redox balance. We speculate that Smyd-2 overexpression compromises anti-lipid peroxidation defenses, leading to the accumulation of toxic lipid peroxides in neurons subjected to OGD/R. This accumulation depletes unsaturated fatty acids, causing brittleness in the neuronal membrane's lipid bilayer, ultimately reducing membrane fluidity and permeability and resulting in neuronal degeneration and death.

We focus on Smyd-2 knockout increasing the content of GSH, a potent inhibitor of the peroxidation of phospholipids in OGD/R-challenged HT-22 cells. GSH, acting as an auxiliary reductant, facilitates the reduction of strongly oxidizing phospholipid hydroperoxides (PLOOHs) to non-toxic PLOOHs in neurons. This anti-reducing detoxification process is mainly associated with glutathione reductase GPX-4 [61]. GPX-4, an emerging ferroptosis regulator, neutralizes excess intracellular lipid peroxides, preserving lipid bilayer homeostasis [13,62–64]. Our immunofluorescence staining showed a negative correlation between Smyd-2 and GPX-4 expression in OGD/R-treated HT-22 cells, indicating that Smyd-2 upregulation may be associated with increased ferroptosis.

In vivo experiments further demonstrated that Smyd-2 (KO) significantly reduces CIR-induced hippocampal ferroptosis by reversing the downregulation of GPX-4, FTH-1, and SLC7A11 while decreasing ACSL-4 expression. FTH-1 has been proven to inhibit ferroptosis through ferritinophagy in the 6-OHDA-induced Parkinson's Disease (PD) model [65]. SLC7A11 is an essential catalytic subunit of the cystine–glutamate transport system (System XC-). Meanwhile, ischemic stroke produces mass thrombin, simultaneously aggravates calcium overload in the neurovascular unit (NVU), and promotes the phosphorylation of Cytoplasmic Phospholipase A-2 (cPLA-2) to release Arachidonic Acid (AA) [66,67]. ACSL-4 then acetylates AA, incorporating it into phosphatidylethanolamine (PE) and phosphatidylcholine (PC), increasing the susceptibility to lipid peroxidation and thereby driving ferroptosis [68–70]. Interestingly, the transcription factor Nrf-2, known for its role in oxidative stress resistance, is upregulated after CIR. The differential expression of Nrf-2 and other anti-ferroptosis proteins in OGD/R-induced HT-22 cells likely reflects variations in the "regulatory time window" of oxidative stress defense mechanisms. Thus, Smyd-2 knockout may alleviate neuronal injury by regulating ferroptosis-related peroxidation reactions and restoring neuronal membrane fluidity.

Another intriguing finding of our research is that Smyd-2 knockout via adenovirus-delivered siRNA in HT-22 cells primarily inhibits Erastin-induced ferroptosis but does not significantly affect tumor necrosis factor-α (TNF-α)/Smac-mimetic-induced apoptosis or necroptosis. Interestingly, although SOD levels correlated well with MDA, we observed a slight increase in SOD content in Smyd-2 knockout HT-22 cells under TNF-α/Smac mimetic induction. This suggests the involvement of other antioxidant stress pathways or a potential temporal lag in lipid oxidation induction relative to ROS removal in this context. A potential alternative explanation for this phenomenon comes from a study by Dhanushka et al., which revealed that ROS can deplete glutathione and oxidize the Cys-13 residue of Smyd-2. This oxidation disrupts Smyd-2's interactions with Hsp90 and N2A, leading to N2A degradation by MMP-2 and calpain-1, ultimately compromising muscle fiber integrity [60]. These findings imply that Smyd-2 may be negatively influenced by ROS accumulation, suggesting the presence of feedforward and feedback loops involving Smyd-2 in ferroptosis pathways.

Interestingly, Smyd-2 has been shown to promote tumor growth in Ovarian Clear-Cell Carcinoma (OCCC) and Colorectal Cancer (CRC) by inhibiting TNF-induced apoptosis [71,72]. Although these findings differ from our observations, this discrepancy may stem from differences in the activation of pro-death or pro-survival signaling pathways between HT-22 cells and tumor cells under TNF-α/Smac mimetic induction. Moreover, in other ischemic diseases, Smyd-2 inhibits p53-dependent cardiomyocyte death by methylating lysine 382 on p53, thereby reducing its transcriptional activity [59]. Thus, it is evident that in different diseases, non-histone trans-methylases can methylate different sites or poly-methylate the same site, forming complex signaling networks to regulate different types of PCD. Indeed, the prooxidative role of Smyd-2 in other pathological conditions induced neuronal apoptosis model and neuronal necroptosis model remains to be further investigated and established.

Erastin was discovered initially as a small-molecule compound that induces ST and RASV-12 tumor cell death and has since been widely used as a ferroptosis-inducing drug in vitro [73,74]. Unlike other ferroptosis inducers, Erastin mediates ferroptosis in multiple ways, including acting on voltage-dependent anion channels to increase mitochondrial outer membrane permeability, activating ACSL-4 to accelerate the process of lipid peroxidation, reducing cysteine uptake by directly crashing System XC^-, and upregulating the expression of GSK-3β to decrease Nrf-2 expression in the nucleus [74–76]. It has been shown that Erastin-induced ferroptosis in HT-22 cells can be blocked by Zileuton, a LOX-5 inhibitor, via a reduction in cytoplasmic ROS accumulation [77]. We demonstrated that the anti-lipid peroxidation ability and mitochondrial protection conferred by Smyd-2 knockout in OGD/R-damaged HT-22 cells could be reversed by Erastin. Thus, Smyd-2 is a crucial regulator with a critical adjusting function in the processes of neuronal ferroptosis.

Triggering neuronal ferroptosis requires both "conditions" and "opportunities". The "conditions" are the formation of PUFA-PL and an increment in the labile iron pool (LIP), both of which provide prooxidant iron components for the endogenous Fenton reaction and induce lipid metabolism disorders in the mitochondria [78,79]. The "opportunities" refer to the incapacity of intracellular anti-lipid peroxidation systems, including the previously mentioned cystine–glutamate (System XC^-), Nrf-2/Keap-1, FSP-1–CoQH-2, and DHODH–CoQH-2 systems. Nrf-2/Keap-1 mainly acts in the cytoplasm and mitochondria [80–82]. Our experiments show that Smyd-2 knockout in OGD/R-induced HT-22 cells displays an extraordinary upregulation of Nrf-2, phosphorylated Nrf-2 (p-Nrf-2), and nucleus Nrf-2, also accompanied by deformed or absent mitochondria and neuronal degeneration. The results above indicate that there is likely a link between Smyd-2 and Nrf-2/Keap-1. ML-385 partially reversed the neuroprotective effect of Smyd-2 knockout in the CIR-created hippocampus lesions, characterized by the significantly increased volume of cerebral infarction, more obvious motor impairment, more outstanding cognitive dysfunction, and more heavy ferrous iron accumulation in the hippocampus. Furthermore, we found that the methylation level of lysine residues on Nrf-2 significantly increased after CIR. The regulatory mechanism of Smyd-2 in CIR involves the Nrf-2-Keap-1-ARE pathway. The administration of the Nrf-2 inhibitor ML-385 decreases the expression of anti-ferroptosis proteins in Smyd-2 knockout hippocampus after CIR, including Nrf-2, p-Nrf-2, nuclear Nrf-2, GPX-4, FTH-1, SLC7A11, and NQO-1, while increasing the expression of ferroptosis-related oxidative stress factors 15-LOX and COX-2.

The Nrf-2-Keap-1 pathway plays a central role in maintaining neuronal redox homeostasis and regulating the antioxidative stress response during CIR [83–87]. Under normal conditions, Nrf-2 levels are kept low by forming a complex with Keap-1-Cul-3-E3 ubiquitin ligase, facilitating continuous ubiquitin-mediated degradation. Keap-1 also functions as a cysteine-based sensor for electrophilic reagents and serves as a target for drugs that promote Nrf-2 dissociation. OGD/R stimulation inhibits the proteasomal degradation of Nrf-2, thereby increasing its abundance to counteract LPO [22,23,88]. Notably, Nrf-2 regulates a substantial number of genes involved in oxidative stress defense, with 244 of its 645 basal target genes and 654 inducible direct target genes counteracting LPO [89]. Additionally, Nrf-2 drives the antioxidant responsive element (ARE)-mediated expression of phase II detoxifying enzymes, including HO-1, COX-2, NQO-1, and 15-LOX [90–93].

Our data indicate that the hippocampal antioxidative stress response is activated after CIR. Smyd-2 knockout in hippocampal neurons further enhances this response, significantly improving both antioxidative capacity and neuronal morphology. Specifically, Smyd-2 knockout favors the Nrf-2-Keap-1-ARE pathway while inhibiting the Nrf-2-Keap-1–ubiquitin pathway, which promotes Nrf-2 degradation via the E3 ubiquitin ligase. By disrupting the "Hinge-and-Latch" model of Nrf-2-Keap-1, Smyd-2 knockout allows more Nrf-2 to escape ubiquitin degradation. Interestingly, Smyd-2 knockout does not affect Keap-1 expression after CIR. This may be attributed to the slower reaction of Keap-1 to electrophiles in the short term. Studies suggest that Keap-1 downregulates its own expression via autophagic degradation mechanisms during prolonged oxidative stress, allowing for sustained Nrf-2 activation [21,94,95]. Despite its critical role in oxidative stress, Keap-1 has limited therapeutic significance in acute disease models such as ischemic stroke, largely due to its bio-refractory properties and long half-life (e.g., 12.7 h in HepG-2 cells) [96,97]. However, these characteristics make Keap-1 a more suitable target for chronic diseases like cancer.

The PTMs of Nrf-2 through phosphorylation or acetylation are adaptable in CIR treatment, which is the essential prerequisite for the nuclear translocation and function of Nrf-2 [98,99]. Keratinocyte growth factor (KGF) phosphorylates Nrf-2 at Ser-40 to promote its activation in endometrial cells [100]. Other cytoplasmic modifiers, including Protein Kinase C (PKC), Mitogen-Activated Protein Kinase (MAPK), Glycogen Synthase Kinase 3β (GSK-3β), and Phosphatidylinositol 3-Kinase (PI3K), also phosphorylate Nrf-2 at serine or threonine residues [101–103]. Even within the nucleus, GSK-3β regulates excessive Nrf-2 levels by phosphorylating it, allowing recognition by β-TrCP, which forms an SCFβ-TRCP E3 ubiquitin ligase complex for proteasomal degradation [104]. This regulation shares similarities with Keap-1-mediated degradation; however, the threshold and coordination mechanisms require further investigation. Therefore, exploring PTMs in different Nrf-2 domains offers great potential for understanding and targeting its regulatory mechanisms.

Our study showed that the adenovirus-mediated overexpression of Smyd-2 does not affect the nuclear translocation of Nrf-2 by downregulating the expression of PKC-δ and p-PKC-δ. Instead, Smyd-2 methylates lysine residues in the Neh-2 domain of Nrf-2, increasing the steric hindrance for phosphorylation at Ser-40, which blocks Nrf-2 phosphorylation signal transduction and inhibits its nuclear translocation. The methylation of Nrf-2 decreases the expression of total Nrf-2 protein, phosphorylated Nrf-2, and nucleus Nrf-2; that is, Smyd-2 stabilizes the forward transduction of the Nrf-2-Keap-1-CUL-3–ubiquitin signaling pathway, continuously promoting the ubiquitin-mediated degradation of Nrf-2 to keep the DGR domain of Keap-1 in an unsaturated state that can capture more newly translated Nrf-2 in the cytosols. Smyd-2 "hijacks" Nrf-2 by methylating it in the cytoplasm, leading to its enzymatic hydrolysis and the suppression of Nrf-2 nuclear translocation.

PKC phosphorylates Nrf-2, participating in nuclear transcriptional regulation, which upregulates the expression of HO-1 and NQO-1 while reducing the expression of 15-LOX and COX-2 [105,106]. Previous studies have demonstrated that the dissociation of Nrf-2 is due to the modification of the active cysteine residue of the electrophilic Keap-1 to reduce the activity of the KEAP-1-Cul-3-E3 ubiquitin ligase complex constantly. In contrast, the direct modification of Nrf-2 in the Hinge-and-Latch system is limited to PKC-associated phosphorylation pathways [107,108]. Unexpectedly, Smyd-2 methylates lysine residues between the DLG and ETGE motifs at the Neh-2 domain of Nrf-2 to block its phosphorylation activation. It also stabilizes Nrf-2 in the neuronal cytoplasm and promotes the targeted binding of the Keap1-Cul3-E3 ubiquitin ligase to lysine residues of Nrf-2. Ubiquitinated Nrf-2 is then delivered to the 26S proteasome for degradation. Significantly, AREs mediate the transcriptional induction of the Rbx-1, Cul-3, and Keap-1 genes [109]. This negative feedback regulation prevents the hyperactivation of the Nrf-2-ARE pathway to some extent, but it delays the recovery of the antioxidative stress function in hippocampal neurons following CIR-induced injury.

5. Conclusions

This paper provides a detailed description of how Smyd-2 methylates the LYS-508 residue of Nrf-2 to regulate its phosphorylation level, effectively "hijacking" Nrf-2 into the Nrf-2-Keap-1–ubiquitin degradation system. This process induces ferroptosis and lipid peroxidation in hippocampal neuronal membranes and mitochondria as a result of CIR. After ruling out the potential effects of Smyd-2 on TNF-α and Smac-mimetic-induced neuronal apoptosis and necroptosis, we further establish the crucial regulatory role of Smyd-2 in neuronal ferroptosis. Given the constitutive expression of Smyd-2 in hippocampal neurons and its quick-acting Nrf-2-dependent curative pathway, the regimen of Smyd-2-related compounds will be viable clinically against ischemic stroke.

Author Contributions: D.L.: review of the relevant literature, writing—original draft—data curation, formal analysis, investigation, methodology, validation, visualization. Y.Z.: writing—review and editing—supervision. All authors have read and agreed to the published version of this manuscript.

Funding: This research was funded by the Macau Science and Technology Development Fund (FDCT(0021/2020/AGJ, 0011/2020/A1, 0012/2021/AMJ, 003/2022/ALC, 0092/2022/A2)) and the National Natural Science Foundation of China (81973320).

Institutional Review Board Statement: The animal study protocol was approved by the Ethical Committee on Animal Experiments of the School of Pharmacy, Fudan University (protocol code 2018-03-YL-MYC-01 and date 1 March 2018).

Informed Consent Statement: Not applicable.

Data Availability Statement: All data needed to evaluate the conclusions in this paper are present in this paper and available on request.

Acknowledgments: We thank Zhonglian Cao for expert animal handling.

Conflicts of Interest: The authors declare no conflicts of interest.

References

1. Cipolla, M.J.; Liebeskind, D.S.; Chan, S.L. The importance of comorbidities in ischemic stroke: Impact of hypertension on the cerebral circulation. *J. Cereb. Blood Flow. Metab.* **2018**, *38*, 2129–2149. [CrossRef] [PubMed]
2. van Sloten, T.T.; Sedaghat, S.; Carnethon, M.R.; Launer, L.J.; Stehouwer, C.D.A. Cerebral microvascular complications of type 2 diabetes: Stroke, cognitive dysfunction, and depression. *Lancet Diabetes Endocrinol.* **2020**, *8*, 325–336. [CrossRef]
3. Krinock, M.J.; Singhal, N.S. Diabetes, stroke, and neuroresilience: Looking beyond hyperglycemia. *Ann. N. Y. Acad. Sci.* **2021**, *1495*, 78–98. [CrossRef]

4. Castro, P.; Azevedo, E.; Sorond, F. Cerebral Autoregulation in Stroke. *Curr. Atheroscler. Rep.* **2018**, *20*, 37. [CrossRef]

5. Maida, C.D.; Norrito, R.L.; Daidone, M.; Tuttolomondo, A.; Pinto, A. Neuroinflammatory Mechanisms in Ischemic Stroke: Focus on Cardioembolic Stroke, Background, and Therapeutic Approaches. *Int. J. Mol. Sci.* **2020**, *21*, 6454. [CrossRef] [PubMed]

6. Loubinoux, I.; Kronenberg, G.; Endres, M.; Schumann-Bard, P.; Freret, T.; Filipkowski, R.K.; Kaczmarek, L.; Popa-Wagner, A. Post-stroke depression: Mechanisms, translation and therapy. *J. Cell. Mol. Med.* **2012**, *16*, 1961–1969. [CrossRef] [PubMed]

7. Andrabi, S.S.; Parvez, S.; Tabassum, H. Ischemic stroke and mitochondria: Mechanisms and targets. *Protoplasma* **2020**, *257*, 335–343. [CrossRef]

8. Kawabori, M.; Shichinohe, H.; Kuroda, S.; Houkin, K. Clinical Trials of Stem Cell Therapy for Cerebral Ischemic Stroke. *Int. J. Mol. Sci.* **2020**, *21*, 7380. [CrossRef] [PubMed]

9. Fan, J.L.; Brassard, P.; Rickards, C.A.; Nogueira, R.C.; Nasr, N.; McBryde, F.D.; Fisher, J.P.; Tzeng, Y.C. Integrative cerebral blood flow regulation in ischemic stroke. *J. Cereb. Blood Flow. Metab.* **2022**, *42*, 387–403. [CrossRef] [PubMed]

10. Toyokuni, S.; Yanatori, I.; Kong, Y.; Zheng, H.; Motooka, Y.; Jiang, L. Ferroptosis at the crossroads of infection, aging and cancer. *Cancer Sci.* **2020**, *111*, 2665–2671. [CrossRef]

11. Zhang, R.; Lei, J.; Chen, L.; Wang, Y.; Yang, G.; Yin, Z.; Luo, L. gamma-Glutamylcysteine Exerts Neuroprotection Effects against Cerebral Ischemia/Reperfusion Injury through Inhibiting Lipid Peroxidation and Ferroptosis. *Antioxidants* **2022**, *11*, 1653. [CrossRef]

12. Cao, J.Y.; Dixon, S.J. Mechanisms of ferroptosis. *Cell. Mol. Life Sci.* **2016**, *73*, 2195–2209. [CrossRef]

13. Yang, W.S.; Stockwell, B.R. Ferroptosis: Death by Lipid Peroxidation. *Trends Cell Biol.* **2016**, *26*, 165–176. [CrossRef]

14. Forcina, G.C.; Dixon, S.J. GPX4 at the Crossroads of Lipid Homeostasis and Ferroptosis. *Proteomics* **2019**, *19*, e1800311. [CrossRef]

15. Loboda, A.; Damulewicz, M.; Pyza, E.; Jozkowicz, A.; Dulak, J. Role of Nrf2/HO-1 system in development, oxidative stress response and diseases: An evolutionarily conserved mechanism. *Cell. Mol. Life Sci.* **2016**, *73*, 3221–3247. [CrossRef] [PubMed]

16. Fan, Z.; Wirth, A.K.; Chen, D.; Wruck, C.J.; Rauh, M.; Buchfelder, M.; Savaskan, N. Nrf2-Keap1 pathway promotes cell proliferation and diminishes ferroptosis. *Oncogenesis* **2017**, *6*, e371. [CrossRef]

17. Karunatilleke, N.C.; Fast, C.S.; Ngo, V.; Brickenden, A.; Duennwald, M.L.; Konermann, L.; Choy, W.Y. Nrf2, the Major Regulator of the Cellular Oxidative Stress Response, is Partially Disordered. *Int. J. Mol. Sci.* **2021**, *22*, 7434. [CrossRef]

18. Wu, J.; Wang, H.; Tang, X. Rexinoid inhibits Nrf2-mediated transcription through retinoid X receptor alpha. *Biochem. Biophys. Res. Commun.* **2014**, *452*, 554–559. [CrossRef] [PubMed]

19. Abed, D.A.; Goldstein, M.; Albanyan, H.; Jin, H.; Hu, L. Discovery of direct inhibitors of Keap1-Nrf2 protein-protein interaction as potential therapeutic and preventive agents. *Acta Pharm. Sin. B* **2015**, *5*, 285–299. [CrossRef]

20. Unoki, T.; Akiyama, M.; Kumagai, Y.; Goncalves, F.M.; Farina, M.; da Rocha, J.B.T.; Aschner, M. Molecular Pathways Associated with Methylmercury-Induced Nrf2 Modulation. *Front. Genet.* **2018**, *9*, 373. [CrossRef] [PubMed]

21. Bellezza, I.; Giambanco, I.; Minelli, A.; Donato, R. Nrf2-Keap1 signaling in oxidative and reductive stress. *Biochim. Biophys. Acta Mol. Cell Res.* **2018**, *1865*, 721–733. [CrossRef]

22. Baird, L.; Yamamoto, M. The Molecular Mechanisms Regulating the KEAP1-NRF2 Pathway. *Mol. Cell Biol.* **2020**, *40*, e00099-20. [CrossRef] [PubMed]

23. Kopacz, A.; Kloska, D.; Forman, H.J.; Jozkowicz, A.; Grochot-Przeczek, A. Beyond repression of Nrf2: An update on Keap1. *Free Radic. Biol. Med.* **2020**, *157*, 63–74. [CrossRef]

24. Ma, Q. Role of nrf2 in oxidative stress and toxicity. *Annu. Rev. Pharmacol. Toxicol.* **2013**, *53*, 401–426. [CrossRef]

25. Tonelli, C.; Chio, I.I.C.; Tuveson, D.A. Transcriptional Regulation by Nrf2. *Antioxid. Redox Signal* **2018**, *29*, 1727–1745. [CrossRef] [PubMed]

26. Fonseca Lopez, F.; Miao, J.; Damjanovic, J.; Bischof, L.; Braun, M.B.; Ling, Y.; Hartmann, M.D.; Lin, Y.S.; Kritzer, J.A. Computational Prediction of Cyclic Peptide Structural Ensembles and Application to the Design of Keap1 Binders. *J. Chem. Inf. Model.* **2023**, *63*, 6925–6937. [CrossRef] [PubMed]

27. Liu, L.; Locascio, L.M.; Dore, S. Critical Role of Nrf2 in Experimental Ischemic Stroke. *Front. Pharmacol.* **2019**, *10*, 153. [CrossRef] [PubMed]

28. Yang, J.; Su, J.; Wan, F.; Yang, N.; Jiang, H.; Fang, M.; Xiao, H.; Wang, J.; Tang, J. Tissue kallikrein protects against ischemic stroke by suppressing TLR4/NF-kappaB and activating Nrf2 signaling pathway in rats. *Exp. Ther. Med.* **2017**, *14*, 1163–1170. [CrossRef] [PubMed]

29. Li, X.N.; Shang, N.Y.; Kang, Y.Y.; Sheng, N.; Lan, J.Q.; Tang, J.S.; Wu, L.; Zhang, J.L.; Peng, Y. Caffeic acid alleviates cerebral ischemic injury in rats by resisting ferroptosis via Nrf2 signaling pathway. *Acta Pharmacol. Sin.* **2023**, *45*, 248–267. [CrossRef]

30. Hwang, J.Y.; Aromolaran, K.A.; Zukin, R.S. The emerging field of epigenetics in neurodegeneration and neuroprotection. *Nat. Rev. Neurosci.* **2017**, *18*, 347–361. [CrossRef] [PubMed]

31. Narne, P.; Pandey, V.; Phanithi, P.B. Role of Nitric Oxide and Hydrogen Sulfide in Ischemic Stroke and the Emergent Epigenetic Underpinnings. *Mol. Neurobiol.* **2019**, *56*, 1749–1769. [CrossRef] [PubMed]

32. Yi, X.; Jiang, X.J.; Fang, Z.M. Histone methyltransferase SMYD2: Ubiquitous regulator of disease. *Clin. Epigenet.* **2019**, *11*, 112. [CrossRef]

33. Abu-Farha, M.; Lambert, J.P.; Al-Madhoun, A.S.; Elisma, F.; Skerjanc, I.S.; Figeys, D. The tale of two domains: Proteomics and genomics analysis of SMYD2, a new histone methyltransferase. *Mol. Cell. Proteom.* **2008**, *7*, 560–572. [CrossRef]

34. Rapanelli, M.; Tan, T.; Wang, W.; Wang, X.; Wang, Z.J.; Zhong, P.; Frick, L.; Qin, L.; Ma, K.; Qu, J.; et al. Behavioral, circuitry, and molecular aberrations by region-specific deficiency of the high-risk autism gene Cul3. *Mol. Psychiatry* **2021**, *26*, 1491–1504. [CrossRef] [PubMed]

35. Jarrell, D.K.; Hassell, K.N.; Crans, D.C.; Lanning, S.; Brown, M.A. Characterizing the Role of SMYD2 in Mammalian Embryogenesis-Future Directions. *Vet. Sci.* **2020**, *7*, 63. [CrossRef] [PubMed]

36. Wang, J.; Zhong, W.; Cheng, Q.; Xiao, C.; Xu, J.; Su, Z.; Su, H.; Liu, X. Histone methyltransferase Smyd2 contributes to blood-brain barrier breakdown in stroke. *Clin. Transl. Med.* **2022**, *12*, e761. [CrossRef] [PubMed]

37. Fluri, F.; Schuhmann, M.K.; Kleinschnitz, C. Animal models of ischemic stroke and their application in clinical research. *Drug Des. Dev. Ther.* **2015**, *9*, 3445–3454. [CrossRef]

38. Bouet, V.; Boulouard, M.; Toutain, J.; Divoux, D.; Bernaudin, M.; Schumann-Bard, P.; Freret, T. The adhesive removal test: A sensitive method to assess sensorimotor deficits in mice. *Nat. Protoc.* **2009**, *4*, 1560–1564. [CrossRef]

39. Liu, H.; Li, C.; Zhang, X.; Chen, H.; Zhang, Q.; Zeng, Y.; Zheng, S.; Zou, J.; Zhao, Y.; Zheng, X.; et al. BMSC-Exosomes attenuate ALP dysfunction by restoring lysosomal function via the mTOR/TFEB Axis to reduce cerebral ischemia-reperfusion injury. *Exp. Neurol.* **2024**, *376*, 114726. [CrossRef]

40. Qian, Y.; Yang, L.; Chen, J.; Zhou, C.; Zong, N.; Geng, Y.; Xia, S.; Yang, H.; Bao, X.; Chen, Y.; et al. SRGN amplifies microglia-mediated neuroinflammation and exacerbates ischemic brain injury. *J. Neuroinflamm.* **2024**, *21*, 35. [CrossRef]

41. Wang, R.; Li, L.; Wang, B. Poncirin ameliorates oxygen glucose deprivation/reperfusion injury in cortical neurons via inhibiting NOX4-mediated NLRP3 inflammasome activation. *Int. Immunopharmacol.* **2022**, *102*, 107210. [CrossRef] [PubMed]

42. MacDonald, M.L.; Murray, I.V.; Axelsen, P.H. Mass spectrometric analysis demonstrates that BODIPY 581/591 C11 overestimates and inhibits oxidative lipid damage. *Free Radic. Biol. Med.* **2007**, *42*, 1392–1397. [CrossRef]

43. Yoshida, Y.; Shimakawa, S.; Itoh, N.; Niki, E. Action of DCFH and BODIPY as a probe for radical oxidation in hydrophilic and lipophilic domain. *Free Radic. Res.* **2003**, *37*, 861–872. [CrossRef] [PubMed]

44. Gong, Y.; Luo, S.; Fan, P.; Zhu, H.; Li, Y.; Huang, W. Growth hormone activates PI3K/Akt signaling and inhibits ROS accumulation and apoptosis in granulosa cells of patients with polycystic ovary syndrome. *Reprod. Biol. Endocrinol.* **2020**, *18*, 121. [CrossRef] [PubMed]

45. Su, X.T.; Wang, L.; Ma, S.M.; Cao, Y.; Yang, N.N.; Lin, L.L.; Fisher, M.; Yang, J.W.; Liu, C.Z. Mechanisms of Acupuncture in the Regulation of Oxidative Stress in Treating Ischemic Stroke. *Oxidative Med. Cell Longev.* **2020**, *2020*, 7875396. [CrossRef] [PubMed]

46. Chen, H.; He, Y.; Chen, S.; Qi, S.; Shen, J. Therapeutic targets of oxidative/nitrosative stress and neuroinflammation in ischemic stroke: Applications for natural product efficacy with omics and systemic biology. *Pharmacol. Res.* **2020**, *158*, 104877. [CrossRef]

47. Li, X.N.; Yang, S.Q.; Li, M.; Li, X.S.; Tian, Q.; Xiao, F.; Tang, Y.Y.; Kang, X.; Wang, C.Y.; Zou, W.; et al. Formaldehyde induces ferroptosis in hippocampal neuronal cells by upregulation of the Warburg effect. *Toxicology* **2021**, *448*, 152650. [CrossRef]

48. Ding, H.; Wang, Z.; Song, W. CTRP3 protects hippocampal neurons from oxygen-glucose deprivation-induced injury through the AMPK/Nrf2/ARE pathway. *Hum. Exp. Toxicol.* **2021**, *40*, 1153–1162. [CrossRef]

49. Zhu, W.; Cui, G.; Li, T.; Chen, H.; Zhu, J.; Ding, Y.; Zhao, L. Docosahexaenoic Acid Protects Traumatic Brain Injury by Regulating NOX2 Generation via Nrf2 Signaling Pathway. *Neurochem. Res.* **2020**, *45*, 1839–1850. [CrossRef] [PubMed]

50. Zhao, X.; Li, S.; Mo, Y.; Li, R.; Huang, S.; Zhang, A.; Ni, X.; Dai, Q.; Wang, J. DCA Protects against Oxidation Injury Attributed to Cerebral Ischemia-Reperfusion by Regulating Glycolysis through PDK2-PDH-Nrf2 Axis. *Oxidative Med. Cell Longev.* **2021**, *2021*, 5173035. [CrossRef] [PubMed]

51. Park, J.S.; Saeed, K.; Jo, M.H.; Kim, M.W.; Lee, H.J.; Park, C.B.; Lee, G.; Kim, M.O. LDHB Deficiency Promotes Mitochondrial Dysfunction Mediated Oxidative Stress and Neurodegeneration in Adult Mouse Brain. *Antioxidants* **2022**, *11*, 261. [CrossRef]

52. Dong, X.; Zhang, Z.; Shu, X.; Zhuang, Z.; Liu, P.; Liu, R.; Xia, S.; Bao, X.; Xu, Y.; Chen, Y. MFG-E8 Alleviates Cognitive Impairments Induced by Chronic Cerebral Hypoperfusion by Phagocytosing Myelin Debris and Promoting Remyelination. *Neurosci. Bull.* **2024**, *40*, 483–499. [CrossRef]

53. Halestrap, A.P. Calcium, mitochondria and reperfusion injury: A pore way to die. *Biochem. Soc. Trans.* **2006**, *34*, 232–237. [CrossRef] [PubMed]

54. Li, C.; Sun, G.; Chen, B.; Xu, L.; Ye, Y.; He, J.; Bao, Z.; Zhao, P.; Miao, Z.; Zhao, L.; et al. Nuclear receptor coactivator 4-mediated ferritinophagy contributes to cerebral ischemia-induced ferroptosis in ischemic stroke. *Pharmacol. Res.* **2021**, *174*, 105933. [CrossRef]

55. Cimarosti, H.; Henley, J.M. Investigating the mechanisms underlying neuronal death in ischemia using in vitro oxygen-glucose deprivation: Potential involvement of protein SUMOylation. *Neuroscientist* **2008**, *14*, 626–636. [CrossRef] [PubMed]

56. Ryou, M.G.; Mallet, R.T. An In Vitro Oxygen-Glucose Deprivation Model for Studying Ischemia-Reperfusion Injury of Neuronal Cells. *Methods Mol. Biol.* **2018**, *1717*, 229–235. [CrossRef] [PubMed]

57. Kerimoglu, C.; Agis-Balboa, R.C.; Kranz, A.; Stilling, R.; Bahari-Javan, S.; Benito-Garagorri, E.; Halder, R.; Burkhardt, S.; Stewart, A.F.; Fischer, A. Histone-methyltransferase MLL2 (KMT2B) is required for memory formation in mice. *J. Neurosci.* **2013**, *33*, 3452–3464. [CrossRef] [PubMed]

58. Chisholm, N.C.; Henderson, M.L.; Selvamani, A.; Park, M.J.; Dindot, S.; Miranda, R.C.; Sohrabji, F. Histone methylation patterns in astrocytes are influenced by age following ischemia. *Epigenetics* **2015**, *10*, 142–152. [CrossRef]
59. Sajjad, A.; Novoyatleva, T.; Vergarajauregui, S.; Troidl, C.; Schermuly, R.T.; Tucker, H.O.; Engel, F.B. Lysine methyltransferase Smyd2 suppresses p53-dependent cardiomyocyte apoptosis. *Biochim. Biophys. Acta* **2014**, *1843*, 2556–2562. [CrossRef]
60. Munkanatta Godage, D.N.P.; VanHecke, G.C.; Samarasinghe, K.T.G.; Feng, H.Z.; Hiske, M.; Holcomb, J.; Yang, Z.; Jin, J.P.; Chung, C.S.; Ahn, Y.H. SMYD2 glutathionylation contributes to degradation of sarcomeric proteins. *Nat. Commun.* **2018**, *9*, 4341. [CrossRef]
61. Soria, F.N.; Perez-Samartin, A.; Martin, A.; Gona, K.B.; Llop, J.; Szczupak, B.; Chara, J.C.; Matute, C.; Domercq, M. Extrasynaptic glutamate release through cystine/glutamate antiporter contributes to ischemic damage. *J. Clin. Investig.* **2014**, *124*, 3645–3655. [CrossRef] [PubMed]
62. Chen, X.; Li, J.; Kang, R.; Klionsky, D.J.; Tang, D. Ferroptosis: Machinery and regulation. *Autophagy* **2021**, *17*, 2054–2081. [CrossRef] [PubMed]
63. Stockwell, B.R.; Jiang, X.; Gu, W. Emerging Mechanisms and Disease Relevance of Ferroptosis. *Trends Cell Biol.* **2020**, *30*, 478–490. [CrossRef]
64. Zheng, J.; Conrad, M. The Metabolic Underpinnings of Ferroptosis. *Cell Metab.* **2020**, *32*, 920–937. [CrossRef] [PubMed]
65. Tian, Y.; Lu, J.; Hao, X.; Li, H.; Zhang, G.; Liu, X.; Li, X.; Zhao, C.; Kuang, W.; Chen, D.; et al. FTH1 Inhibits Ferroptosis Through Ferritinophagy in the 6-OHDA Model of Parkinson's Disease. *Neurotherapeutics* **2020**, *17*, 1796–1812. [CrossRef] [PubMed]
66. Strokin, M.; Sergeeva, M.; Reiser, G. Docosahexaenoic acid and arachidonic acid release in rat brain astrocytes is mediated by two separate isoforms of phospholipase A2 and is differently regulated by cyclic AMP and Ca2+. *Br. J. Pharmacol.* **2003**, *139*, 1014–1022. [CrossRef]
67. Hartz, A.M.S.; Rempe, R.G.; Soldner, E.L.B.; Pekcec, A.; Schlichtiger, J.; Kryscio, R.; Bauer, B. Cytosolic phospholipase A2 is a key regulator of blood-brain barrier function in epilepsy. *FASEB J.* **2019**, *33*, 14281–14295. [CrossRef]
68. Kagan, V.E.; Mao, G.; Qu, F.; Angeli, J.P.; Doll, S.; Croix, C.S.; Dar, H.H.; Liu, B.; Tyurin, V.A.; Ritov, V.B.; et al. Oxidized arachidonic and adrenic PEs navigate cells to ferroptosis. *Nat. Chem. Biol.* **2017**, *13*, 81–90. [CrossRef]
69. He, S.; Li, R.; Peng, Y.; Wang, Z.; Huang, J.; Meng, H.; Min, J.; Wang, F.; Ma, Q. ACSL4 contributes to ferroptosis-mediated rhabdomyolysis in exertional heat stroke. *J. Cachexia Sarcopenia Muscle* **2022**, *13*, 1717–1730. [CrossRef]
70. Kuwata, H.; Nakatani, E.; Shimbara-Matsubayashi, S.; Ishikawa, F.; Shibanuma, M.; Sasaki, Y.; Yoda, E.; Nakatani, Y.; Hara, S. Long-chain acyl-CoA synthetase 4 participates in the formation of highly unsaturated fatty acid-containing phospholipids in murine macrophages. *Biochim. Biophys. Acta Mol. Cell Biol. Lipids* **2019**, *1864*, 1606–1618. [CrossRef] [PubMed]
71. Kojima, M.; Sone, K.; Oda, K.; Hamamoto, R.; Kaneko, S.; Oki, S.; Kukita, A.; Kawata, A.; Honjoh, H.; Kawata, Y.; et al. The histone methyltransferase SMYD2 is a novel therapeutic target for the induction of apoptosis in ovarian clear cell carcinoma cells. *Oncol. Lett.* **2020**, *20*, 153. [CrossRef]
72. Lai, Y.; Yang, Y. SMYD2 facilitates cancer cell malignancy and xenograft tumor development through ERBB2-mediated FUT4 expression in colon cancer. *Mol. Cell. Biochem.* **2020**, *477*, 2149–2159. [CrossRef] [PubMed]
73. Yang, Y.; Luo, M.; Zhang, K.; Zhang, J.; Gao, T.; Connell, D.O.; Yao, F.; Mu, C.; Cai, B.; Shang, Y.; et al. Nedd4 ubiquitylates VDAC2/3 to suppress erastin-induced ferroptosis in melanoma. *Nat. Commun.* **2020**, *11*, 433. [CrossRef]
74. Zhao, Y.; Li, Y.; Zhang, R.; Wang, F.; Wang, T.; Jiao, Y. The Role of Erastin in Ferroptosis and Its Prospects in Cancer Therapy. *Onco Targets Ther.* **2020**, *13*, 5429–5441. [CrossRef]
75. Yagoda, N.; von Rechenberg, M.; Zaganjor, E.; Bauer, A.J.; Yang, W.S.; Fridman, D.J.; Wolpaw, A.J.; Smukste, I.; Peltier, J.M.; Boniface, J.J.; et al. RAS-RAF-MEK-dependent oxidative cell death involving voltage-dependent anion channels. *Nature* **2007**, *447*, 864–868. [CrossRef] [PubMed]
76. Sato, M.; Kusumi, R.; Hamashima, S.; Kobayashi, S.; Sasaki, S.; Komiyama, Y.; Izumikawa, T.; Conrad, M.; Bannai, S.; Sato, H. The ferroptosis inducer erastin irreversibly inhibits system xc- and synergizes with cisplatin to increase cisplatin's cytotoxicity in cancer cells. *Sci. Rep.* **2018**, *8*, 968. [CrossRef]
77. Liu, Y.; Wang, W.; Li, Y.; Xiao, Y.; Cheng, J.; Jia, J. The 5-Lipoxygenase Inhibitor Zileuton Confers Neuroprotection against Glutamate Oxidative Damage by Inhibiting Ferroptosis. *Biol. Pharm. Bull.* **2015**, *38*, 1234–1239. [CrossRef]
78. Conrad, M.; Proneth, B. Selenium: Tracing Another Essential Element of Ferroptotic Cell Death. *Cell Chem. Biol.* **2020**, *27*, 409–419. [CrossRef] [PubMed]
79. Zou, Y.; Henry, W.S.; Ricq, E.L.; Graham, E.T.; Phadnis, V.V.; Maretich, P.; Paradkar, S.; Boehnke, N.; Deik, A.A.; Reinhardt, F.; et al. Plasticity of ether lipids promotes ferroptosis susceptibility and evasion. *Nature* **2020**, *585*, 603–608. [CrossRef] [PubMed]
80. Liu, M.; Kong, X.Y.; Yao, Y.; Wang, X.A.; Yang, W.; Wu, H.; Li, S.; Ding, J.W.; Yang, J. The critical role and molecular mechanisms of ferroptosis in antioxidant systems: A narrative review. *Ann. Transl. Med.* **2022**, *10*, 368. [CrossRef]
81. Bridges, R.J.; Natale, N.R.; Patel, S.A. System xc(-) cystine/glutamate antiporter: An update on molecular pharmacology and roles within the CNS. *Br. J. Pharmacol.* **2012**, *165*, 20–34. [CrossRef]
82. Lei, G.; Mao, C.; Yan, Y.; Zhuang, L.; Gan, B. Ferroptosis, radiotherapy, and combination therapeutic strategies. *Protein Cell* **2021**, *12*, 836–857. [CrossRef] [PubMed]

83. Al-Mubarak, B.R.; Bell, K.F.S.; Chowdhry, S.; Meakin, P.J.; Baxter, P.S.; McKay, S.; Dando, O.; Ashford, M.L.J.; Gazaryan, I.; Hayes, J.D.; et al. Non-canonical Keap1-independent activation of Nrf2 in astrocytes by mild oxidative stress. *Redox Biol.* **2021**, *47*, 102158. [CrossRef]

84. Zhang, T.; Wu, P.; Budbazar, E.; Zhu, Q.; Sun, C.; Mo, J.; Peng, J.; Gospodarev, V.; Tang, J.; Shi, H.; et al. Mitophagy Reduces Oxidative Stress Via Keap1 (Kelch-Like Epichlorohydrin-Associated Protein 1)/Nrf2 (Nuclear Factor-E2-Related Factor 2)/PHB2 (Prohibitin 2) Pathway After Subarachnoid Hemorrhage in Rats. *Stroke* **2019**, *50*, 978–988. [CrossRef]

85. Alfieri, A.; Srivastava, S.; Siow, R.C.; Modo, M.; Fraser, P.A.; Mann, G.E. Targeting the Nrf2-Keap1 antioxidant defence pathway for neurovascular protection in stroke. *J. Physiol.* **2011**, *589*, 4125–4136. [CrossRef]

86. Zhang, X.; Wu, Q.; Wang, Z.; Li, H.; Dai, J. Keap1-Nrf2/ARE signal pathway activated by butylphthalide in the treatment of ischemic stroke. *Am. J. Transl. Res.* **2022**, *14*, 2637–2646. [PubMed]

87. Chu, S.F.; Zhang, Z.; Zhou, X.; He, W.B.; Chen, C.; Luo, P.; Liu, D.D.; Ai, Q.D.; Gong, H.F.; Wang, Z.Z.; et al. Ginsenoside Rg1 protects against ischemic/reperfusion-induced neuronal injury through miR-144/Nrf2/ARE pathway. *Acta Pharmacol. Sin.* **2019**, *40*, 13–25. [CrossRef] [PubMed]

88. Suzuki, T.; Yamamoto, M. Molecular basis of the Keap1-Nrf2 system. *Free Radic. Biol. Med.* **2015**, *88*, 93–100. [CrossRef] [PubMed]

89. Malhotra, D.; Portales-Casamar, E.; Singh, A.; Srivastava, S.; Arenillas, D.; Happel, C.; Shyr, C.; Wakabayashi, N.; Kensler, T.W.; Wasserman, W.W.; et al. Global mapping of binding sites for Nrf2 identifies novel targets in cell survival response through ChIP-Seq profiling and network analysis. *Nucleic Acids Res.* **2010**, *38*, 5718–5734. [CrossRef]

90. Wu, Y.; Zhao, Y.; Yang, H.Z.; Wang, Y.J.; Chen, Y. HMGB1 regulates ferroptosis through Nrf2 pathway in mesangial cells in response to high glucose. *Biosci. Rep.* **2021**, *41*, BSR20202924. [CrossRef]

91. Subedi, L.; Lee, J.H.; Yumnam, S.; Ji, E.; Kim, S.Y. Anti-Inflammatory Effect of Sulforaphane on LPS-Activated Microglia Potentially through JNK/AP-1/NF-kappaB Inhibition and Nrf2/HO-1 Activation. *Cells* **2019**, *8*, 194. [CrossRef] [PubMed]

92. Banning, A.; Brigelius-Flohe, R. NF-kappaB, Nrf2, and HO-1 interplay in redox-regulated VCAM-1 expression. *Antioxid. Redox Signal* **2005**, *7*, 889–899. [CrossRef]

93. Abu-Halaka, D.; Shpaizer, A.; Zeigerman, H.; Kanner, J.; Tirosh, O. DMF-Activated Nrf2 Ameliorates Palmitic Acid Toxicity While Potentiates Ferroptosis Mediated Cell Death: Protective Role of the NO-Donor S-Nitroso-N-Acetylcysteine. *Antioxidants* **2023**, *12*, 512. [CrossRef] [PubMed]

94. Ulasov, A.V.; Rosenkranz, A.A.; Georgiev, G.P.; Sobolev, A.S. Nrf2/Keap1/ARE signaling: Towards specific regulation. *Life Sci.* **2022**, *291*, 120111. [CrossRef] [PubMed]

95. Ichimura, Y.; Waguri, S.; Sou, Y.S.; Kageyama, S.; Hasegawa, J.; Ishimura, R.; Saito, T.; Yang, Y.; Kouno, T.; Fukutomi, T.; et al. Phosphorylation of p62 activates the Keap1-Nrf2 pathway during selective autophagy. *Mol. Cell* **2013**, *51*, 618–631. [CrossRef] [PubMed]

96. Ding, X.; Jian, T.; Wu, Y.; Zuo, Y.; Li, J.; Lv, H.; Ma, L.; Ren, B.; Zhao, L.; Li, W.; et al. Ellagic acid ameliorates oxidative stress and insulin resistance in high glucose-treated HepG2 cells via miR-223/keap1-Nrf2 pathway. *Biomed. Pharmacother.* **2019**, *110*, 85–94. [CrossRef]

97. Taguchi, K.; Fujikawa, N.; Komatsu, M.; Ishii, T.; Unno, M.; Akaike, T.; Motohashi, H.; Yamamoto, M. Keap1 degradation by autophagy for the maintenance of redox homeostasis. *Proc. Natl. Acad. Sci. USA* **2012**, *109*, 13561–13566. [CrossRef] [PubMed]

98. Bloom, D.A.; Jaiswal, A.K. Phosphorylation of Nrf2 at Ser40 by protein kinase C in response to antioxidants leads to the release of Nrf2 from INrf2, but is not required for Nrf2 stabilization/accumulation in the nucleus and transcriptional activation of antioxidant response element-mediated NAD(P)H:quinone oxidoreductase-1 gene expression. *J. Biol. Chem.* **2003**, *278*, 44675–44682. [CrossRef] [PubMed]

99. Ganner, A.; Pfeiffer, Z.C.; Wingendorf, L.; Kreis, S.; Klein, M.; Walz, G.; Neumann-Haefelin, E. The acetyltransferase p300 regulates NRF2 stability and localization. *Biochem. Biophys. Res. Commun.* **2020**, *524*, 895–902. [CrossRef] [PubMed]

100. Hu, H.; Hao, L.; Tang, C.; Zhu, Y.; Jiang, Q.; Yao, J. Activation of KGFR-Akt-mTOR-Nrf2 signaling protects human retinal pigment epithelium cells from Ultra-violet. *Biochem. Biophys. Res. Commun.* **2018**, *495*, 2171–2177. [CrossRef] [PubMed]

101. Huang, H.C.; Nguyen, T.; Pickett, C.B. Phosphorylation of Nrf2 at Ser-40 by protein kinase C regulates antioxidant response element-mediated transcription. *J. Biol. Chem.* **2002**, *277*, 42769–42774. [CrossRef]

102. Banerjee, N.; Wang, H.; Wang, G.; Boor, P.J.; Khan, M.F. Redox-sensitive Nrf2 and MAPK signaling pathways contribute to trichloroethene-mediated autoimmune disease progression. *Toxicology* **2021**, *457*, 152804. [CrossRef] [PubMed]

103. Reddy, N.M.; Potteti, H.R.; Vegiraju, S.; Chen, H.J.; Tamatam, C.M.; Reddy, S.P. PI3K-AKT Signaling via Nrf2 Protects against Hyperoxia-Induced Acute Lung Injury, but Promotes Inflammation Post-Injury Independent of Nrf2 in Mice. *PLoS ONE* **2015**, *10*, e0129676. [CrossRef]

104. Cuadrado, A.; Kugler, S.; Lastres-Becker, I. Pharmacological targeting of GSK-3 and NRF2 provides neuroprotection in a preclinical model of tauopathy. *Redox Biol.* **2018**, *14*, 522–534. [CrossRef]

105. Song, X.; Long, D. Nrf2 and Ferroptosis: A New Research Direction for Neurodegenerative Diseases. *Front. Neurosci.* **2020**, *14*, 267. [CrossRef] [PubMed]

106. Chang, L.C.; Chiang, S.K.; Chen, S.E.; Yu, Y.L.; Chou, R.H.; Chang, W.C. Heme oxygenase-1 mediates BAY 11-7085 induced ferroptosis. *Cancer Lett.* **2018**, *416*, 124–137. [CrossRef]

107. Niture, S.K.; Khatri, R.; Jaiswal, A.K. Regulation of Nrf2-an update. *Free Radic. Biol. Med.* **2014**, *66*, 36–44. [CrossRef] [PubMed]

108. Canning, P.; Bullock, A.N. New strategies to inhibit KEAP1 and the Cul3-based E3 ubiquitin ligases. *Biochem. Soc. Trans.* **2014**, *42*, 103–107. [CrossRef] [PubMed]
109. Kaspar, J.W.; Jaiswal, A.K. An autoregulatory loop between Nrf2 and Cul3-Rbx1 controls their cellular abundance. *J. Biol. Chem.* **2010**, *285*, 21349–21358. [CrossRef] [PubMed]

Article

Chlorine-Induced Toxicity on Murine Cornea: Exploring the Potential Therapeutic Role of Antioxidants

Seungwon An [1,2,*], Khandaker Anwar [1], Mohammadjavad Ashraf [1,3], Kyu-Yeon Han [1] and Ali R. Djalilian [1,*]

[1] Department of Ophthalmology and Visual Sciences, University of Illinois at Chicago, Chicago, IL 60612, USA; kanwar@uic.edu (K.A.); mashra5@uic.edu (M.A.); biohan72@uic.edu (K.-Y.H.)
[2] Clinical Stem Cell Laboratory, UI Blood & Marrow Transplant Program, University of Illinois Hospital and Health Sciences System, Chicago, IL 60612, USA
[3] Department of Pathology, Shiraz University of Medical Sciences, Shiraz 71348-14336, Iran
* Correspondence: seunga1@uic.edu (S.A.); adjalili@uic.edu (A.R.D.); Tel.: +1-(312)-996-3767 (S.A. & A.R.D.)

Abstract: Chlorine (Cl_2) exposure poses a significant risk to ocular health, with the cornea being particularly susceptible to its corrosive effects. Antioxidants, known for their ability to neutralize reactive oxygen species (ROS) and alleviate oxidative stress, were explored as potential therapeutic agents to counteract chlorine-induced damage. In vitro experiments using human corneal epithelial cells showed decreased cell viability by chlorine-induced ROS production, which was reversed by antioxidant incubation. The mitochondrial membrane potential decreased due to both low and high doses of Cl_2 exposure; however, it was recovered through antioxidants. The wound scratch assay showed that antioxidants mitigated impaired wound healing after Cl_2 exposure. In vivo and ex vivo, after Cl_2 exposure, increased corneal fluorescein staining indicates damaged corneal epithelial and stromal layers of mice cornea. Likewise, Cl_2 exposure in human ex vivo corneas led to corneal injury characterized by epithelial fluorescein staining and epithelial erosion. However, antioxidants protected Cl_2-induced damage. These results highlight the effects of Cl_2 on corneal cells using in vitro, ex vivo, and in vivo models while also underscoring the potential of antioxidants, such as vitamin A, vitamin C, resveratrol, and melatonin, as protective agents against acute chlorine toxicity-induced corneal injury. Further investigation is needed to confirm the antioxidants' capacity to alleviate oxidative stress and enhance the corneal healing process.

Keywords: chlorine; ROS; JC-1; antioxidants; wound healing

Citation: An, S.; Anwar, K.; Ashraf, M.; Han, K.-Y.; Djalilian, A.R. Chlorine-Induced Toxicity on Murine Cornea: Exploring the Potential Therapeutic Role of Antioxidants. *Cells* **2024**, *13*, 458. https://doi.org/10.3390/cells13050458

Academic Editors: Dimitrios Karamichos and Ritva Tikkanen

Received: 26 December 2023
Revised: 1 March 2024
Accepted: 3 March 2024
Published: 5 March 2024

1. Introduction

The cornea, as the transparent outermost layer of the eye, is susceptible to injuries and damage from various factors [1–7]. Corneal wound healing is a crucial process for restoring its integrity and visual function. However, oxidative stress resulting from reactive oxygen species (ROS) can hinder the healing process, leading to delayed recovery and potential complications, such as photokeratoconjunctivities, photokeratitis, pingueculae and pterygia, cataracts, glaucoma, and macular degeneration [2–10]. Corneal injuries trigger an inflammatory response, leading to the release of ROS, which can cause cellular damage and impede the healing process [11–13]. Oxidative stress disrupts cellular functions, delays the migration and proliferation of corneal cells, and interferes with the extracellular matrix remodeling necessary for effective wound closure [13–15]. Antioxidants are molecules that neutralize ROS, preventing cellular damage and maintaining a healthy redox balance [16–19]. They scavenge free radicals, stabilize cell membranes, and modulate signaling pathways involved in inflammation and tissue repair. In recent years, research has focused on the potential benefits of antioxidants in supporting corneal wound healing [20–24]. The enzymatic antioxidants that have been documented to be present in the cornea include superoxide dismutases (SODs), catalase (CAT), glutathione peroxidases (GPXs), reductase (GR), and glucose-6-phosphate dehydrogenase (G6PD) [25,26].

Common antioxidants include vitamins C and E, superoxide dismutase (SOD), glutathione, and various plant-derived compounds that also have antioxidant effects [27,28]. The therapeutic effects of antioxidants are investigated on corneal epithelial and stromal cells exposed to oxidative stress, resulting in wound healing [20–24]. Antioxidants, such as vitamin A, vitamin C, melatonin, NAC (N-acetylcysteine), and resveratrol, play essential roles in neutralizing harmful reactive oxygen species (ROS) and protecting cells from oxidative damage [25]. Vitamin A, or beta-carotene, is an antioxidant and can be effective in enhancing wound strength in rats [29]. Vitamins C, D, and E and acetylcysteine helped corneal wound healing [30–32]. Especially, vitamin C prevents lipid peroxidation and apoptosis in corneal endothelial cells and improves the antioxidant enzyme activity in rat eyes [33–35]. Melatonin is a hormone that regulates the sleep–wake cycle and has strong antioxidant properties. Research suggests that melatonin may help protect ocular tissues, including the cornea, from oxidative damage caused by environmental factors such as UV radiation. Melatonin ameliorates oxidative stress in granular corneal dystrophy, dry eye, and diabetic models [36–41]. N-Acetylcysteine (NAC) exhibits various beneficial effects, such as rescuing oxidative stress-induced angiogenesis in a mouse corneal alkali-burn model, increasing corneal endothelial cell survival in a mouse model of Fuchs endothelial corneal dystrophy, reducing oxidative stress for cytosine arabinoside in a rat model, and promoting the long-term survival of cones in a model of retinitis pigmentosa [42–45]. Resveratrol demonstrates protective effects on human corneal epithelial cells, safeguarding them from inflammation, oxidative stress damage, cytotoxicity induced by moxifloxacin and benzalkonium chloride, hyperosmolar conditions, and enhancing wound healing through the attenuation of oxidative stress-induced impairment of cell proliferation and migration, showcasing the potential for the treatment of dry eye disease and various ocular diseases [46–51]. Of note, some of the antioxidants noted above, such as resveratrol, may not be strictly antioxidants and may affect other pathways other than ROS activity.

Several studies have highlighted the significance of antioxidant activities within the cornea. Tsao et al. explored the effect of total antioxidant capacity (TAC) in aqueous humor on corneal endothelial health, discovering that both TAC and ascorbic acid (AA) independently safeguarded against low endothelial cell density [52]. Additionally, Higuchi et al. conducted research into the role of antioxidants in the treatment of corneal disorders, pinpointing selenoprotein P as a substance that imparts antioxidative effects on corneal epithelial cells [53]. In their study, Koskela et al. delved into oxidative stress and protein accumulation in different corneal diseases, identifying that oxidative stress and the activation of the molecular chaperone response were prevalent in keratoconus, macular corneal dystrophy, and Fuchs endothelial corneal dystrophy [54]. Stoddard et al. evaluated the bioavailability and effectiveness of antioxidants in human corneal limbal epithelial cells, ascertaining that quercetin, epigallocatechin gallate, n-propyl gallate, and gallic acid all demonstrated antioxidant activity [55]. These collective studies underscore the vital role of antioxidants in preserving corneal health and their potential in the therapeutic treatment of corneal disorders.

Acute chlorine toxicity on the cornea refers to the harmful effects of exposure to chlorine gas or chlorine-containing substances on the eye's corneal tissue [56,57]. Chlorine is a highly reactive and corrosive chemical commonly used in industrial settings, swimming pools, and household cleaning products [57–60]. When chlorine gas or chlorine-based compounds come into contact with the cornea, they can cause severe damage, leading to various ocular symptoms and potential long-term consequences [57,61,62]. Exposure to chlorine gas can lead to symptom onset at concentrations of 1–3 ppm, which is characterized by mucus membrane irritation. Eye irritation becomes evident at 5–15 ppm, accompanied by moderate upper respiratory tract irritation [56]. Higher concentrations of chlorine gas, such as 430 ppm, can result in death within 30 min, and concentrations exceeding 1000 ppm can lead to death within just a few minutes [56]. The severity of symptoms tends to increase with higher concentrations of chlorine gas [63]. Following chlorine exposure, the eyes may show signs of infections, abrasions, and corrosions in the conjunctiva [64]. The symptoms from chlorine gas exposure can occur immediately or be delayed, appearing 24 h after exposure [56,63]. Chlorine-related corneal injuries typically heal within one to two days and are characterized by a burning sensation and superficial disruption of the corneal epithelium. The cornea is highly sensitive and vulnerable to chemical exposure, and acute chlorine toxicity on the cornea can lead to a range of symptoms reported, including tearing, soreness, severe discomfort, conjunctiva edema, conjunctivitis, excessive tearing, blurred vision, a sensation of having a foreign object in the eye, photophobia, corneal abrasions, and superficial punctate keratopathy [65–68], as well as foreign body sensation in the eye, pterygium, chronic conjunctivitis, and premature presbyopia [56,61]. The affected eye may also become swollen, and vision may be temporarily blurred or reduced [69,70]. Chlorine exposure can cause direct injury to the corneal epithelial cells, leading to the loss of the epithelial layer and the formation of corneal ulcers [56,71]. Chlorine, a disinfectant used in swimming pools and tap water, can damage the corneal epithelium. As a result, frequent swimmers may experience symptoms such as redness, itching, ocular surface epithelial damage, and eye irritation [71]. An ophthalmic examination may reveal ciliary injection and superficial punctate keratitis, which can be attributed to chlorine's presence in swimming pools [72].

Our study demonstrates that antioxidants have protective effects on corneal cells, shielding them from oxidative damage caused by Cl_2 exposure. Furthermore, these antioxidants promote cell migration and accelerate wound closure, indicating their potential to enhance the healing process of corneal injuries. Combining antioxidant treatment with standard care or other regenerative approaches may result in synergistic effects, further augmenting corneal wound healing. Future research should focus on (1) exploring the underlying mechanisms of Cl_2 injury to the cornea and the protective role of antioxidants in observed changes, such as fluorescein staining, corneal thickness, and epithelial edema; (2) evaluating antioxidants using in vivo and ex vivo models; and (3) optimizing antioxidant formulations, dosages, and delivery methods to maximize their therapeutic potential. Antioxidants show significant promise in supporting corneal wound healing by combating oxidative stress and creating an environment conducive to tissue repair. As our understanding of their mechanisms deepens and more clinical evidence emerges, antioxidant-based therapies could become valuable tools in ophthalmology, contributing to the recovery of corneal injuries and overall improvement of ocular health.

2. Materials and Methods

2.1. Materials

The following materials were used for this study: vitamin A (#11017, Cayman, CO, USA), vitamin C (#A4403, Sigma-Aldrich, St. Louis, MO, USA), melatonin (#M 5250, Sigma-Aldrich, St. Louis, MO, USA), N-Acetyl Cysteine (NAC; #A9165, Sigma-Aldrich, St. Louis, MO, USA), chlorine (#198016, Thermo Fisher Scientific, Waltham, MA, USA), and resveratrol (#554325, Sigma-Aldrich, St. Louis, MO, USA).

2.2. Cell Culture

Human corneal epithelial cell (HCEC) cultures were initiated from cadaver corneas kindly provided by the Illinois (Chicago, IL, USA) and Midwest eye banks (Ann Arbor, MI, USA) [73,74]. Human corneal epithelial cells were cultured in 10% fetal bovine serum (FBS; # F2442, Sigma-Aldrich, St. Louis, MO, USA), 1X L-glutamine (#MT25005CI, Corning, NY, USA), 1X NEAA (#11140050, Gibco, Bilings, MT, USA), and 1% penicillin-streptomycin (P/S; #MT30002CI, Corning, NY, USA) in 5% CO_2 at 37 °C [73,74].

2.3. LDH Toxicity Assay

HCECs were precisely dispensed into 96-well culture plates at a density of 3×10^4 cells per well, utilizing a complete growth medium optimized for this cell type. Following an incubation period of 12 h to permit cell adhesion and stabilization, the cells underwent a single washing step with 200 µL of phosphate-buffered saline (PBS) at isotonic concentration.

Subsequently, the cells were subjected to varying concentrations of antioxidant compounds, which were administered in a basal Dulbecco's Modified Eagle Medium (DMEM) and incubated for 24 h to assess the protective efficacy against oxidative stress. In parallel, to evaluate the cellular response to Cl_2, a similar protocol was employed wherein post-wash, HCECs were incubated with Cl_2 at concentrations ranging from 1 to 3000 ppm, dissolved in DMEM, for 24 h. Upon the completion of the exposure period, a volume of 50 μL of the cell culture supernatant was carefully combined with an equal volume of the LDH reaction mixture, prepared by the stipulated guidelines provided by the manufacturer (#C2030, Thermo Fisher Scientific, Waltham, MA, USA). This mixture was then transferred to a 96-well flat-bottom plate and allowed to incubate at ambient temperature for 30 min, facilitating the development of the enzymatic reaction. The resultant chromogenic substrate conversion was quantitatively measured by recording the optical density at dual wavelengths, specifically 490 nm and 680 nm, utilizing a Cytation5 microplate reader. The reliability and reproducibility of the data were ensured by conducting the assays in triplicate across six independent experimental replicates, as substantiated by references [73,74].

2.4. Cell Proliferation

HCECs were cultured on a 4-well chamber slide and incubated for 12 h to promote cell adhesion and stabilization. After this, the cells were treated with a different culture medium for 2 h, washed, and then cultured for another 24 h to encourage proliferation. The proliferation of HCECs was assessed by measuring DNA content using the CyQuant® NF Cell Proliferation Assay (#C35006, Invitrogen, Waltham, MA, USA). Following a total of 36 h of incubation, the supernatant was removed, and cells were incubated with a $1\times$ CyQuant dye solution for fluorescence development. Fluorescence intensity, reflecting cell proliferation, was measured with an excitation wavelength of 485 nm and an emission wavelength of 530 nm using a Gen5 plate reader and conducted in triplicate for six samples to ensure data reliability [74].

2.5. In Vitro Scratch Assay

HCECs were plated in 6-well culture plates at a density of 5×10^6 cells per well, using media supplemented with 10% FBS to ensure optimal growth conditions. Following a 12-h incubation period to establish confluent monolayers, a sterile 200 μL pipette tip was employed to introduce a standardized scratch, simulating a wound. After this wounding procedure, monolayers were rinsed twice with $1\times$ PBS to eliminate any detached cells. Prior to the administration of antioxidant treatments, cells were subjected to a 30-min exposure to Cl_2, which was followed by another two $1\times$ PBS washes to remove any residual Cl_2. The migration and closure of the scratch wound were monitored at designated time intervals. This was accomplished by capturing sequential images of the scratch area with a high-resolution spinning disk confocal microscope (Z1; Carl Zeiss Meditec, Jena, Germany). Quantitative analysis of the wound healing process was facilitated by utilizing ImageJ software to measure the area of the scratch that remained unhealed over time. To ensure the reproducibility and accuracy of the results, these assays were conducted in triplicate with five independent experimental replicates [74].

2.6. Mitochondria Membrane Potential Assay

The evaluation of mitochondrial membrane potential in human corneal epithelial cells (HCECs) was conducted utilizing the JC-1 Mitochondrial Membrane Potential Assay Kit (Catalog #ab113850, Abcam, Cambridge, MA, USA), in strict adherence to the supplier's protocol. HCECs were seeded into 96-well opaque culture plates at a density of 2×10^5 cells per well and allowed to adhere and grow for 24 h. After incubation, the cells underwent a 30-min treatment with Cl_2, after which they were washed and further incubated with antioxidants for an additional 24 h to assess the protective effects on mitochondrial integrity. The JC-1 assay was then performed by incubating the cells with a 1 μM JC-1 staining solution for 30 min at 37 °C. A parallel set of wells received dilution buffer alone, serving as the control condition. Post incubation, imaging was carried out in the dilution buffer to evaluate the mitochondrial membrane potential. Fluorometric detection was executed employing a Cytation5 plate reader, with an excitation wavelength set at 475 nm and dual emission wavelengths of 530 $\pm$ 15 nm and 590 $\pm$ 17.5 nm to distinguish between the monomeric and aggregated states of JC-1, respectively, indicative of the mitochondrial membrane potential status [74].

2.7. Chlorine Treatment on Naïve Murine Eyes

All in vivo procedures were meticulously executed in strict accordance with the ARVO Statement for the Use of Animals in Ophthalmic and Vision Research, ensuring the highest standards of ethical conduct. The experimental protocol received full endorsement from the University of Illinois at Chicago's Committee on the Ethics of Animal Experiments (UIC) and the Biosafety Committee, affirming the commitment to ethical research practices. C57BL/6J mice, aged between six and ten weeks, were anesthetized via an intraperitoneal administration of a ketamine–xylazine solution, at dosages of 100 mg/kg and 5 mg/kg, respectively, as referenced in a previous publication [74]. These wild-type mice served as the biological model for assessing ocular toxicity attributable to chlorine exposure. The experimental regimen involved the application of graded concentrations of Cl_2 (ranging from 1 to 2000 ppm; 10 µL for 30 s) to the murine corneas, administered daily over two weeks. Repetitive measurements were carried out in triplicates with four independent subjects per group. Post-treatment, corneal integrity was evaluated using a 1 mg/mL fluorescein solution (BioGlo; HUB Pharmaceuticals, Plymouth, CA, USA), which was applied to the corneal surface for one minute. After the application, any residual staining solution was carefully blotted away with Kimwipes. The extent of corneal injury was then assessed by examining and capturing images of the fluorescein staining under a Nikon FS-2 slit lamp at 30X magnification. Quantitative analysis of the fluorescein staining intensity was conducted using MetaMorph software (Molecular Devices, Version 7.8.13.0), enabling precise data acquisition on corneal damage following chlorine exposure [74].

2.8. Ex vivo Model of Human and Murine Cornea Culture

For murine eyeballs, wild-type mice were used to take basal images of bright-field and fluorescein (1 mg/mL) for basal and followed by treatment with chlorine injury (100–500 ppm, 2 mL, 30–60 min) and then with antioxidants (or vehicle control) for up to 7 days (n = 12 per group). The eyes were examined and imaged with a slit lamp every day for 3 days along with fluorescein staining visualized under a cobalt blue light. The outcome measures in the ex vivo human corneas include (i) corneal damage after chlorine exposure (bright-field image and fluorescein staining), (ii) histopathologic examination of the corneal structure and corneal epithelial cells (H&E), and (iii) corneal epithelial/stromal cell apoptosis (TUNEL staining).

Donated human corneas from an Eversight eye bank facility (Michigan, Ohio, Illinois, New Jersey, and Connecticut) were used. For the human cornea, intact human corneas were selected and washed with $1\times$ PBS containing antibiotics. Human corneas were imaged with bright-field and fluorescein staining (1 mg/mL) for basal and followed by treatment with chlorine injury (100–500 ppm, 2 mL, 30–60 min) and then with treatment with antioxidants (or the vehicle control) for up to 7 days (n = 12 per group). The eyes were examined and imaged with a slit lamp every day for 3 days along with fluorescein staining visualized under a cobalt blue light. The outcome measures in the ex vivo human corneas include (i) corneal damage after chlorine exposure (bright-field image and fluorescein staining), (ii) histopathologic examination of the corneal structure and corneal epithelial cells (H&E), and (iii) corneal epithelial/stromal cell apoptosis (TUNEL staining).

2.9. Histology

For hematoxylin and eosin (H&E) staining, cryo-sections were fixed in neutral buffered 10% formaldehyde (Sigma-Aldrich, St. Louis, MO, USA) for 20 min and followed the protocol as previously described [73,74].

2.10. Detection of ROS (O_2^- and H_2O_2) in Tissues

For IF staining, two cryo-sections from each group (total of 14 slides, seven different groups) were fixed in neutral buffered 10% formaldehyde (Sigma-Aldrich, St. Louis, MO, USA) for 20 min, following the previously described protocol [73,74]. The final concentrations of DCF-DA (20 µM, Fisher, #50-187-4597) and DHE (5 µM, Cayman, #NC2189794) were freshly prepared and applied to the slides for 30–60 min in a moisture chamber without light. After incubation, the slides were washed twice with $1\times$ PBS and prepared with 1 µL of DAPI for confocal microscopy (Z1; Carl Zeiss Meditec, Jena, Germany). DCF-DA staining was used to detect reactive oxygen species (ROS) in green fluorescence (Ex:Em = 502/523 nm), while DHE staining was specifically used to detect superoxide (O_2^-) in red fluorescence (Ex:Em = 490/595 nm).

2.11. Statistical Analysis

Statistical analysis was performed utilizing GraphPad Prism 5 software (Version 5.01, GraphPad Software, Inc., San Diego, CA, USA). The data are expressed as mean $\pm$ standard deviation (SD), derived from three independent experimental runs. To ascertain the significance of differences between groups, two-tailed nonparametric *t*-tests were employed, with the analyses facilitated by both GraphPad Prism and Microsoft Excel software (Version 2019, Microsoft Corp., Redmond, WA, USA). A P value of less than 0.05 was predetermined as the threshold for statistical significance.

3. Results

3.1. Cytotoxicity Assay of Antioxidants and Cl_2 on HCECs

In this study, we aimed to determine the optimal range of antioxidants and the cytotoxicity concentration (CC50) of Cl_2 in HCECs using a lactate dehydrogenase (LDH) cytotoxicity assay. In order to determine the non-toxic range for the antioxidant, a range of concentration was tested, which showed a dose-dependent decrease in cell viability, with noticeable effects observed at the following concentrations: 0.5 µM for vitamin A, 0.4 µM for vitamin C, 10 µM for resveratrol, 1 mM for melatonin, and 1 mM for NAC (Figure 1a–e). Chlorine exposure demonstrated cytotoxic effects on HCECs, with a reduction in cell viability starting at 1 ppm and approximately 50% loss of viability at 100 ppm (Figure 1f). Based on these findings, we identified the optimal dose of antioxidants to be 100 ppm Cl_2 for our model of chlorine-mediated cell injury. This concentration strikes a balance between antioxidant protection and chlorine-induced injury, making it suitable for further investigation in our experimental setup.

3.2. Cell Proliferation of Antioxidants on Cl_2-Treated HCECs

We investigated the impact of antioxidants on HCECs after exposure to Cl_2. At 100 ppm Cl_2 exposure, cell viability decreased, but treatment with antioxidants (vitamin A, vitamin C, resveratrol, melatonin, and NAC) effectively reversed the Cl_2-induced cell damage (Figure 2a). Moreover, incubation with antioxidants alone, without Cl_2 exposure, led to a significant increase in cell proliferation compared to the untreated control (vitamin A: 1.45 ± 0.17, vitamin C: 1.99 ± 0.26, resveratrol: 1.83 ± 0.20, melatonin: 2.59 ± 0.21, NAC: 2.55 ± 0.16 vs. control: 1.0 ± 0.1) in HCECs (Figure 2b). However, when HCECs were exposed to 100 ppm Cl_2 before antioxidant incubation, the fold change in cell proliferation significantly decreased compared to the untreated control (vitamin A: 0.62 ± 0.08, vitamin C: 0.93 ± 0.13, resveratrol: 0.77 ± 0.10, melatonin: 0.92 ± 0.13, NAC: 1.05 ± 0.11, Cl_2: 0.12 ± 0.01 vs. control: 1.0 ± 0.1) in HCECs (Figure 2b). These results strongly suggest that antioxidants have the potential to reverse Cl_2-mediated inhibition of cell proliferation in

HCECs. They not only enhance cell proliferation when applied alone but also counteract the negative effects of Cl_2 exposure, thus offering a promising avenue for mitigating chlorine-induced damage and promoting cell recovery.

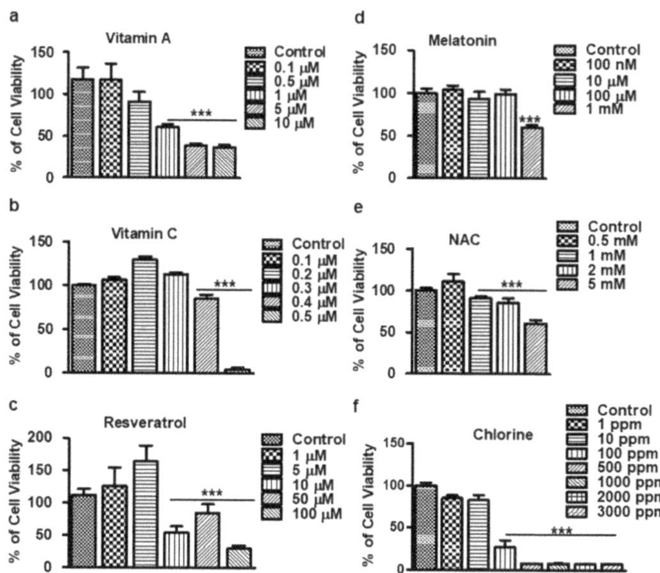

Figure 1. Cytotoxicity assay of antioxidants in the HCECs. (**a–f**) The cells were treated for 24 h with varying concentrations. The cell viability is reported as the percentage of the control group (100%). All data are presented as the mean $\pm$ SEM (n = 6). A significant difference *** $p < 0.001$ using one-way ANOVA analysis witsh Tukey's post hoc analysis was observed in the percentage of cell viability vs. the control group (untreated). NAC: N-Acetyl Cysteine.

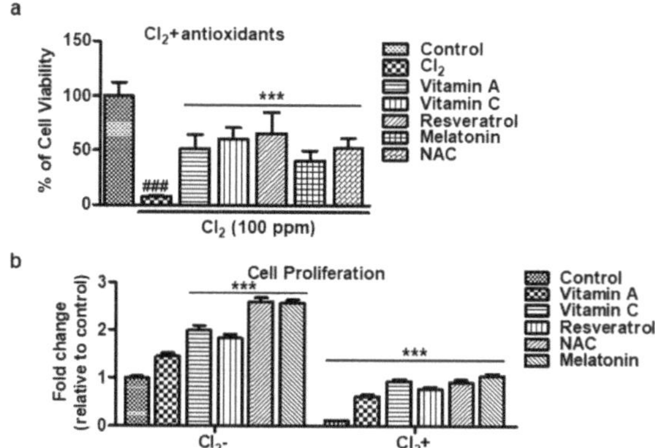

Figure 2. Cell proliferation of antioxidants in Cl_2-treated HCECs. (**a,b**) The HCECs are exposed to 100 ppm Cl_2 for 30 min and are followed by treatment with antioxidants for 24 h. The results indicate the percentage of cell proliferation vs. the control cells (untreated). Values are the mean $\pm$ SEM (n = 6). The data were analyzed by one-way ANOVA analysis with Tukey's post hoc analysis. A significant difference, ### $p < 0.001$ was observed in the percentage of cell viability vs. untreated cells and Cl_2-treated cells. A significant difference, *** $p < 0.001$ was observed in the percentage of cell viability vs. antioxidant-treated cells. Vitamin A: 100 µM, vitamin C: 300 µM, resveratrol: 5 µM, melatonin: 100 µM, NAC: 500 µM.

3.3. Effect of Antioxidant on Cl₂-Induced ROS Production in HCECs

To assess the impact of antioxidants after Cl_2 exposure on HCECs, we measured cellular ROS levels. As depicted in Figure 3a, various concentrations of Cl_2 induced a significant four~six-fold increase in cellular ROS accumulation compared to the untreated control (Cl_2 1 ppm: 1.53 ± 0.18, 10 ppm: 1.72 ± 0.30, 100 ppm: 3.49 ± 0.36, 500 ppm: 3.04 ± 0.84, 1000 ppm: 4.63 ± 0.67, 2000 ppm: 4.89 ± 1.02, 3000 ppm: 6.14 ± 0.94 vs. control: 0.89 ± 0.04). In contrast, treatment with antioxidants did not show a significant difference compared to the untreated control (vitamin A: 1.01 ± 0.13, vitamin C: 1.47 ± 0.07, resveratrol: 1.61 ± 0.04, melatonin: 1.12 ± 0.10, NAC: 0.85 ± 0.07 vs. control: 0.89 ± 0.04). Moreover, when cells were exposed to 100 ppm Cl_2, ROS levels surged by four to five times; however, incubating antioxidants mitigated the cellular ROS accumulation to one~three-fold compared to the Cl_2-treated group. These results highlight that Cl_2 exposure induces mitochondrial damage and elevates oxidative stress, resulting in a significant increase in cellular ROS levels (Figure 3b). However, the administration of antioxidants effectively combats these deleterious effects, offering a potential therapeutic approach to ameliorate oxidative stress and its detrimental consequences induced by Cl_2 exposure in HCECs (Figure 3b).

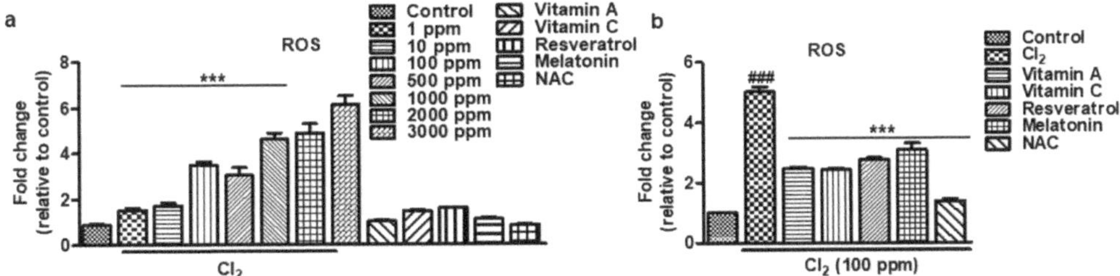

Figure 3. Effect of antioxidants on Cl2-induced ROS production in HCECs. (**a**,**b**) The HCECs were pretreated with Cl_2 for 30 min followed by 24 h of antioxidant treatment. The results indicate the fold change of ROS level vs. the control cells (untreated). Values are the mean $\pm$ SEM ($n = 6$). A significant difference, *** $p < 0.001$ was observed in the fold change of ROS vs. untreated cells and Cl_2-treated cells. ### $p < 0.001$ was observed in the fold change of ROS vs. untreated cells and antioxidant-treated cells. Vitamin A: 100 µM, vitamin C: 300 µM, resveratrol: 5 µM, melatonin: 100 µM, NAC: 500 µM.

3.4. Mitochondrial Membrane Potential in HCECs

To assess the effects of antioxidants on maintaining high mitochondrial membrane potential (MMP) after Cl_2 exposure, HCECs were treated with antioxidants following Cl_2 exposure. As MMP is a critical indicator of mitochondrial activity, the use of antioxidants appears to preserve mitochondrial function in the presence of Cl_2. As depicted, exposure to 100 ppm Cl_2 for 30 min resulted in a significant decrease in the percentage of MMP compared to the control group (100 ppm: 28.59% vs. control: 100%). However, when antioxidants were incubated after Cl_2 exposure, the decline in MMP induced by Cl_2 was ameliorated compared to the Cl_2-exposed group (Figure 4). These findings collectively indicate that Cl_2 exposure impairs mitochondrial function, leading to a decrease In MMP. However, antioxidant therapy effectively restores the compromised mitochondrial function induced by Cl_2 exposure, highlighting the potential of antioxidants in preserving mitochondrial activity and mitigating the adverse effects of Cl_2 on HCECs.

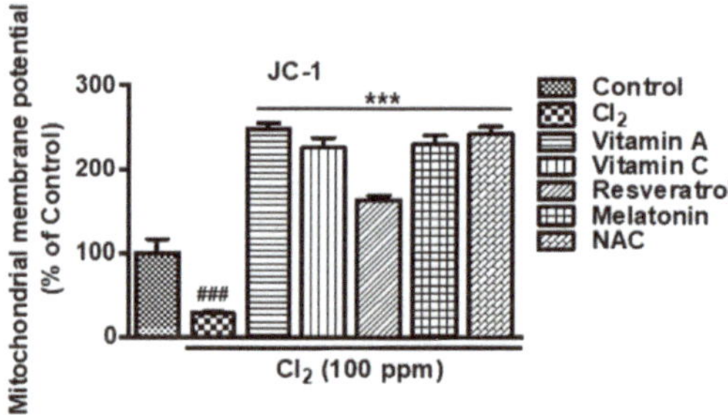

Figure 4. Mitochondrial membrane potential in HCECs. Cells are exposed to 100 ppm Cl_2 for 30 min prior to treating antioxidants. The results indicate the percentage of mitochondrial membrane potential vs. the control cells (untreated). Values are the mean $\pm$ SEM ($n = 6$). A significant difference, ### $p < 0.001$ was observed in the percentage of cell viability vs. untreated cells. A significant difference, *** $p < 0.001$ was observed in the percentage of cell viability vs. Cl_2-treated cells. Vitamin A: 100 µM, vitamin C: 300 µM, resveratrol: 5 µM, melatonin: 100 µM, NAC: 500 µM.

3.5. Wound-Healing Assay to Detect Cell Migration of HCECs

To investigate whether antioxidants can promote wound healing in HCECs delayed by Cl_2 exposure, we incubated the optimal doses of antioxidants (vitamin A: 100 µM, vitamin C: 300 µM, resveratrol: 5 µM, melatonin: 100 µM, NAC: 500 µM) after Cl_2 exposure. As depicted in Figure 5, Cl_2 exposure resulted in cell damage and delayed wound healing, as evidenced by the reduced percentage of wound closure (Cl_2: 16.6 $\pm$ 1.78% vs. control group: 35.4 $\pm$ 9.96%). However, when vitamin C and NAC were incubated after Cl_2 exposure, they significantly promoted wound healing compared to the control group (vitamin C: 68.8 $\pm$ 13.09%, NAC: 60.6 $\pm$ 14.61%). In contrast, vitamin A, resveratrol, and melatonin did not show a significant improvement in wound healing compared to the control group (vitamin A: 12.6 $\pm$ 4.20%, resveratrol: 16 $\pm$ 4.14%, melatonin: 17 $\pm$ 4.02%). These findings indicate that antioxidant therapy, particularly with vitamin C and NAC, can effectively counteract the delay in cornea cell wound healing caused by Cl_2 exposure. These antioxidants demonstrate the potential to promote wound closure and may hold promise as therapeutic agents to facilitate the recovery of corneal tissue after Cl_2-induced damage. Further investigations are warranted to understand the specific mechanisms underlying the wound healing-promoting effects of these antioxidants in HCECs.

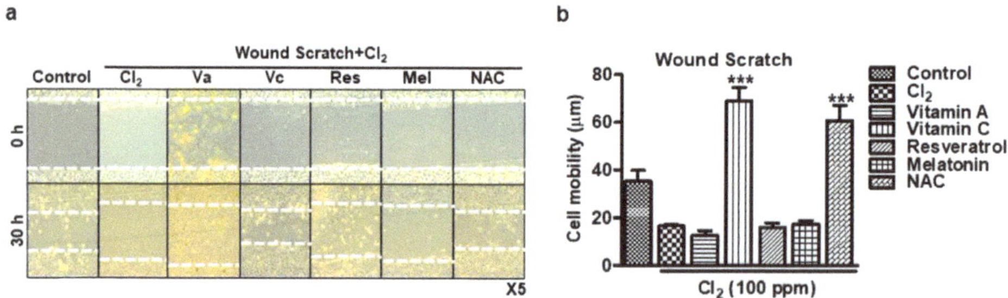

Figure 5. Wound-healing assay to detect cell migration of HCECs. (**a,b**) Wound scratch in HCECs exposed to 100 ppm Cl_2 for 30 min prior to treating antioxidants. (**a**) Representative images showing scratch wound assay in HCLE cells. White dot: wound area. (**b**) Graph showing wound healing rate for different conditions in epithelial scratch wounds ($n = 5$/group) at 30 h. *** $p < 0.001$ was observed in the cell mobility (µm) vs. untreated cells and Cl_2-treated cells. Va: vitamin A, Vc: vitamin C, Res: resveratrol, Mel: melatonin, vitamin A: 100 µM, vitamin C: 300 µM, resveratrol: 5 µM, melatonin: 100 µM, NAC: 500 µM.

3.6. In Vivo Evaluation of Cl_2 Impact on Mice Eyes

Next, we performed experiments in a murine model to determine whether Cl_2 exposure can cause corneal epitheliopathy in vivo. We applied freshly prepared Cl_2 and exposure to naïve murine corneas for 1 min per topical application per day for up to 2 weeks. The fluorescein staining was greatly increased at 1 week and 2 weeks compared to day 0 (Figure 6a). Subsequently, Cl_2 exposure induced damage to the central cornea and stromal layer (asterisk) in murine corneas (Figure 6b). Cl_2 exposure at 1000 ppm and 2000 ppm resulted in significantly higher corneal fluorescein staining after 1 week compared to baseline (1000 ppm: 1.56-fold, 2000 ppm: 2.23-fold, vs. control). Starting from 2 weeks, Cl_2 exposure at 10 ppm to 2000 ppm showed a significant increase in corneal fluorescein staining compared to the control group at 2 weeks (10 ppm: 1.9-fold, 100 ppm: 1.79-fold, 500 ppm:1.92-fold, 1000 ppm: 1.4-fold, 2000 ppm: 2.1-fold, vs. control) (Figure 6c). This result indicated that Cl_2 exposure can cause dose-dependent corneal epitheliopathy in vivo.

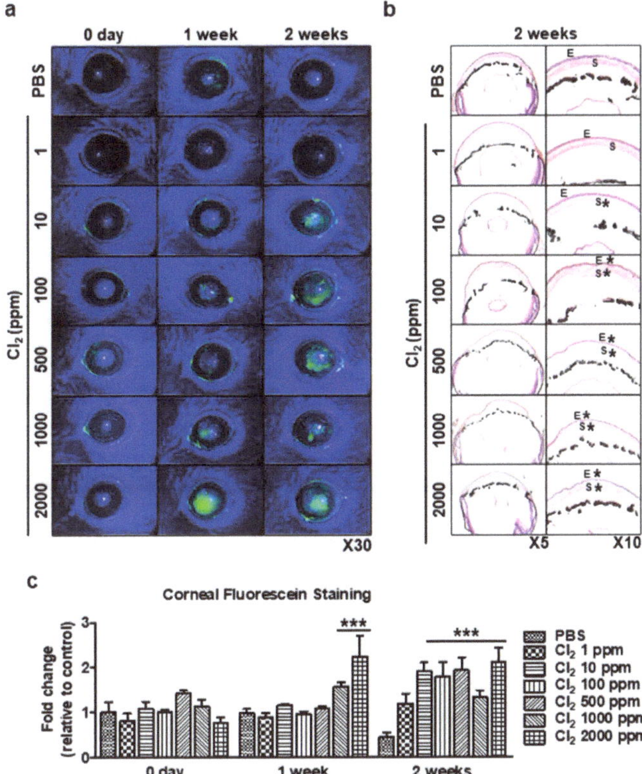

Figure 6. In vivo evaluation of chlorine's impact on mice eyes using corneal fluorescein staining. Mice corneas were applied to various doses of Cl_2 (1, 10, 100, 500, 1000, and 2000 ppm; 10 μL, 30 s) once a day for 2 weeks. (**a**) Representative images of murine corneas showing fluorescein staining with Cl_2 treatment. (**b**) H&E staining of various doses of Cl_2-treated murine corneas. E: epithelium, S: stroma. *: damaged area. (**c**) Graph showing the intensity fold change of corneal fluorescein staining after application of Cl_2 treatment ($n = 4$/group) for 2 weeks. Values are the mean $\pm$ SEM ($n = 4$). The results indicate that corneal fluorescein staining was greatly increased in a dose-dependent manner compared to the control group (PBS-treated). A significant difference, *** $p < 0.001$ was observed in the fold change of fluorescein staining vs. the control groups (PBS-treated on 1 week or 2 weeks).

3.7. Ex Vivo Evaluation of Cl_2 Effects

In parallel experiments, we investigated whether the effect of Cl_2 exposure on naïve murine cornea was determined by an ex vivo model. Significantly greater corneal fluorescein staining was observed on day 2 following the application of 500 ppm Cl_2 compared to day 0 (Cl_2: 1.71-fold vs. PBS: 1.09-fold) (Figure 7a,b). H&E and microscopic analysis of murine eyes showed corneal epithelial loss and stromal edema upon Cl_2 exposure compared with the control (asterisk, Cl_2: 20.99 $\pm$ 11.16 μm vs. PBS: 60.67 $\pm$ 14.23 μm) (Figures 7c,d and S1a,b). Therefore, Cl_2 exposure to the eye causes severe ocular toxicity, corneal epithelial damage, and abnormal stroma structure ex vivo. As shown in Figure 7e–h, antioxidant-treated groups showed less corneal fluorescein staining (PBS: 1.27-fold, Cl_2: 14.64-fold, Cl_2 + vitamin A: 5.23-fold, Cl_2 + vitamin C: 1.35-fold, Cl_2 + resveratrol: 1.91-fold, Cl_2 + melatonin: 5.89-fold, Cl_2 + NAC: 1.27-fold on day 2) and prevented corneal epithelial loss compared to the Cl_2-exposed group (PBS: 80.00 $\pm$ 2.00 μm, Cl_2: 14.67 $\pm$ 5.51 μm, Cl_2 + vitamin A: 71.67 $\pm$ 15.28 μm, Cl_2 + vitamin C: 73.33 $\pm$ 33.39 μm, Cl_2 + resveratrol: 66.00 $\pm$ 1.73 μm, Cl_2 + melatonin: 77.33 $\pm$ 4.04 μm, Cl_2 + NAC: 84.00 $\pm$ 8.54 μm on day 2). In parallel in vitro experiments, ROS and superoxide were measured to determine the effect of antioxidants after Cl_2 exposure on the murine cornea. Fluorescence microscopy images showed increasing ROS and superoxide generation after Cl_2 exposure, while antioxidant treatment ameliorated compared to the Cl_2-exposed group (Figures 7i and S1c).

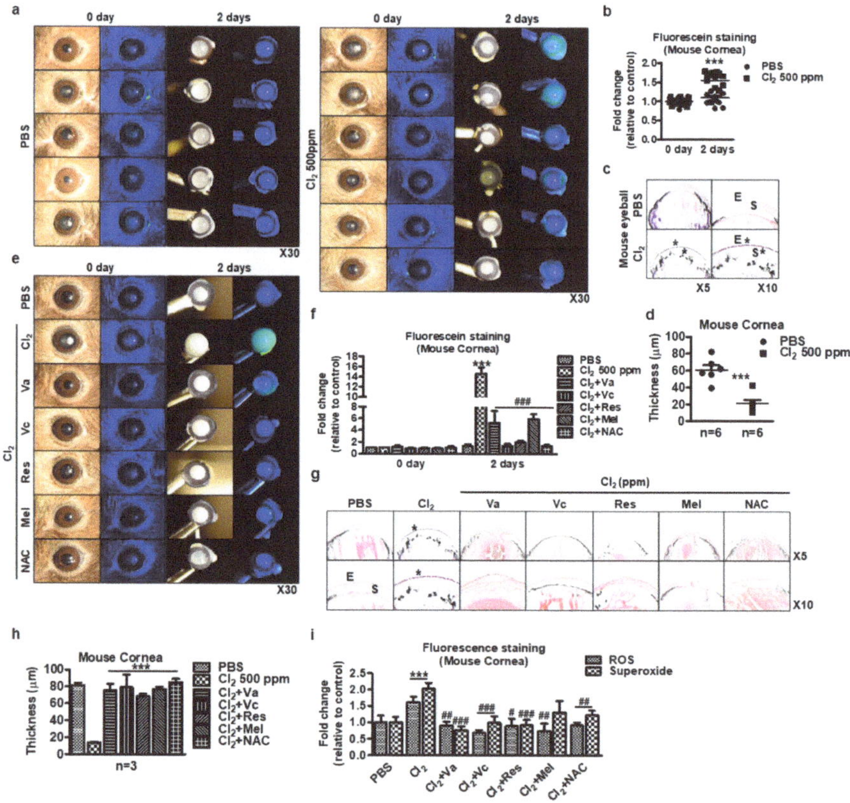

Figure 7. Ex vivo evaluation of chlorine's effects on mice eyeballs. (**a–c**) Mouse eyeballs were exposed to 500 ppm Cl_2 for 2 h. Subsequently, the eyeballs were washed two times and then incubated in

1x PBS for 2 days. (**a**) Representative images of murine whole eyeballs showing fluorescein staining with or without Cl_2 treatment. (**b**) Graph showing the intensity fold change of corneal fluorescein staining after application of Cl_2 treatment ($n = 12$/group) for 2 days. Values are the mean $\pm$ SEM ($n = 11$). The results indicate that corneas with 500 ppm Cl_2 became opaque and hazy with more fluorescein staining of the cornea and conjunctiva than the PBS-treated group. (**c**) H&E staining on murine whole eyeballs after application of PBS or chlorine 500 ppm for 2 days. E: epithelium, S: stroma. *: damaged area. (**d**) Murine cornea thickness after 500 ppm Cl_2 exposure. *** $p < 0.001$ was observed in the cornea thickness vs. the control group (PBS treatment). Values are the mean $\pm$ SEM (PBS: $n = 11$, Cl_2: $n = 12$). (**e**) Representative images of murine whole eyeballs showed fluorescein staining. (**f**) Graph showing the intensity fold change of corneal fluorescein staining after application of antioxidants ($n = 4$/group) for 2 days. *** $p < 0.001$ was observed in the corneal fluorescein staining vs. the control group (PBS-treated on day 2). ### $p < 0.001$ was observed in the corneal fluorescein staining vs. Cl_2-treated group on day 2. (**g**) H&E staining on murine whole eyeballs with antioxidants after application of Cl_2 500 ppm for 2 days. E: epithelium, S: stroma. *: damaged area. Vitamin A (Va): 100 μM, vitamin C (Vc): 300 μM, resveratrol (Res): 5 μM, melatonin (Mel): 100 μM, and NAC: 500 μM for 2 days. (**h**) Murine cornea thickness. *** $p < 0.001$ was observed in the cornea thickness vs. the control group (PBS-treated). (**i**) Relative fold changes of immunofluorescence intensity from Figure S1c. *** $p < 0.001$ was observed in the ROS and superoxide groups vs. the control group (PBS-treated). # $p < 0.05$, ## $p < 0.01$, ### $p < 0.001$ were observed in the ROS and superoxide groups vs. the Cl_2 treatment group.

To investigate the effects of Cl_2 on human cornea ex vivo, we selected intact human corneas for Cl_2 exposure. As shown in Figure 8, 500 ppm Cl_2 exposure was not shown to change corneal fluorescein staining compared to the PBS control group at day 1 (1.03 ± 0.01 vs. 0.98 ± 0.01). Interestingly, Cl_2 exposure significantly increased corneal fluorescein staining compared to the PBS control group at day 2 (1.45 ± 0.18 vs. 1.07 ± 0.18) (Figures 8b and S2). Subsequently, Cl_2 exposure induced damage to the central cornea (asterisk), more so than the peripheral cornea or corneal–limbus area in human corneas (Figure 8c). Moreover, H&E and microscopic analysis showed that Cl_2 exposure resulted in a 1.2~1.6-fold increase in human corneal thickness compared with the control corneas (Figures 8d and S3). In a parallel experiment, antioxidants were treated after Cl_2 exposure decreased corneal fluorescein staining compared to the PBS control group on day 2 (Figure 8e,f) (PBS: 1.40-fold, Cl_2: 2.56-fold, Cl_2 + vitamin A: 1.03-fold, Cl_2 + vitamin C: 1.50-fold, Cl_2 + resveratrol: 1.00-fold, Cl_2 + melatonin: 1.27-fold, Cl_2 + NAC: 0.78-fold). As a result, Cl_2 exposure to the cornea showed (1) a rough surface of the epithelial layer on the corneal surface, (2) a loose epithelial layer (corneal epithelial erosion), (3) separation of the epithelial–stromal layer, (4) and corneal edema; however, (5) antioxidant treatment protects Cl_2-induced epithelial–stromal damage.

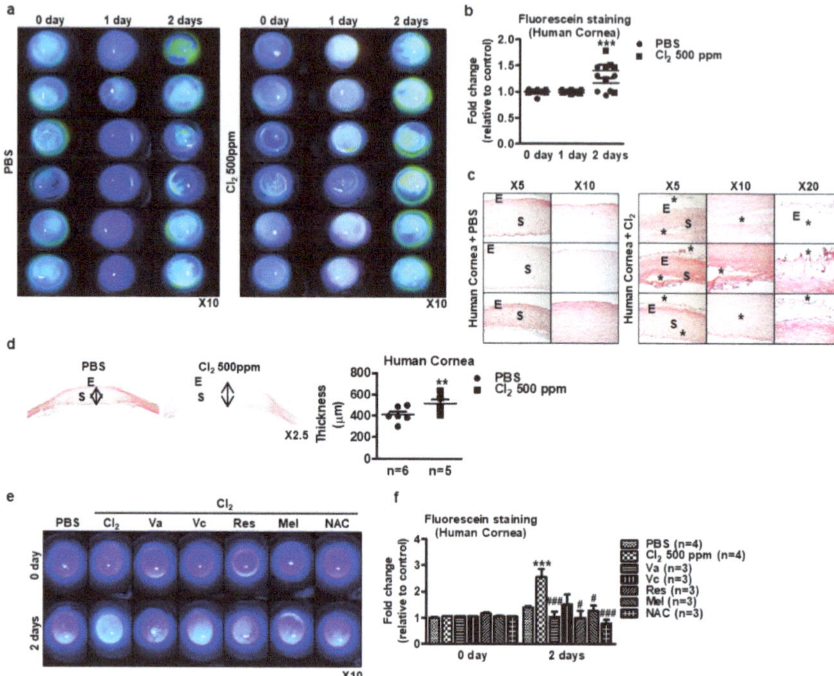

Figure 8. Ex vivo evaluation of Cl_2 effects on human corneas. (**a–d**) Human corneas were exposed to 500 ppm Cl_2 for 3 days. (**a**) Representative images of human corneas showing fluorescein staining with or without Cl_2 treatment. (**b**) Graph showing the intensity fold change of human cornea fluorescein staining after Cl_2 exposure (n = 12/group) for 3 days. *** $p < 0.001$ was observed in the fold change of fluorescein staining vs. the control group (PBS). The results indicate that corneas exposure to 500 ppm Cl_2 showed higher levels of fluorescein staining on both the cornea and conjunctival compared to the PBS-exposed group. (**c**) H&E staining of Cl_2-treated human corneas. Black star: damaged area, E: epithelium, S: stroma. (**d**) Human cornea thickness after 500 ppm Cl_2 exposure. ** $p < 0.05$ was observed in the cornea thickness vs. the control group (PBS). Values are the mean ± SEM (PBS: n = 12, Cl_2: n = 12). (**e**) Representative images of human corneas showed fluorescein staining by Cl_2 treatment and followed by antioxidants (Va, Vc, Res, Mel, and NAC). (**f**) Graph showing the intensity fold change of corneal fluorescein staining after application of antioxidants (n = 4/group) for 2 days. *** $p < 0.001$ was observed in the corneal fluorescein staining vs. the control group (PBS treatment on day 2). # $p < 0.05$, ### $p < 0.001$ were observed in the corneal fluorescein staining vs. Cl_2-treated group on day 2.

4. Discussion

In this study, we tested the effects of Cl_2 in in vitro (human corneal epithelial cells), ex vivo (mouse eyeballs and human corneas), and in vivo mouse models. The main findings of our study are as follows: Cl_2 exposure significantly (1) decreased cell viability, (2) increased ROS generation, (3) decreased MMP, and (4) delayed in vitro wound healing. However, known antioxidants (vitamin A, vitamin C, resveratrol, melatonin, and NAC) could reverse Cl_2-mediated damages. Moreover, ex vivo and in vivo studies showed that Cl_2 exposure showed (5) corneal epithelial damage, (6) separation of the epithelial–stromal layer, and (7) corneal edema. Therefore, we proposed that antioxidants have therapeutic potential to protect against Cl_2 eye injury and could be used in the development of targeted ocular therapies.

Previous studies have established the toxic effects of chlorine gas exposure. Cl_2 gas, a highly reactive and toxic substance, is classified as a pulmonary irritant. It finds widespread use in various industries and household applications, including water treatment, disinfection, and cleaning products [59,60,66]. Exposure to chlorine gas, being water-soluble, can result in a range of health issues, contingent upon the dose and duration of exposure [57,59,60]. Acute exposure to high doses of Cl_2 gas may cause dyspnea, violent cough, nausea, vomiting, lightheadedness, headache, chest pain, abdominal discomfort, and corneal burns. Moreover, even low doses of Cl_2 gas can lead to chest pain, cough, sore throat, and hemoptysis [59,60,66].

We have previously reported that nitrogen mustard similarly induced ROS, change in MMP, and delay in wound healing in corneal epithelial cells, which was mitigated by the mesenchymal stem cell secretome [74,75]. Our current findings revealed that Cl_2 decreased cell viability (Figure 1f) and cell proliferation (Figure 2), increased intracellular ROS generation (Figure 3), decreased MMP in HCECs (Figure 4), and delayed wound healing (Figure 5). However, the optimized dose of antioxidants (vitamin A, vitamin C, resveratrol, and melatonin) incubation showed significantly reversed Cl_2-mediated cellular damages, such as increased cell viability and cell proliferation, downregulated ROS accumulation, and stabilized MMP levels. While we did not investigate the mechanisms underlying the protective effects of antioxidants in our current study, considering our current focus on the therapeutic potential of antioxidants on Cl_2 exposure, it will be of great interest to investigate the Cl_2-involved mechanism by dose and exposure duration in the future.

Our in vitro similarly showed that Cl_2 exposure delayed corneal wound healing (Figure 5) and increased corneal fluorescein staining in an in vivo model (Figure 6). Interestingly, more than 10 ppm Cl_2 gradually increased corneal fluorescein staining over a period of two weeks (Figure 6a,c), which indicates a more long-lasting effect. These results suggest that continuous Cl_2 exposure causes ocular surface epithelial damage and likely deeper layers of the epithelium, including the more basal cells, which include stem/progenitor cells. Only vitamin C and NA incubation after Cl_2 exposure promotes wound healing in vitro (Figure 5), and other tested antioxidants did not have the same effect on wound healing. It is interesting to note that both vitamin C and NAC are mostly known as antioxidants, while the tested chemicals (vitamin A, melatonin, resveratrol) are also known to affect other pathways. Future studies are needed to determine the effects of antioxidants and the specific mechanisms after Cl_2 injury.

In murine corneas, Cl_2 exposure significantly increased epithelial edema (Figure 7c,d). The effect of antioxidants after Cl_2 exposure in murine corneas in vivo will be studied in the future. Also, our previous study demonstrated that induced loss of membrane integrity of surface epithelium and corneal stromal matrix by nitrogen mustard exposure resulted in epithelial and stromal inflammation and apoptosis [74]. In our study, we employed both in vivo and ex vivo models using human and mouse corneas. To evaluate in vitro data, we conducted ex vivo experiments using both human cornea and mouse cornea. Cl_2 injury increased mouse corneal fluorescein staining (epithelial damage); however, the antioxidant treatment showed significantly less staining than the Cl_2-treated group (Figure 7e,f). In correlation with staining data, H&E staining data suggested that Cl_2 injury significantly damaged the corneal epithelial layer, but antioxidant (Va, Vc, Res, Mel, and NAC) treatment prevented Cl_2-induced corneal epithelial damage (Figure 7g,h). According to the data, there is a correlation between ROS accumulation in human corneal epithelial cells (Figure 3b) and the effect of antioxidants on MMP (Figure 4). IF staining data further supports the impact of antioxidants on Cl_2-induced ROS and superoxide detection in in vivo tissues (Figures 7i and S1c). The provided data confirms that staining with ROS and superoxide can ascertain the injury of murine corneal epithelium and the therapeutic effect of antioxidants.

In parallel experiments, murine eyeballs and donated intact human corneas were used to determine the effects of Cl_2 in ex vivo conditions. The results showed that corneas exposed to Cl_2 exhibited increased fluorescein staining on both the cornea and conjunctiva areas (Figure 8a,b). Additionally, H&E staining revealed damaged areas in the Cl_2-treated human corneas, specifically in the epithelium and stroma. Furthermore, the corneas exposed to 500 ppm Cl_2 showed increased epithelial edema compared to the control group (Figure 8c,d). Statistical analysis indicated significant differences in fluorescein staining intensity and cornea thickness between the Cl_2-exposed group and the control group (Figure 8d). More interestingly, antioxidant-treated groups (Va, Vc, Res, Mel, and NAC) decreased fluorescein staining intensity means preventing human corneal epithelial damage compared to Cl_2-treated human corneas (Figure 8e). It suggests that antioxidants can protect against chlorine exposure to mouse corneas. These samples can be indirectly assessed for antioxidant effects through oxidative stress and superoxide staining [74]. Direct assessment can be achieved using the Trolox Equivalent Antioxidant Capacity method [52,76,77]. Furthermore, we hypothesize that Cl_2-induced ROS triggers mitochondrial dysfunction. To investigate this, we propose using cellular oxygen consumption rate (OCR) and extracellular acidification rate (ECAR) as functional assays. These tools will allow us to evaluate the capacity of antioxidants to mitigate potential mitochondrial dysfunction caused by Cl_2-induced ROS. Based on the findings, further investigations and future plans could include as follows: (1) extending the exposure time beyond 3 days to evaluate the long-term impact of Cl_2 on human corneal tissues; (2) conducting experiments with varying concentrations of Cl_2 to assess the dose-dependent effects on human corneas; (3) exploring the underlying mechanisms of Cl_2 injury on the corner and the protection mechanism of antioxidants responsible for the observed changes in fluorescein staining, corneal thickness, and epithelial edema; (4) evaluation of antioxidants in the in vivo and ex vivo models (relating to oxidative stress/superoxide), MMP, Trolox Equivalent Antioxidant Capacity, cellular oxygen consumption rates and extracellular acidification rates, (5) influence of Cl_2 on the corneal limbal stem cells using 2 mm or limbus-to-limbus corneal wound model with an especially low dose of Cl_2 treatment; and (6) assessing the effectiveness of potential therapeutic agents or treatments aimed at mitigating Cl_2-induced corneal damage. It is essential to continue research in this area to enhance our understanding of Cl_2-induced corneal toxicity and develop strategies to protect and treat affected individuals effectively.

In summary, our study demonstrates the protective role of antioxidants in preventing Cl_2-induced corneal injury. This protection is associated with enhanced corneal epithelial cell migration, proliferation, and maintenance of mitochondrial dynamic balance in human cornea cells. Furthermore, we successfully replicated Cl_2-induced corneal injury in both murine eyeballs and donated human corneas using ex vivo models. However, to gain a comprehensive understanding of the effects of antioxidants on Cl_2-induced injury in the ex vivo model, further studies are warranted.

5. Conclusions

Our study sheds light on the potential benefits of antioxidant therapy in countering acute chlorine-induced corneal injury. These findings hold promise for developing effective treatments to safeguard ocular health and mitigate the harmful effects of Cl_2 exposure to the cornea.

Supplementary Materials: The following supporting information can be downloaded at: https://www.mdpi.com/article/10.3390/cells13050458/s1, Figure S1: H&E staining on Cl_2-exposed murine eyeball. (a,b) H&E staining of Cl_2-exposed murine eyeballs for 3 days. (c) IF staining of ROS (DCF-DA: Green), superoxide (DHE: Red), and DAPI (Blue). Scale bar, 50 μm; Figure S2: Fluorescein staining on the human cornea after Cl_2 exposure. (a,b) Representative images of human corneas showing fluorescein staining with or without 500ppm Cl_2 treatment for 3 days. BF: Bright field, Flu: Fluorescein; Figure S3: H&E staining on Cl_2-exposed human corneas. (a,b) H&E staining of Cl_2-treated human corneas. E: Epithelium, S: Stroma, En: Endothelium.

Author Contributions: Conceptualization, A.R.D. and S.A.; methodology, S.A.; software, S.A.; validation, S.A., K.A. and M.A.; formal analysis, S.A. and M.A.; data curation, S.A.; writing—original draft preparation, S.A.; writing—review and editing, S.A., K.-Y.H. and A.R.D.; visualization, S.A. and M.A.; supervision, A.R.D.; project administration, A.R.D. and S.A. All authors have read and agreed to the published version of the manuscript.

Funding: This research was funded by R01-EY024349 (A.R.D.) and a core grant EY01792 from NEI/NIH, an unrestricted grant to UIC Department of Ophthalmology and Physician–Scientist Award (A.R.D.) from Research to Prevent Blindness, and Vision Research Program—Congressionally Directed Medical Research Program VR170180 from the Department of Defense.

Institutional Review Board Statement: The study was conducted according to the guidelines of the Declaration of Helsinki and approved by the Committee on the Ethics of Animal Experiments of University of Illinois at Chicago (UIC). In addition, the protocol was approved by the Biosafety Committee (protocol code #20-079, 1 December 2020).

Informed Consent Statement: Not applicable.

Data Availability Statement: Data are contained within the article and Supplementary Materials.

Acknowledgments: We thank Khandaker Anwar and Hyungjo Lee for their contributions to mouse breeding and maintenance. We appreciate the technical assistance from Khandaker Anwar and Mohammadjavad Ashraf in histology and surgical operations, respectively. We are grateful to Kyu-Yeon Han for the technical support, discussing the data, and revision.

Conflicts of Interest: The authors declare no conflicts of interest.

References

1. Sridhar, M.S. Anatomy of cornea and ocular surface. *Indian J. Ophthalmol.* **2018**, *66*, 190–194. [CrossRef]
2. Wilson, S.E. Corneal wound healing. *Exp. Eye Res.* **2020**, *197*, 108089. [CrossRef] [PubMed]
3. Andley, U.P.; Rhim, J.S.; Chylack, L.T., Jr.; Fleming, T.P. Propagation and immortalization of human lens epithelial cells in culture. *Investig. Ophthalmol. Vis. Sci.* **1994**, *35*, 3094–3102.
4. Balasubramanian, D. Ultraviolet radiation and cataract. *J. Ocul. Pharmacol. Ther.* **2000**, *16*, 285–297. [CrossRef] [PubMed]
5. Zigman, S. Lens UVA photobiology. *J. Ocul. Pharmacol. Ther.* **2000**, *16*, 161–165. [CrossRef] [PubMed]
6. Van Kuijk, F.J. Effects of ultraviolet light on the eye: Role of protective glasses. *Environ. Health Perspect.* **1991**, *96*, 177–184. [CrossRef] [PubMed]
7. Tenkate, T.D. Occupational exposure to ultraviolet radiation: A health risk assessment. *Environ. Health Perspect.* **1999**, *14*, 187–209. [CrossRef] [PubMed]
8. Roberts, J.E. Ocular phototoxicity. *J. Photochem. Photobiol. B* **2001**, *64*, 136–143. [CrossRef] [PubMed]
9. Brennan, L.A.; Kantorow, M. Mitochondrial function and redox control in the aging eye: Role of MsrA and other repair systems in cataract and macular degenerations. *Exp. Eye Res.* **2009**, *88*, 195–203. [CrossRef]
10. Taylor, H.R.; Munoz, B.; West, S.; Bressler, N.M.; Bressler, S.B.; Rosenthal, F.S. Visible light and risk of age-related macular degeneration. *Trans. Am. Ophthalmol. Soc.* **1990**, *88*, 163–173, discussion 173–178.
11. Morescalchi, F.; Duse, S.; Gambicorti, E.; Romano, M.R.; Costagliola, C.; Semeraro, F. Proliferative vitreoretinopathy after eye injuries: An overexpression of growth factors and cytokines leading to a retinal keloid. *Mediat. Inflamm.* **2013**, *2013*, 269787. [CrossRef]
12. Shoham, A.; Hadziahmetovic, M.; Dunaief, J.L.; Mydlarski, M.B.; Schipper, H.M. Oxidative stress in diseases of the human cornea. *Free Radic. Biol. Med.* **2008**, *45*, 1047–1055. [CrossRef] [PubMed]
13. Ljubimov, A.V.; Saghizadeh, M. Progress in corneal wound healing. *Prog. Retin. Eye Res.* **2015**, *49*, 17–45. [CrossRef] [PubMed]
14. Wang, G.; Yang, F.; Zhou, W.; Xiao, N.; Luo, M.; Tang, Z. The initiation of oxidative stress and therapeutic strategies in wound healing. *Biomed. Pharmacother.* **2023**, *157*, 114004. [CrossRef] [PubMed]
15. Kunkemoeller, B.; Kyriakides, T.R. Redox Signaling in Diabetic Wound Healing Regulates Extracellular Matrix Deposition. *Antioxid. Redox Signal.* **2017**, *27*, 823–838. [CrossRef]
16. He, L.; He, T.; Farrar, S.; Ji, L.; Liu, T.; Ma, X. Antioxidants Maintain Cellular Redox Homeostasis by Elimination of Reactive Oxygen Species. *Cell. Physiol. Biochem.* **2017**, *44*, 532–553. [CrossRef]
17. Rose, R.C.; Richer, S.P.; Bode, A.M. Ocular oxidants and antioxidant protection. *Proc. Soc. Exp. Biol. Med.* **1998**, *217*, 397–407. [CrossRef]
18. Tan, B.L.; Norhaizan, M.E.; Liew, W.P.; Sulaiman Rahman, H. Antioxidant and Oxidative Stress: A Mutual Interplay in Age-Related Diseases. *Front. Pharmacol.* **2018**, *9*, 1162. [CrossRef]
19. Babizhayev, M.A.; Yegorov, Y.E. Reactive Oxygen Species and the Aging Eye: Specific Role of Metabolically Active Mitochondria in Maintaining Lens Function and in the Initiation of the Oxidation-Induced Maturity Onset Cataract—A Novel Platform of Mitochondria-Targeted Antioxidants with Broad Therapeutic Potential for Redox Regulation and Detoxification of Oxidants in Eye Diseases. *Am. J. Ther.* **2016**, *23*, e98–e117. [CrossRef] [PubMed]

20. Hallberg, C.K.; Trocme, S.D.; Ansari, N.H. Acceleration of corneal wound healing in diabetic rats by the antioxidant trolox. *Res. Commun. Mol. Pathol. Pharmacol.* **1996**, *93*, 3–12.

21. Tsai, C.Y.; Woung, L.C.; Yen, J.C.; Tseng, P.C.; Chiou, S.H.; Sung, Y.J.; Liu, K.T.; Cheng, Y.H. Thermosensitive chitosan-based hydrogels for sustained release of ferulic acid on corneal wound healing. *Carbohydr. Polym.* **2016**, *135*, 308–315. [CrossRef] [PubMed]

22. Shukla, A.; Rasik, A.M.; Dhawan, B.N. Asiaticoside-induced elevation of antioxidant levels in healing wounds. *Phytother. Res.* **1999**, *13*, 50–54. [CrossRef]

23. Chen, J.; Lan, J.; Liu, D.; Backman, L.J.; Zhang, W.; Zhou, Q.; Danielson, P. Ascorbic Acid Promotes the Stemness of Corneal Epithelial Stem/Progenitor Cells and Accelerates Epithelial Wound Healing in the Cornea. *Stem. Cells Transl. Med.* **2017**, *6*, 1356–1365. [CrossRef] [PubMed]

24. Kant, V.; Gopal, A.; Pathak, N.N.; Kumar, P.; Tandan, S.K.; Kumar, D. Antioxidant and anti-inflammatory potential of curcumin accelerated the cutaneous wound healing in streptozotocin-induced diabetic rats. *Int. Immunopharmacol.* **2014**, *20*, 322–330. [CrossRef] [PubMed]

25. Chen, Y.; Mehta, G.; Vasiliou, V. Antioxidant defenses in the ocular surface. *Ocul. Surf.* **2009**, *7*, 176–185. [CrossRef]

26. Fecondo, J.V.; Augusteyn, R.C. Superoxide dismutase, catalase and glutathione peroxidase in the human cataractous lens. *Exp. Eye Res.* **1983**, *36*, 15–23. [CrossRef]

27. Imam, M.U.; Zhang, S.; Ma, J.; Wang, H.; Wang, F. Antioxidants Mediate Both Iron Homeostasis and Oxidative Stress. *Nutrients* **2017**, *9*, 671. [CrossRef] [PubMed]

28. Oresajo, C.; Pillai, S.; Manco, M.; Yatskayer, M.; McDaniel, D. Antioxidants and the skin: Understanding formulation and efficacy. *Dermatol. Ther.* **2012**, *25*, 252–259. [CrossRef]

29. Gerber, L.E.; Erdman, J.W., Jr. Effect of dietary retinyl acetate, beta-carotene and retinoic acid on wound healing in rats. *J. Nutr.* **1982**, *112*, 1555–1564. [CrossRef]

30. Gujral, G.S.; Askari, S.N.; Ahmad, S.; Zakir, S.M.; Saluja, K. Topical vitamin C, vitamin E, and acetylcysteine as corneal wound healing agents: A comparative study. *Indian J. Ophthalmol.* **2020**, *68*, 2935–2939. [CrossRef]

31. Reins, R.Y.; Hanlon, S.D.; Magadi, S.; McDermott, A.M. Effects of Topically Applied Vitamin D during Corneal Wound Healing. *PLoS ONE* **2016**, *11*, e0152889. [CrossRef]

32. Ringsdorf, W.M., Jr.; Cheraskin, E. Vitamin C and human wound healing. *Oral Surg. Oral Med. Oral Pathol.* **1982**, *53*, 231–236. [CrossRef] [PubMed]

33. Serbecic, N.; Beutelspacher, S.C. Anti-oxidative vitamins prevent lipid-peroxidation and apoptosis in corneal endothelial cells. *Cell Tissue Res.* **2005**, *320*, 465–475. [CrossRef] [PubMed]

34. Jelodar, G.; Akbari, A.; Nazifi, S. The prophylactic effect of vitamin C on oxidative stress indexes in rat eyes following exposure to radiofrequency wave generated by a BTS antenna model. *Int. J. Radiat. Biol.* **2013**, *89*, 128–131. [CrossRef] [PubMed]

35. Serbecic, N.; Beutelspacher, S.C. Vitamins inhibit oxidant-induced apoptosis of corneal endothelial cells. *Jpn. J. Ophthalmol.* **2005**, *49*, 355–362. [CrossRef] [PubMed]

36. Choi, S.I.; Dadakhujaev, S.; Ryu, H.; Im Kim, T.; Kim, E.K. Melatonin protects against oxidative stress in granular corneal dystrophy type 2 corneal fibroblasts by mechanisms that involve membrane melatonin receptors. *J. Pineal Res.* **2011**, *51*, 94–103. [CrossRef] [PubMed]

37. Siu, A.W.; Maldonado, M.; Sanchez-Hidalgo, M.; Tan, D.X.; Reiter, R.J. Protective effects of melatonin in experimental free radical-related ocular diseases. *J. Pineal Res.* **2006**, *40*, 101–109. [CrossRef] [PubMed]

38. Wang, B.; Zuo, X.; Peng, L.; Wang, X.; Zeng, H.; Zhong, J.; Li, S.; Xiao, Y.; Wang, L.; Ouyang, H.; et al. Melatonin ameliorates oxidative stress-mediated injuries through induction of HO-1 and restores autophagic flux in dry eye. *Exp. Eye Res.* **2021**, *205*, 108491. [CrossRef] [PubMed]

39. Jiang, T.; Chang, Q.; Cai, J.; Fan, J.; Zhang, X.; Xu, G. Protective Effects of Melatonin on Retinal Inflammation and Oxidative Stress in Experimental Diabetic Retinopathy. *Oxidative Med. Cell. Longev.* **2016**, *2016*, 3528274. [CrossRef]

40. Lundmark, P.O.; Pandi-Perumal, S.R.; Srinivasan, V.; Cardinali, D.P. Role of melatonin in the eye and ocular dysfunctions. *Vis. Neurosci.* **2006**, *23*, 853–862. [CrossRef]

41. Gul, M.; Emre, S.; Esrefoglu, M.; Vard, N. Protective effects of melatonin and aminoguanidine on the cornea in streptozotocin-induced diabetic rats. *Cornea* **2008**, *27*, 795–801. [CrossRef]

42. Kubota, M.; Shimmura, S.; Kubota, S.; Miyashita, H.; Kato, N.; Noda, K.; Ozawa, Y.; Usui, T.; Ishida, S.; Umezawa, K.; et al. Hydrogen and N-acetyl-L-cysteine rescue oxidative stress-induced angiogenesis in a mouse corneal alkali-burn model. *Investig. Ophthalmol. Vis. Sci.* **2011**, *52*, 427–433. [CrossRef]

43. Kim, E.C.; Meng, H.; Jun, A.S. N-Acetylcysteine increases corneal endothelial cell survival in a mouse model of Fuchs endothelial corneal dystrophy. *Exp. Eye Res.* **2014**, *127*, 20–25. [CrossRef] [PubMed]

44. Balci, Y.I.; Acer, S.; Yagci, R.; Kucukatay, V.; Sarbay, H.; Bozkurt, K.; Polat, A. N-acetylcysteine supplementation reduces oxidative stress for cytosine arabinoside in rat model. *Int. Ophthalmol.* **2017**, *37*, 209–214. [CrossRef]

45. Lee, S.Y.; Usui, S.; Zafar, A.B.; Oveson, B.C.; Jo, Y.J.; Lu, L.; Masoudi, S.; Campochiaro, P.A. N-Acetylcysteine promotes long-term survival of cones in a model of retinitis pigmentosa. *J. Cell. Physiol.* **2011**, *226*, 1843–1849. [CrossRef] [PubMed]

46. Tsai, T.Y.; Chen, T.C.; Wang, I.J.; Yeh, C.Y.; Su, M.J.; Chen, R.H.; Tsai, T.H.; Hu, F.R. The effect of resveratrol on protecting corneal epithelial cells from cytotoxicity caused by moxifloxacin and benzalkonium chloride. *Investig. Ophthalmol. Vis. Sci.* **2015**, *56*, 1575–1584. [CrossRef]
47. Shetty, R.; Subramani, M.; Murugeswari, P.; Anandula, V.R.; Matalia, H.; Jayadev, C.; Ghosh, A.; Das, D. Resveratrol Rescues Human Corneal Epithelial Cells Cultured in Hyperosmolar Conditions: Potential for Dry Eye Disease Treatment. *Cornea* **2020**, *39*, 1520–1532. [CrossRef] [PubMed]
48. Lancon, A.; Frazzi, R.; Latruffe, N. Anti-Oxidant, Anti-Inflammatory and Anti-Angiogenic Properties of Resveratrol in Ocular Diseases. *Molecules* **2016**, *21*, 304. [CrossRef] [PubMed]
49. Zhou, X.; Ruan, Q.; Ye, Z.; Chu, Z.; Xi, M.; Li, M.; Hu, W.; Guo, X.; Yao, P.; Xie, W. Resveratrol accelerates wound healing by attenuating oxidative stress-induced impairment of cell proliferation and migration. *Burns* **2021**, *47*, 133–139. [CrossRef]
50. Bryl, A.; Falkowski, M.; Zorena, K.; Mrugacz, M. The Role of Resveratrol in Eye Diseases-A Review of the Literature. *Nutrients* **2022**, *14*, 2974. [CrossRef] [PubMed]
51. Galiniak, S.; Aebisher, D.; Bartusik-Aebisher, D. Health benefits of resveratrol administration. *Acta Biochim. Pol.* **2019**, *66*, 13–21. [CrossRef]
52. Tsao, Y.T.; Wu, W.C.; Chen, K.J.; Yeh, L.K.; Hwang, Y.S.; Hsueh, Y.J.; Chen, H.C.; Cheng, C.M. Analysis of aqueous humor total antioxidant capacity and its correlation with corneal endothelial health. *Bioeng. Transl. Med.* **2021**, *6*, e10199. [CrossRef] [PubMed]
53. Higuchi, A. Development of New Pharmaceutical Candidates With Antioxidant Activity for the Treatment of Corneal Disorders. *Cornea* **2019**, *38* (Suppl. S1), S45–S49. [CrossRef]
54. Vottonen, L.; Koskela, A.; Felszeghy, S.; Wylegala, A.; Kryszan, K.; Gurubaran, I.S.; Kaarniranta, K.; Wylegala, E. Oxidative Stress and Cellular Protein Accumulation Are Present in Keratoconus, Macular Corneal Dystrophy, and Fuchs Endothelial Corneal Dystrophy. *J. Clin. Med.* **2023**, *12*, 4332. [CrossRef]
55. Stoddard, A.R.; Koetje, L.R.; Mitchell, A.K.; Schotanus, M.P.; Ubels, J.L. Bioavailability of antioxidants applied to stratified human corneal epithelial cells. *J. Ocul. Pharmacol. Ther.* **2013**, *29*, 681–687. [CrossRef] [PubMed]
56. Al-Shami, K.; Almurabi, S.; Shatnawi, J.; Qasagsah, K.; Shatnawi, G.; Nashwan, A.J. Ophthalmic Manifestations of Chlorine Gas Exposure: What Do We Know So Far? *Cureus* **2023**, *15*, e35590. [CrossRef] [PubMed]
57. Nodelman, V.; Ultman, J.S. Longitudinal distribution of chlorine absorption in human airways: A comparison to ozone absorption. *J. Appl. Physiol.* **1999**, *87*, 2073–2080. [CrossRef]
58. Na, W.; Wang, Y.; Li, A.; Zhu, X.; Xue, C.; Ye, Q. Acute chlorine poisoning caused by an accident at a swimming pool. *Toxicol. Ind. Health* **2021**, *37*, 513–519. [CrossRef]
59. Morim, A.; Guldner, G.T. *Chlorine Gas Toxicity*; StatPearls: Treasure Island, FL, USA, 2023.
60. Das, R.; Blanc, P.D. Chlorine gas exposure and the lung: A review. *Toxicol. Ind. Health* **1993**, *9*, 439–455. [CrossRef]
61. D'Alessandro, A.; Kuschner, W.; Wong, H.; Boushey, H.A.; Blanc, P.D. Exaggerated responses to chlorine inhalation among persons with nonspecific airway hyperreactivity. *Chest* **1996**, *109*, 331–337. [CrossRef]
62. Kanikowska, A.; Napiorkowska-Baran, K.; Graczyk, M.; Kucharski, M.A. Influence of chlorinated water on the development of allergic diseases—An overview. *Ann. Agric. Environ. Med.* **2018**, *25*, 651–655. [CrossRef]
63. Morris, J.B.; Wilkie, W.S.; Shusterman, D.J. Acute respiratory responses of the mouse to chlorine. *Toxicol. Sci.* **2005**, *83*, 380–387. [CrossRef]
64. Jurkuvenaite, A.; Benavides, G.A.; Komarova, S.; Doran, S.F.; Johnson, M.; Aggarwal, S.; Zhang, J.; Darley-Usmar, V.M.; Matalon, S. Upregulation of autophagy decreases chlorine-induced mitochondrial injury and lung inflammation. *Free Radic. Biol. Med.* **2015**, *85*, 83–94. [CrossRef] [PubMed]
65. Mackie, E.; Svendsen, E.; Grant, S.; Michels, J.E.; Richardson, W.H. Management of chlorine gas-related injuries from the Graniteville, South Carolina, train derailment. *Disaster Med. Public Health Prep.* **2014**, *8*, 411–416. [CrossRef] [PubMed]
66. Govier, P.; Coulson, J.M. Civilian exposure to chlorine gas: A systematic review. *Toxicol. Lett.* **2018**, *293*, 249–252. [CrossRef] [PubMed]
67. Racioppi, F.; Daskaleros, P.A.; Besbelli, N.; Borges, A.; Deraemaeker, C.; Magalini, S.I.; Martinez Arrieta, R.; Pulce, C.; Ruggerone, M.L.; Vlachos, P. Household bleaches based on sodium hypochlorite: Review of acute toxicology and poison control center experience. *Food Chem. Toxicol.* **1994**, *32*, 845–861. [CrossRef]
68. Ingram, T.A., 3rd. Response of the human eye to accidental exposure to sodium hypochlorite. *J. Endod.* **1990**, *16*, 235–238. [CrossRef] [PubMed]
69. Jones, E.; Palmer, I.; Wessely, S. Enduring beliefs about effects of gassing in war: Qualitative study. *BMJ* **2007**, *335*, 1313–1315. [CrossRef]
70. Slaughter, R.J.; Watts, M.; Vale, J.A.; Grieve, J.R.; Schep, L.J. The clinical toxicology of sodium hypochlorite. *Clin. Toxicol.* **2019**, *57*, 303–311. [CrossRef]
71. Ishioka, M.; Kato, N.; Kobayashi, A.; Dogru, M.; Tsubota, K. Deleterious effects of swimming pool chlorine on the corneal epithelium. *Cornea* **2008**, *27*, 40–43. [CrossRef]
72. Regalado Farreras, D.C.; Puente, C.G.; Estrela, C. Sodium hypochlorite chemical burn in an endodontist's eye during canal treatment using operating microscope. *J. Endod.* **2014**, *40*, 1275–1279. [CrossRef]
73. An, S.; Anwar, K.; Ashraf, M.; Lee, H.; Jung, R.; Koganti, R.; Ghassemi, M.; Djalilian, A.R. Wound-Healing Effects of Mesenchymal Stromal Cell Secretome in the Cornea and the Role of Exosomes. *Pharmaceutics* **2023**, *15*, 1486. [CrossRef]

74. An, S.; Shen, X.; Anwar, K.; Ashraf, M.; Lee, H.; Koganti, R.; Ghassemi, M.; Djalilian, A.R. Therapeutic Potential of Mesenchymal Stem Cell-Secreted Factors on Delay in Corneal Wound Healing by Nitrogen Mustard. *Int. J. Mol. Sci.* **2022**, *23*, 11510. [CrossRef] [PubMed]
75. Gidfar, S.; Milani, F.Y.; Milani, B.Y.; Shen, X.; Eslani, M.; Putra, I.; Huvard, M.J.; Sagha, H.; Djalilian, A.R. Rapamycin Prolongs the Survival of Corneal Epithelial Cells in Culture. *Sci. Rep.* **2017**, *7*, 40308. [CrossRef] [PubMed]
76. Bilgihan, K.; Bilgihan, A.; Adiguzel, U.; Sezer, C.; Yis, O.; Akyol, G.; Hasanreisoglu, B. Keratocyte apoptosis and corneal antioxidant enzyme activities after refractive corneal surgery. *Eye* **2002**, *16*, 63–68. [CrossRef] [PubMed]
77. Jurja, S.; Negreanu-Pirjol, T.; Vasile, M.; Hincu, M.; Ciuluvica, R.; Negreanu-Pirjol, B.S. Comparative antioxidant activity of various ophthalmic product types for artificial tears under different experimental conditions. *Exp. Ther. Med.* **2022**, *23*, 330. [CrossRef]

Article

Allyl Isothiocianate Induces Ca^{2+} Signals and Nitric Oxide Release by Inducing Reactive Oxygen Species Production in the Human Cerebrovascular Endothelial Cell Line hCMEC/D3

Roberto Berra-Romani [1], Valentina Brunetti [2], Giorgia Pellavio [3], Teresa Soda [4], Umberto Laforenza [3], Giorgia Scarpellino [2] and Francesco Moccia [2,*]

[1] Department of Biomedicine, School of Medicine, Benemérita Universidad Autónoma de Puebla, Puebla 72410, Mexico; rberra001@hotmail.com
[2] Laboratory of General Physiology, Department of Biology and Biotechnology "L. Spallanzani", University of Pavia, 27100 Pavia, Italy; valentina.brunetti01@universitadipavia.it (V.B.); giorgia.scarpellino@unipv.it (G.S.)
[3] Department of Molecular Medicine, University of Pavia, 27100 Pavia, Italy; giorgia.pellavio@unipv.it (G.P.); lumberto@unipv.it (U.L.)
[4] Department of Health Sciences, University of Magna Graecia, 88100 Catanzaro, Italy; teresa.soda@unicz.it
* Correspondence: francesco.moccia@unipv.it; Tel.: +39-382-987-613

check for updates

Citation: Berra-Romani, R.; Brunetti, V.; Pellavio, G.; Soda, T.; Laforenza, U.; Scarpellino, G.; Moccia, F. Allyl Isothiocianate Induces Ca^{2+} Signals and Nitric Oxide Release by Inducing Reactive Oxygen Species Production in the Human Cerebrovascular Endothelial Cell Line hCMEC/D3. *Cells* **2023**, *12*, 1732. https://doi.org/10.3390/cells12131732

Academic Editors: Rossana Morabito and Alessia Remigante

Received: 29 May 2023
Revised: 20 June 2023
Accepted: 24 June 2023
Published: 27 June 2023

Abstract: Nitric oxide (NO) represents a crucial mediator to regulate cerebral blood flow (CBF) in the human brain both under basal conditions and in response to somatosensory stimulation. An increase in intracellular Ca^{2+} concentrations ([Ca^{2+}]$_i$) stimulates the endothelial NO synthase to produce NO in human cerebrovascular endothelial cells. Therefore, targeting the endothelial ion channel machinery could represent a promising strategy to rescue endothelial NO signalling in traumatic brain injury and neurodegenerative disorders. Allyl isothiocyanate (AITC), a major active constituent of cruciferous vegetables, was found to increase CBF in non-human preclinical models, but it is still unknown whether it stimulates NO release in human brain capillary endothelial cells. In the present investigation, we showed that AITC evoked a Ca^{2+}-dependent NO release in the human cerebrovascular endothelial cell line, hCMEC/D3. The Ca^{2+} response to AITC was shaped by both intra- and extracellular Ca^{2+} sources, although it was insensitive to the pharmacological blockade of transient receptor potential ankyrin 1, which is regarded to be among the main molecular targets of AITC. In accord, AITC failed to induce transmembrane currents or to elicit membrane hyperpolarization, although NS309, a selective opener of the small- and intermediate-conductance Ca^{2+}-activated K$^+$ channels, induced a significant membrane hyperpolarization. The AITC-evoked Ca^{2+} signal was triggered by the production of cytosolic, but not mitochondrial, reactive oxygen species (ROS), and was supported by store-operated Ca^{2+} entry (SOCE). Conversely, the Ca^{2+} response to AITC did not require Ca^{2+} mobilization from the endoplasmic reticulum, lysosomes or mitochondria. However, pharmacological manipulation revealed that AITC-dependent ROS generation inhibited plasma membrane Ca^{2+}-ATPase (PMCA) activity, thereby attenuating Ca^{2+} removal across the plasma membrane and resulting in a sustained increase in [Ca^{2+}]$_i$. In accord, the AITC-evoked NO release was driven by ROS generation and required ROS-dependent inhibition of PMCA activity. These data suggest that AITC could be exploited to restore NO signalling and restore CBF in brain disorders that feature neurovascular dysfunction.

Keywords: allyl isothiocianate; hCMEC/D3; reactive oxygen species; Ca^{2+} signalling; nitric oxide; plasma membrane Ca^{2+}-ATPase; store-operated Ca^{2+} entry

1. Introduction

The central nervous system (CNS) is not able to store energy, and therefore, neuronal activity is maintained by the continuous supply of oxygen and nutrients through the capillary bed [1], which accounts for ~90% of brain vasculature [2]. The mechanisms by which an increase in the firing rate leads to an increase in cerebral blood flow (CBF) to active neurons is known as neurovascular coupling (NVC) or functional hyperemia [1]. Capillary endothelial cells (cECs), which represent the core component of the blood–brain barrier (BBB), are also able to sense neuronal activity and thereby signal upstream arterioles to dilate and increase CBF to firing neurons [3–6]. In mouse brain cortical vasculature, synaptically released neurotransmitters cause an increase in endothelial Ca^{2+} concentration by stimulating inositol-1,4,5-trisphosphate ($InsP_3$) receptors ($InsP_3Rs$) and transient receptor potential vanilloid 4 (TRPV4) channels, which are, respectively, located on the endoplasmic reticulum (ER) and the plasma membrane (PM) [4,7,8]. The ensuing Ca^{2+}-dependent recruitment of endothelial nitric oxide (NO) synthase (eNOS) is instrumental to generating the gasotransmitter NO [9], which dilates upstream arterioles and redirects local CBF through the active capillary branches [4]. NO is also fundamental in maintaining basal CBF [10] and initiating NVC upon somatosensory stimulation in the human brain [11]. Human brain cECs can release NO in response to a variety of neurotransmitters and neuromodulators [12–16], which evoke an increase in intracellular Ca^{2+} concentration to engage the eNOS. The Ca^{2+} response to neurohumoral mediators is triggered by ER Ca^{2+} mobilization through $InsP_3Rs$ and lysosomal Ca^{2+} release through two-pore channels (TPCs) and is sustained over time by store-operated Ca^{2+} entry (SOCE) [12–16]. Conversely, ryanodine receptors are absent in hCMEC/D3 cells and do not contribute to agonist-induced Ca^{2+} signals [12]. Targeting the endothelial Ca^{2+}-handling machinery could, therefore, provide an alternative approach to rescue NO signalling, CBF and NVC in cerebrovascular disorders [17,18].

Allyl isothiocyanate (AITC) is an organosulfur phytochemical compound that is an abundant constituent of common cruciferous vegetables, such as broccoli, cabbage, mustard, brussels sprouts and cauliflower [19,20]. AITC induced vasodilation in rat cerebral arteries in both ex vivo brain slices [20,21] and the brain of living animals [22] via an endothelial-dependent mechanism. The hemodynamic response to AITC in rat cortical vasculature requires extracellular Ca^{2+} entry via Transient Receptor Potential Ankyrin 1 (TRPA1) channels [20]. Nevertheless, AITC has also been found to induce intracellular Ca^{2+} signals via a TRPA1-independent mechanism, which involves the production of reactive oxygen species (ROS). However, the sources(s) of AITC-induced ROS-dependent Ca^{2+} signals remain(s) to be identified [23–25]. Similarly, it is still unknown whether ROS may support the endothelial Ca^{2+} response to AITC and whether this Ca^{2+} signal leads to NO release in human brain cECs.

In the present investigation, we sought to assess whether and how AITC evokes Ca^{2+}-dependent NO release in hCMEC/D3 cells, which represent the most widespread model of human cerebrovascular cECs [26–29]. We found that AITC induced NO production through the ROS-dependent inhibition of plasma membrane Ca^{2+}-ATPase (PMCA) activity, which resulted in the accumulation of intracellular Ca^{2+} and eNOS engagement. These findings support the view that AITC administration represents a promising strategy to rescue NO signalling in cerebral disorders associated to endothelial dysfunction.

2. Materials and Methods

2.1. Cell Culture

Human cerebral microvascular endothelial cells (hCMEC/D3) were obtained from Institut National de la Santé et de la Recherche Médicale (INSERM, Paris, France). hCMEC/D3 cells cultured between passage 25 and 35 were used. As described in [30], the cells were seeded at a concentration of 27.000 cells/cm^2 and grown in tissue culture flasks coated with 0.1 mg/mL rat tail collagen type 1, in the following medium: EBM-2 medium (Lonza, Basel, Switzerland) supplemented with 5% fetal bovine serum, 1% Penicillin-Streptomycin, 1.4 μM hydrocortisone, 5 μg/mL ascorbic acid, 1/100 chemically defined lipid concentrate (Life Technologies, Milan, Italy), 10 mM HEPES and 1 ng/mL basic fibroblast growth factor. The cells were cultured at 37 °C, 5% CO_2-saturated humidity.

2.2. Solutions

Physiological salt solution (PSS) had the following composition (in mM): 150 NaCl, 6 KCl, 1.5 CaCl$_2$, 1 MgCl$_2$, 10 glucose, 10 HEPES. In Ca^{2+}-free solution (0Ca^{2+}), Ca^{2+} was substituted with 2 mM NaCl, and 0.5 mM EGTA was added. Solutions were titrated to pH 7.4 with NaOH. The osmolality of PSS as measured with an osmometer (Wescor 5500, Logan, UT, USA) was 300–310 mOsm/L.

2.3. [Ca^{2+}]$_i$, NO and ROS Imaging

AITC-TRPA1-mediated changes in [Ca^{2+}]$_i$ were monitored in hCMEC/D3 cells loaded with the selective Ca^{2+}-fluorophore, loaded with 4 μM fura-2 acetoxymethyl ester (Fura-2/AM; 1 mM stock in dimethyl sulfoxide) in PSS for 30 min at 37 °C and 5% CO$_2$, as described in [12]. After washing in PSS, the coverslip was fixed to the bottom of a Petri dish and the cells were observed by an upright epifluorescence Axiolab microscope (Carl Zeiss, Oberkochen, Germany), usually equipped with a Zeiss ×40 Achroplan objective (water-immersion, 2.0 mm working distance, 0.9 numerical aperture). The cells were excited alternately at 340 and 380 nm, and the emitted light was detected at 510 nm. A first neutral density filter (1 or 0.3 optical density) reduced the overall intensity of the excitation light, and a second neutral density filter (optical density = 0.3) was coupled to the 380 nm filter to approach the intensity of the 340 nm light. A round diaphragm was used to increase the contrast. The excitation filters were mounted on a filter wheel (Lambda 10, Sutter Instrument, Novato, CA, USA). Custom software, working in the LINUX environment, was used to drive the camera (Extended-ISIS Camera, Photonic Science, Millham, UK) and the filter wheel, and to measure and plot online the fluorescence from 10 up to 40 rectangular "regions of interest" (ROIs). Each ROI was identified by a number. Since cell borders were not clearly identifiable, a ROI may not include the whole cell or may include part of an adjacent cell. Adjacent ROIs never superimposed. [Ca^{2+}]$_i$ was monitored by measuring, for each ROI, the ratio of the mean fluorescence emitted at 510 nm when exciting alternatively at 340 and 380 nm (F$_{340}$/F$_{380}$). An increase in [Ca^{2+}]$_i$ causes an increase in the ratio [31]. Ratio measurements were performed and plotted online every 3 s. The experiments were performed at room temperature (22 °C).

To evaluate NO release, hCMEC/D3 cells were loaded with 4-Amino-5-methylamino-2′,7′-difluorofluorescein diacetate (DAF-FM) (1 μM) for 60 min in PSS at 22 °C, as illustrated in [30]. DAF-FM fluorescence was measured by using the same imaging setup described above for Ca^{2+} measurements but with a different filter set, i.e., excitation at 480 nm and emission at 535 nm wavelength (emission intensity denoted as NO$_i$ (F$_{535}$/F$_0$)). The changes in DAF-FM fluorescence evoked by extracellular stimulation were recorded and plotted online every 5 s. Measurements of NO were performed at 22 °C.

To measure ROS production, hCMEC/D3 cells were loaded with 2′,7′-dichlorodihydrofluorescein diacetate (H$_2$DCF–DA) (2 μM) in PSS for 30 min at 22 °C, as recently described [32,33]. After washout of the excess fluorophore from the extracellular solution, H2DCF-DA fluorescence of the probes was recorded (excitation/emission wavelengths, 490/520 nm) by using the same imaging setup described above (denoted as ROS (F$_{491}$/F$_0$)). ROS production was measured at 22 °C.

The cellular production of NO and ROS was reported as relative fluorescence (F/F$_0$) of DAF (for NO) and H$_2$DCF-Da (for ROS), where F is the fluorescence intensity obtained during recordings and F$_0$ is the basal fluorescence intensity.

2.4. Electrophysiological Recordings

The presence of AITC-evoked membrane current was assessed by using a Port-a-patch planar patch-clamp system (Nanion Technologies, Munich, Germany) in the whole-cell, voltage-clamp or current-clamp configurations, at room temperature (22 °C), as described in [16]. Cultured cells (2–3 days after plating) were detached with Detachin and suspended at a cell density of 1–5 × 10^6 cells/mL in external recording solution containing (in mM): 145 NaCl, 2.8 KCl, 2 MgCl$_2$, 10 CaCl$_2$, 10 HEPES and 10 D-glucose (pH = 7.4). Suspended cells were placed on the NPC© chip surface, and the whole-cell configuration was achieved. Internal recording solution, containing (in mM) 10 CsCl, 110 CsF, 10 NaCl, 10 HEPES and 10 EGTA (pH = 7.2, adjusted with CsOH), was deposited in recording chips, having resistances of 3–5 MΩ. The bioelectrical response to agonist (AITC or NS309) stimulation was recorded in the current-clamp mode or in the voltage-clamp mode by using an EPC-10 patch-clamp amplifier (HEKA, Munich, Germany). Changes in the membrane potential were measured in the current-clamp mode, as shown in [34], whereas the current-to-voltage relationship of agonist-evoked membrane current was evaluated in the voltage-clamp mode by applying, every 1 s, voltage ramps ranging from -100 mV to $+100$ mV from a holding potential of -70 mV, as described in [35,36]. Immediately after the whole-cell configuration was established, the cell capacitance and the series resistances (<10 MΩ) were measured. During recordings, these two parameters were measured, and if exceeding $\geq 10\%$ with respect to the initial value, the experiment was discontinued [37]. Liquid junction potential and capacitive currents were cancelled using the automatic compensation of the EPC-10. Data were filtered at 10 kHz and sampled at 5 kHz [38].

2.5. Immunoblotting

Total proteins from primary hCMEC/D3 cells were treated with a RIPA buffer (150 mM NaCl, 0.5% sodium deoxycholate, 0.1% SDS, 0.1% Triton X-100 and 50 mM Tris-HCl, pH 8) supplemented with the protease inhibitor cocktail cOmplete (cOmplete Tablets EASYpack, 04693116001; Merck, Milan, Italy). Total proteins were solubilized in a Laemmli buffer, and 20 µg of proteins was separated by SDS-PAGE using precast gel electrophoresis (4–20% Mini-PROTEAN TGX Stain-Free Gels, Bio-Rad, Segrate, Italy) [12,39] and blotted onto the PVDF Membrane (Trans-Blot Turbo Transfer Pack, #1704156, Bio-Rad, Segrate, Italy) with the Trans-Blot Turbo Transfer System (#1704150, Bio-Rad, Segrate, Italy). After blocking for 1 h at RT in Tris-buffered saline containing 5% nonfat dry milk and 0.1% Tween (blocking solution), membranes were incubated overnight with anti-TRPA1 rabbit antibody (PA146159, 1:500 dilution; Thermo Fisher Scientific, Monza, Italy) in a blocking solution. After washing, the membranes were incubated for 1 h with goat anti-rabbit IgG antibody, peroxidase-conjugated (1:100,000; AP132P; Millipore part of Merck S.p.a., Vimodrone, Italy). Chemiluminescence detection of the bands was performed by incubating the blots with the Westar Supernova Western blotting detection system (CYANAGEN, Bologna, Italy), and the molecular weights were identified using pre-stained molecular weight markers (#161-0376, Bio-Rad Laboratories, Hercules, CA, USA). Anti-β2-microglobulin (B2M) rabbit monoclonal (ab75853, 1:10,000; Abcam, Cambridge, UK) was used on the stripped membrane [12]. Protein bands were detected with the iBright™ CL1000 Imaging System (Thermo Fisher Scientific, Monza, Italy).

2.6. Statistics

All the data have been obtained from hCMEC/D3 cells from three independent experiments. The amplitude of agonist-evoked Ca^{2+} signals was measured as the difference between the ratio at the Ca^{2+} peak and the mean ratio of 30 s baseline before the peak. The dose–response data in Figure 2B were fitted by using the following equation:

$$Y = \frac{100}{1 + \dfrac{EC_{50}}{[AITC]}} \tag{1}$$

where Y is the response (relative to the amplitude of the Ca^{2+} response), [AITC] is the AITC concentration and EC_{50} is the half-maximal effective concentration. The decay time of cyclopiazonic acid (CPA)-evoked Ca^{2+} signals was computed as the time in which $[Ca^{2+}]_i$ decayed from 80% to 20% of its peak amplitude (t_{80-20}) [40,41]. The amplitude of AITC-induced NO production was evaluated as the difference between the maximal increase in DAF-FM fluorescence and the average of 1 min baseline before the peak. The amplitude of AITC-induced ROS production was evaluated as the difference between the maximal increase in H_2DCF-DA fluorescence and the average of 1 min baseline before the peak. Statistical significance ($p < 0.05$) was evaluated by two-tailed Student's *t*-test for unpaired observations or one-way ANOVA analysis followed by the post hoc Dunnett's test, as appropriate. Data relative to Ca^{2+}, NO and ROS signals are presented as mean $\pm$ SE, whereas the number of cells analyzed is indicated in the corresponding bar histograms. The tracings shown in each figure are representative of the Ca^{2+}, NO and ROS signals evoked by each agonist in three independent experiments for each condition.

2.7. Chemicals

Fura-2/AM and DAF-FM/AM were purchased from Molecular Probes (Molecular Probes Europe BV, Leiden, The Netherlands). BTP-2 was obtained from Tocris (Bristol, UK). All other chemicals were of analytical grade and obtained from Sigma Chemical Co. (St. Louis, MO, USA).

3. Results

3.1. AITC Evokes Ca^{2+}-Dependent NO Release in hCMEC/D3 Cells

AITC (30 μM) evoked a slowly rising increase in DAF-FM fluorescence (Figure 1A), which reflected NO release and developed after a latency of 299.3 $\pm$ 10.8 s ($n = 134$). AITC-evoked NO release was significantly ($p < 0.05$) reduced by L-NIO di-hydrochloride (L-NIO; 50 μM), a selective blocker of eNOS activity [15], (Figure 1A,B). Furthermore, AITC-evoked NO release was prevented by buffering intracellular Ca^{2+} with BAPTA-AM (20 μM) (Figure 1A,B). Overall, these findings indicate that AITC stimulates NO production in a Ca^{2+}-dependent manner in hCMEC/D3 cells.

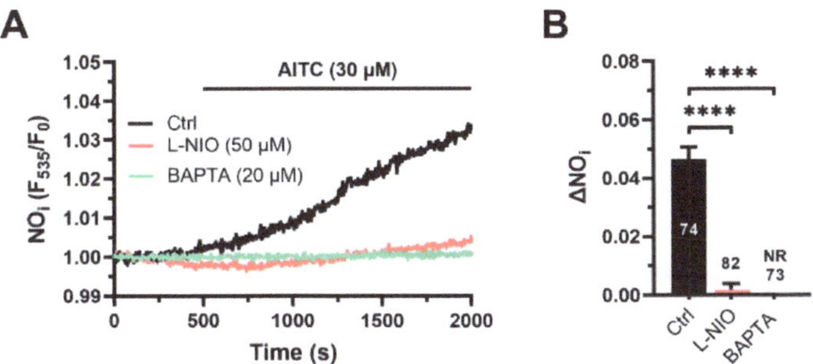

Figure 1. AITC-evoked Ca^{2+}-dependent NO release in hCMEC/D3 cells. (**A**) AITC (30 μM) evoked a slow-rising increase in DAF-FM fluorescence, which reflected NO release and was inhibited by L-NIO (50 μM, 1 h) and BAPTA (20 μM, 2 h). In this and the following figures, AITC has been administered at the time indicated by the black bar that has been drawn above the NO, Ca^{2+}, or ROS tracings. (**B**) Mean $\pm$ SE of the amplitude of AITC-evoked NO release in the absence (Ctrl) or in the presence of L-NIO or BAPTA. One-way ANOVA followed by the post hoc Dunnett's test: **** $p < 0.0001$. Ctrl: control. NR: no response.

3.2. AITC Evokes TRPA1-Independent Ca^{2+} Signals in hCMEC/D3 Cells

In order to assess whether AITC evokes an increase in $[Ca^{2+}]_i$, hCMEC/D3 cells were loaded with the Ca^{2+}-sensitive fluorophore Fura-2/AM. AITC induced a slow increase in $[Ca^{2+}]_i$ that started at 3 μM (Figure 2A,B), reached the maximal response at between 300 μM and 1 mM (Figure 2A,B) and displayed an EC_{50} of 13.77 μM (Figure 2B). The latency of the Ca^{2+} response to 30 μM AITC was 125.2 $\pm$ 5.2 s ($n = 105$), which was significantly ($p < 0.0001$, Student's *t*-test) faster than the latency of the accompanying NO signal (see above). Extensive washout of the agonist from the perfusate did not promote the recovery of the $[Ca^{2+}]_i$ to the baseline at each concentration probed (Figure 2A). This observation suggests that AITC promotes an irreversible modification either of the target channel or of the Ca^{2+}-clearing mechanisms. AITC induces an increase in $[Ca^{2+}]_i$ by activating

TRPA1 channels on the plasma membrane [20,21] or via cytosolic ROS production [23–25]. Immunoblotting identified a major band of ~140 kDa, which is the predicted molecular weight of TRPA1 protein [42], in hCMEC/D3 cells (Figure 2C). However, the Ca^{2+} response to AITC (30 µM) was not affected by HC-030031 (30 µM) (Figure 2D), which is regarded to be among the most selective TRPA1 channel inhibitors [20,42,43]. To further investigate whether the TRPA1 channel was sensitive to AITC in hCMEC/D3 cells, we exploited the planar patch-clamp technique [16]. AITC-dependent Ca^{2+} entry has been shown to hyperpolarize mouse brain microvascular endothelial cells by recruiting the small-and intermediated conductance Ca^{2+}-activated K^+ channels (respectively, SK_{Ca} and IK_{Ca}) [20,44]. However, AITC (30 µM) failed either to activate transmembrane currents (Figure S1A) or to induce membrane hyperpolarization (Figure S1B) in hCMEC/D3 cells. Nevertheless, NS309 (10 µM), a selective SK_{Ca}/IK_{Ca} opener [45,46], activated an outwardly rectifying current reverting at ~ -70 mV (Figure S1A), which is the predicted equilibrium potential for K^+ under our conditions. Furthermore, NS309 (10 µM) caused a rapid and reversible membrane hyperpolarization (Figure S1C), which is consistent with the expression of SK_{Ca}/IK_{Ca} in hCMEC/D3 cells. Therefore, the lack of effect of AITC is not due to the absence of SK_{Ca}/IK_{Ca} channels in these cells. Nevertheless, 4-Hydroxy-nonenal (4-HNE), a metabolite of lipid peroxidation that may serve as an endogenous agonist of TRPA1 in mouse brain microcirculation [5,21], evoked a rapid increase in $[Ca^{2+}]_i$ in hCMEC/D3 cells that was inhibited by HC-030031 (30 µM) (Figure S2). Overall, these findings indicate that TRPA1 is functional but is unlikely to support AITC-evoked Ca^{2+} signals in hCMEC/D3 cells.

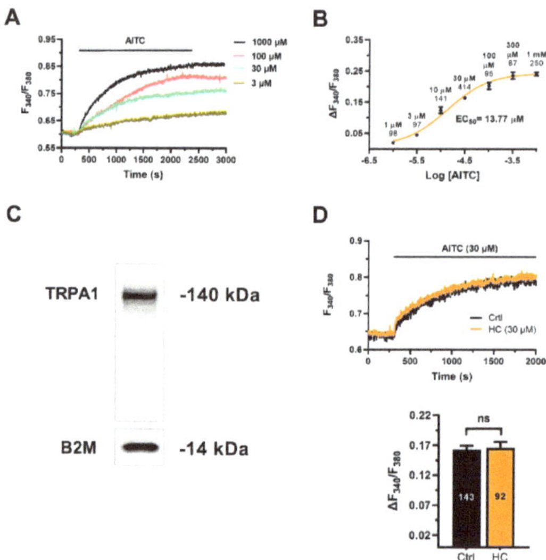

Figure 2. AITC elicits TRPA1-independent Ca^{2+} signals in hCMEC/D3 cells. (**A**) Intracellular Ca^{2+} signals evoked by increasing concentrations of AITC in hCMEC/D3 cells. The $[Ca^{2+}]_i$ never returned to the baseline upon agonist washout. The baseline of the Ca^{2+} responses has been slightly shifted to avoid tracing overlapping. (**B**) Dose-response relationship of the amplitude of AITC-evoked Ca^{2+} signals. The sigmoidal line that fits the dose -response curve was obtained by using Equation (1). (**C**) TRPA1 protein expression in hCMEC/D3 cells. Blots representative of four independent experiments are shown. Major bands of the expected molecular weights are indicated. (**D**) Upper panel, AITC-evoked Ca^{2+} signals were not inhibited by blocking TRPA1 with HC-030031 (30 µM). Lower panel, mean $\pm$ SE of the amplitude of AITC-evoked NO in the absence (Ctrl) and presence of HC-030031 (HC). NS: not significant, Student's *t*-test.

3.3. Cytosolic ROS Mediate the Ca²⁺ Response to AITC in hCMEC/D3 Cells

The role of cytosolic ROS in the Ca^{2+} response to AITC (30 µM) was evaluated in hCMEC/D3 cells pretreated with n-acetylcysteine (NAC; 1 mM, 1 h), a thiol-containing compound that serves as a direct ROS scavenger [47]. The AITC-evoked Ca^{2+} signal was significantly ($p < 0.05$) reduced in the presence of NAC, although removal of the synthetic antioxidant from the perfusate fully restored the amplitude of the increase in $[Ca^{2+}]_i$ (Figure 3A,B). Conversely, the Ca^{2+} response to AITC was not impaired by MitoTEMPO (10 µM) (Figure 3C,D), a selective scavenger of mitochondrial ROS [47]. In accord, AITC (30 µM) caused an immediate and long-lasting increase in cytosolic ROS production (Figure 3E), which was evaluated in hCMEC/D3 cells loaded with the ROS-sensitive fluorescent indicator H_2DCF–DA and exhibited a latency of 20.7 ± 1.8 s ($n = 72$). This delay was significantly ($p < 0.0001$, One-way ANOVA followed by the post hoc Dunnett's test) faster than the latency of the accompanying Ca^{2+} and NO signals (see above). AITC-induced ROS production was abolished by NAC (1 mM) (Figure 3E,F), but not MitoTEMPO (10 µM) (Figure 3E,F). Overall, these findings demonstrate that the Ca^{2+} response to AITC is triggered by cytosolic ROS production.

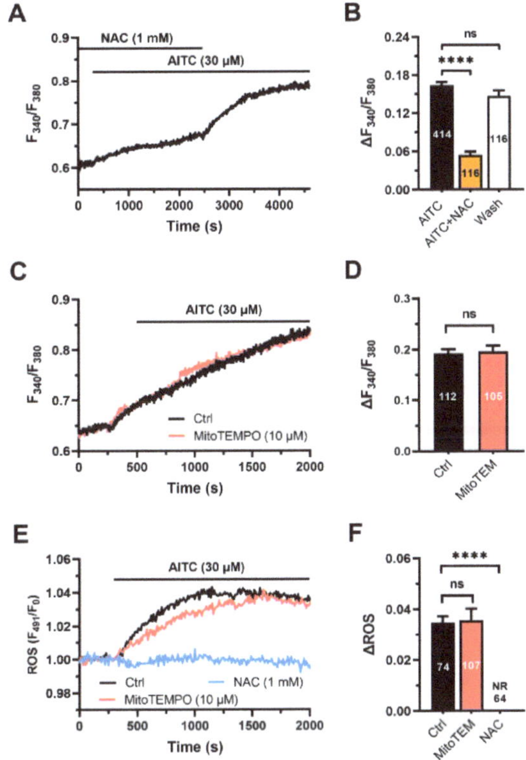

Figure 3. Cytosolic ROS drive AITC-evoked Ca^{2+} signals in hCMEC/D3 cells. (**A**) Intracellular Ca^{2+} signals evoked by AITC (30 µM) were dampened by the antioxidant NAC (1 mM, 1 h). NAC washout from the perfusate caused a further AITC-dependent elevation in $[Ca^{2+}]_i$. (**B**) Mean $\pm$ SE of the amplitude of the Ca^{2+} signals evoked by AITC in the absence (AITC) or in the presence of NAC (AITC+NAC) and upon AITC washout from the perfusate (Wash). One-way ANOVA followed by the post hoc Dunnett's test: **** $p < 0.0001$. NS: not significant. (**C**) Intracellular Ca^{2+} signals evoked by AITC (30 µM) were not affected by scavenging mitochondrial ROS with MitoTEMPO (MitoTEM;

10 μM, 1 h). (**D**) Mean $\pm$ SE of the amplitude of AITC-evoked Ca^{2+} signals in the absence (Ctrl) or in the presence of MitoTEMPO (MitoTEM). NS: not significant, Student's *t*-test. Ctrl: control. (**E**) AITC (30 μM) evoked robust ROS production in hCMEC/D3 cells that was sensitive to NAC (1 mM), but not to MitoTEMPO (MitoTEM; 10 μM). (**F**) Mean $\pm$ SE of the amplitude of AITC-evoked Ca^{2+} signals in the absence (Ctrl) or in the presence of MitoTEMPO (MitoTEM) or NAC. One-way ANOVA followed by the post hoc Dunnett's test: **** $p < 0.0001$. NR: no response.

3.4. The Ca^{2+} Response to AITC Requires Cytosolic Ca^{2+} and Is Sustained by SOCE

In hCMEC/D3 cells, intracellular Ca^{2+} signals leading to NO release are shaped by ER Ca^{2+} release and lysosomal Ca^{2+} mobilization and sustained over time by SOCE [12–14]. Under $0Ca^{2+}$ conditions, AITC (30 μM) was still able to evoke a slow increase in $[Ca^{2+}]_i$ (Figure 4A), which attained a significantly ($p < 0.05$) lower amplitude as compared to that measured in the presence of Ca^{2+} (Figure 4B). Thus, both cytosolic and extracellular Ca^{2+} sources are required to shape the Ca^{2+} response to AITC. Intriguingly, also in the absence of extracellular Ca^{2+}, the increase in $[Ca^{2+}]_i$ evoked by AITC did not decline to the baseline, as otherwise observed in response to physiological neurotransmitters [12,13].

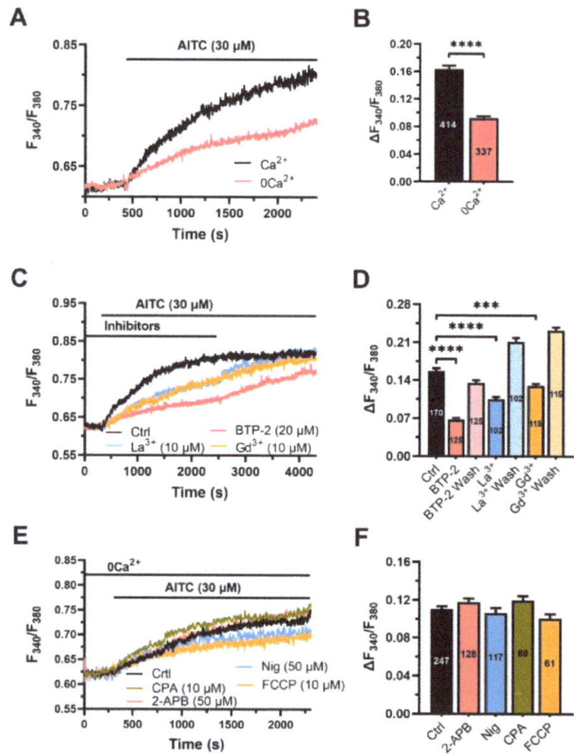

Figure 4. AITC-evoked Ca^{2+} signals require SOCE but not intracellular Ca^{2+} release. (**A**) Intracellular Ca^{2+} signals evoked by AITC (30 μM) were dampened in the absence of extracellular Ca^{2+} ($0Ca^{2+}$). (**B**) Mean $\pm$ SE of the amplitude of AITC-evoked Ca^{2+} signals in the presence (Ca^2) or in the absence of extracellular Ca^2 ($0Ca^2$). Student's *t*-test: *** $p < 0.001$. (**C**) Intracellular Ca^{2+} signals evoked by AITC (30 μM) were dampened upon SOCE inhibition with BTP-2 (20 μM, 20 min), La^{3+} (10 μM, 20 min) and Gd^{3+} (10 μM, 20 min). Removal of each inhibitor from the perfusate caused a further AITC-dependent elevation in $[Ca^{2+}]_i$. (**D**) Mean $\pm$ SE of the amplitude of AITC-evoked Ca^{2+} signals

under the following conditions: in the absence of any inhibitor (Ctrl); in the presence of BTP-2 (BTP-2) or after BTP-2 washout from the perfusate (Wash); in the presence of La^{3+} (La^{3+}) or after La^{3+} washout from the perfusate (Wash); and in the presence of Gd^{3+} (Gd^{3+}) or after Gd^{3+} washout from the perfusate (Wash). One-way ANOVA followed by the post hoc Dunnett's test: **** $p < 0.0001$; *** $p < 0.001$. (**E**) AITC (30 µM) still evoked robust intracellular Ca^{2+} signals in the presence of CPA (10 µM, 30 min), nigericin (Nig; 50 µM, 30 min), 2-APB (50 µM, 30 min) and FCCP (10 µM, 30 min). (**F**) Mean ± SE of the amplitude of AITC-evoked Ca^{2+} signals in the absence (Ctrl) or in the presence of 2-APB, nigericin (Nig), CPA or FCCP.

SOCE represents the main Ca^{2+}-entry pathway that sustains neurotransmitters-induced NO release in hCMEC/D3 cells [12,13]. The Ca^{2+} response to AITC (30 µM) was reversibly reduced by three distinct SOCE inhibitors (Figure 4C,D): BTP-2 (20 µM), La^{3+} (10 µM) and Gd^{3+} (10 µM) [48]. Thus, SOCE also sustains AITC-evoked Ca^{2+} signals in these cells. Note that removal of the SOCE inhibitors from the perfusate caused an increase in the amplitude of the increase in $[Ca^{2+}]_i$ (Figure 4C).

We then focused our attention on the intracellular Ca^{2+} pool targeted by AITC (30 µM). AITC-evoked intracellular Ca^{2+} mobilization was not attenuated by depleting either the ER Ca^{2+} reservoir with CPA (10 µM) or the lysosomal Ca^{2+} store with nigericin (50 µM) (Figure 4E,F). CPA depletes the ER Ca^{2+} store by inhibiting SERCA activity and preventing cytosolic Ca^{2+} sequestration into ER lumen [48], whereas nigericin is a protonophore that prevents cytosolic Ca^{2+} accumulation into lysosomal lumen by inhibiting the H^+-dependent Ca^{2+} uptake mechanism [49,50]. Additionally, AITC-evoked Ca^{2+} mobilization was not affected by blocking inositol-1,4,5-trisphosphate receptors (InsP$_3$Rs), which represent the sole ER Ca^{2+}-releasing channel in hCMEC/D3 cells [12], with 2-Aminoethyl Diphenylborinate (2-APB; 50 µM) (Figure 4E,F). Likewise, AITC-evoked Ca^{2+} release was insensitive to depletion of the mitochondrial Ca^{2+} pool with the protonophore FCCP (10 µM) (Figure 4E,F). Thus, AITC is unlikely to increase the $[Ca^{2+}]_i$ by targeting the main cytosolic Ca^{2+} reservoirs involved in endothelial Ca^{2+} signalling. It turns out that AITC-evoked SOCE activation is not driven by ER Ca^{2+} depletion.

3.5. AITC Inhibits PMCA Activity to Increase the $[Ca^{2+}]_i$ in hCMEC/D3 Cells

The observation that the Ca^{2+} response to AITC does not decline to the baseline either upon removal of the agonist or during prolonged stimulation under $0Ca^{2+}$ conditions suggests that AITC inhibits the Ca^{2+}-clearing mechanisms that maintain the resting $[Ca^{2+}]_i$ in hCMEC/D3 cells. These include sarco-endoplasmic Ca^{2+}-ATPase (SERCA) in the ER, the plasma membrane Ca^{2+}-ATPase (PMCA) and the Na^+/Ca^{2+} exchanger (NCX) on the plasma membrane [18]. AITC-induced inhibition of the Ca^{2+}-clearing mechanisms would result in the accumulation of cytosolic Ca^{2+} and thereby in an increase in $[Ca^{2+}]_i$. Nevertheless, SERCA inhibition does not account for the Ca^{2+} response to AITC since this fully occurs upon inhibition of SERCA activity with CPA (see Figure 4E,F).

To assess whether AITC targets NCX or PMCA or both, we exploited the strategy described to evaluate the H_2O_2-dependent blockade of PMCA activity in mouse parotid acinar cells [41]. CPA evokes a robust increase in $[Ca^{2+}]_i$ by inhibiting SERCA and thereby inducing passive ER Ca^{2+} efflux in the cytoplasm [12]. Under $0Ca^{2+}$ conditions, the Ca^{2+} response to CPA (10 µM) declines to the baseline because of cytosolic Ca^{2+} removal by NCX, PMCA or mitochondrial Ca^{2+} uniporter (MCU) (Figure 5A). Therefore, we first measured the rate of decline of the CPA-evoked increase in $[Ca^{2+}]_i$ in the absence and presence of SEA0400 (10 µM), vanadate ($VO3^-$; 500 µM) and Ru360 (10 µM), which, respectively, block NCX [51], PMCA [51] and MCU [52] (Figure 5A). Preliminary recordings showed that the green fluorescence of carboxyeosin, which also inhibits endothelial PMCA activity [53], was too bright and interfered with Fura-2 fluorescence in hCMEC/D3 cells. The decay rate of CPA-evoked intracellular Ca^{2+} release was slightly, although significantly ($p < 0.05$), slowed by either SEA0400 or Ru360 (Figure 5A,B). Conversely, in the presence of $VO3^-$, the increase in $[Ca^{2+}]_i$ evoked by CPA did not even return to the baseline and resulted in a long-lasting plateau (Figure 5A). The value of t_{80-20}, therefore, could not be measured under these conditions (Figure 5B). Once assessed that PMCA is the main factor responsible for the decay of the CPA-evoked intracellular Ca^{2+} release, we evaluated the effect of AITC (30 μM). AITC mimicked the effect of $VO3^-$ by converting the transient increase in $[Ca^{2+}]_i$ evoked by CPA under $0Ca^{2+}$ conditions into a biphasic elevation in $[Ca^{2+}]_i$ that did not enable us to measure the τ_{80-20} (Figure 5B). The amplitude of the long-lasting plateau caused by $VO3^-$ and AITC in hCMEC/D3 cells stimulated with CPA is displayed in Figure 5C. Furthermore, pretreatment with AITC significantly ($p < 0.05$) reduced the amplitude of the Ca^{2+} response to CPA (Figure 5D). These findings suggest that AITC increases $[Ca^{2+}]_i$ by inhibiting PMCA activity. Consistently, $VO3^-$ (500 μM) caused a slow increase in $[Ca^{2+}]_i$ (Figure 5E) and prevented the intracellular Ca^{2+} response to AITC (Figure 5E). In contrast, neither SEA0400 (10 μM) nor Ru360 (10 μM) increased the resting $[Ca^{2+}]_i$ (Figure 5E) or impaired AITC-induced Ca^{2+} signals under $0Ca^{2+}$ conditions (Figure 5E). The statistical analysis of these data is displayed in Figure 5F. hCMEC/D3 cells were then challenged with hydrogen peroxide (H_2O_2; 100 μM) under $0Ca^{2+}$ conditions. Exogenous administration of H_2O_2 caused a slow increase in $[Ca^{2+}]_i$ that mimicked and prevented the subsequent Ca^{2+} response to AITC (Figure 5G,H). In addition, AITC failed to slow down and convert into a long-lasting plateau the decay phase of CPA-evoked Ca^{2+} transient in hCMEC/D3 cells pretreated with NAC (1 mM) (Figure S3). Finally, exposure of hCMEC/D3 cells with H_2O_2 converted the transient increase in $[Ca^{2+}]_i$ induced by CPA (10 μM) into a biphasic Ca^{2+} signal that presented a long-lasting plateau (Figure S4), as observed with VO3- and AITC.

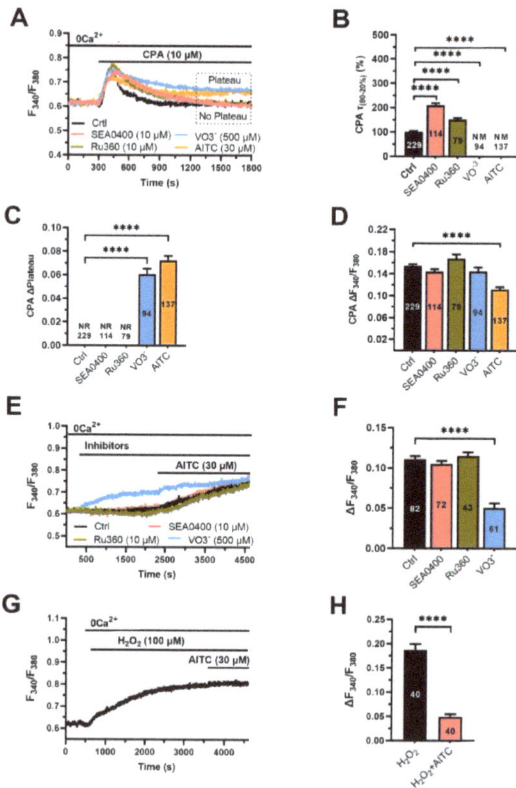

Figure 5. AITC-induced ROS production inhibits PMCA activity in hCMEC/D3 cells. (**A**) Intracellular Ca^{2+} release evoked by CPA (10 μM) in the absence (Ctrl) and presence of SEA0400 (10 μM, 30 min),

$VO3^-$ (500 µM, 30 min), Ru360 (10 µM, 30 min) and AITC (30 µM). In the presence of $VO3^-$ or AITC, the $[Ca^{2+}]_i$ did not return to the baseline, but rather decayed to a sustained plateau level. The baselines of the Ca^{2+} tracings have been overlapped for representative purposes. (**B**) Mean $\pm$ SE of the percentage change in the value of $\tau_{80\text{-}20}$ of CPA-evoked Ca^{2+} release in the absence (Ctrl) or in the presence of SEA0400, Ru360 or $VO3^-$. One-way ANOVA followed by the post hoc Dunnett's test: **** $p < 0.0001$. NM: not measurable. (**C**) Mean $\pm$ SE of the amplitude of the long-lasting plateau evoked by CPA in the presence of $VO3^-$ or AITC. One-way ANOVA followed by the post hoc Dunnett's test: **** $p < 0.0001$. NR: no response (i.e., no plateau arising). (**D**) Mean $\pm$ SE of the amplitude of CPA-evoked intracellular Ca^{2+} release in the absence (Ctrl) or in the presence of SEA0400, Ru360 or $VO3^-$. One-way ANOVA followed by the post hoc Dunnett's test: **** $p < 0.0001$. (**E**) Exogenous administration of $VO3^-$ (500 µM), but not of SEA0400 (10 µM) or Ru360 (10 µM), caused a slow increase in $[Ca^{2+}]_i$ that dampened the subsequent Ca^{2+} response to AITC (30 µM). (**F**) Mean $\pm$ SE of the amplitude of the Ca^{2+} signals evoked by AITC in the absence (Ctrl) or in the presence of SEA0400, Ru360 or $VO3^-$. One-way ANOVA followed by the post hoc Dunnett's test: **** $p < 0.0001$. (**G**) Exogenous administration of H_2O_2 (100 µM) caused a slow increase in $[Ca^{2+}]_i$ that dampened the subsequent Ca^{2+} response to AITC (30 µM). (**H**) Mean $\pm$ SE of the amplitude of the Ca^{2+} signals evoked by H_2O_2 (H_2O_2) or by AITC after H_2O_2 stimulation (AITC+ H_2O_2); Student's *t*-test: **** $p < 0.0001$.

Overall, these findings confirm that the Ca^{2+} response evoked by AITC in the absence of extracellular Ca^{2+} is due to the ROS-dependent inhibition of PMCA activity. This in turn interferes with the extrusion of cytosolic Ca^{2+} across the plasma membrane and results in a progressive elevation in $[Ca^{2+}]_i$. PMCA can be tightly coupled to store-operated channels and is the main responsible for clearing the incoming Ca^{2+} across the plasma membrane [54,55]. SOCE is constitutively activated in hCMEC/D3 cells [12]. Therefore, ROS-dependent PMCA inhibition could also unmask this background Ca^{2+} entry route and thereby explain the blocking effects of BTP-2 and trivalent cations on AITC-evoked Ca^{2+} signals.

3.6. AITC-Evoked NO Release Requires ROS-Dependent Inhibition of PMCA Activity and SOCE

In agreement with the model described above, AITC (30 µM) failed to cause a detectable NO signal in hCMEC/D3 cells pretreated with NAC (1 mM) (Figure 6A). Furthermore, AITC-evoked NO release was significantly ($p < 0.05$) reduced in the presence of either $VO3^-$ (500 µM) (Figure 6A) or BTP-2 (20 µM) (Figure 6A), which, respectively, inhibit PMCA and SOCE. The statistical analysis of these findings is reported in Figure 6B. Notably, the acute addition of $VO3^-$ caused a slow increase in DAF-FM fluorescence (Figure 6C), which was reminiscent of that caused by AITC (30 µM) and was inhibited by L-NIO (50 µM) (Figure 6C,D) and BAPTA (20 µM) (Figure 6C,D). Finally, exogenous administration of H_2O_2 (100 µM) was also able to induce NO release in hCMEC/D3 cells (Figure 6E) and, as reported for AITC and $VO3^-$, H_2O_2-induced NO production was inhibited by L-NIO (50 µM) (Figure 6C,D) and BAPTA (20 µM). These findings confirm that AITC evokes NO release in hCMEC/D3 cells by causing cytosolic ROS production, which inhibits PMCA to increase the $[Ca^{2+}]_i$ and engage eNOS.

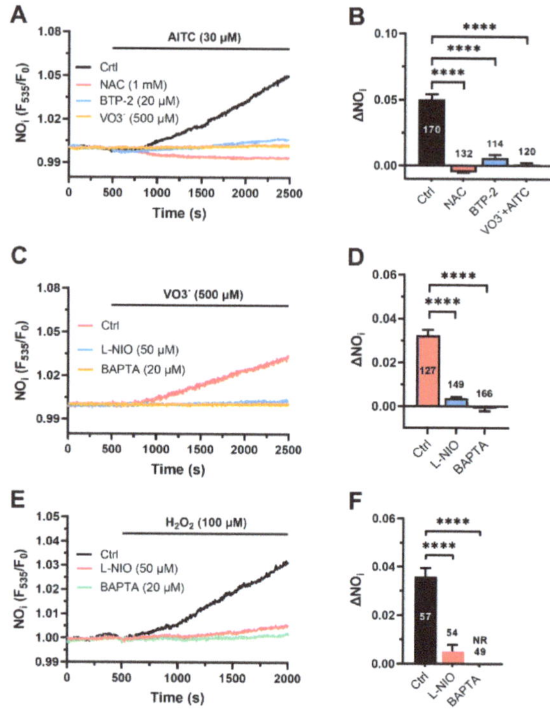

Figure 6. The role of AITC-evoked Ca^{2+} signals in AITC-induced NO release. (**A**) AITC (30 μM) evokes robust NO production in the absence (Ctrl), but not in the presence, of NAC (1 mM, 1 min), BTP-2 (20 μM, 20 min) and $VO3^-$ (500 μM, 30 min). (**B**) Mean ± SE of the amplitude of NO release induced by AITC in the absence (Ctrl) or in the presence of NAC (NAC), BTP-2 (BTP-2) or $VO3^-$. One-way ANOVA followed by the post hoc Dunnett's test: **** $p < 0.0001$. (**C**) $VO3^-$ (500 μM) evokes robust NO production in the absence (Ctrl), but not in the presence, of BAPTA (20 μM, 2 h) or L-NIO (50 μM, 1 h). (**D**) Mean ± SE of the amplitude of NO release induced by $VO3^-$ in the absence (Ctrl) or in the presence of L-NIO (L-NIO) or BAPTA (BAPTA). One-way ANOVA followed by the post hoc Dunnett's test: **** $p < 0.0001$. (**E**) Exogenous administration of H_2O_2 (100 μM) induces NO release in the absence (Ctrl), but not in the presence, of BAPTA (20 μM, 2 h) or L-NIO (50 μM, 1 h). (**F**) Mean ± SE of the amplitude of NO release induced by $VO3^-$ in the absence (Ctrl) or in the presence of L-NIO (L-NIO) or BAPTA (BAPTA). One-way ANOVA followed by the post hoc Dunnett's test: **** $p < 0.0001$.

4. Discussion

Herein, we demonstrated that AITC, a major active constituent of cruciferous vegetables, elicits NO release in the human cerebrovascular endothelial cell line hCMEC/D3 by generating intracellular ROS that inhibit PMCA activity and thereby cause a slow increase in $[Ca^{2+}]_i$. This is the first demonstration of the signalling pathway whereby AITC-dependent ROS production stimulates intracellular Ca^{2+} activity in a TRPA1-independent manner [42]. We further showed that the AITC-induced increase in $[Ca^{2+}]_i$ leads to NO release, thereby providing the proof-of-concept that AITC administration could represent a promising strategy to prevent deficits in CBF.

It has long been known that AITC elicits vasodilation in rat cerebral arteries in ex vivo brain slices [20,21] and increases meningeal blood flow in the brain of living animals [22]. These studies showed that the hemodynamic response to AITC in rat brain vasculature was accomplished by TRPA1-mediated Ca^{2+} entry in cECs, which triggered a regenerative Ca^{2+} wave propagating back to upstream arterioles and recruiting endothelial Ca^{2+}-dependent

K$^+$ channels to hyperpolarize adjacent vascular smooth muscle cells [20,21]. AITC-induced vasodilation in rat brain vasculature was insensitive to eNOS inhibition [20]. Nevertheless, the ion signalling machinery that regulates CBF can slightly differ between human and mice cECs [18]. Herein, we found that AITC causes a slow production of NO in hCMEC/D3 cells loaded with DAF-FM, a selective NO-sensitive fluorescent probe that is widely employed to measure NO release in both cultured cECs [12,13] and in brain microvessels [56]. It has previously been shown that AITC stimulates NO production in adult mouse ventricular cardiomyocytes [57], while this is the first demonstration that AITC evokes NO release in endothelial cells. The AITC-induced increase in DAF-FM fluorescence was suppressed by selectively inhibiting eNOS activity with L-NIO and by buffering intracellular Ca^{2+} levels with BAPTA. Therefore, we hypothesized that AITC was also able to induce NO release through an increase in the [Ca^{2+}]$_i$ in hCMEC/D3 cells.

Consistently, we showed that AITC caused a dose-dependent increase in [Ca^{2+}]$_i$ that was insensitive to agonist removal from the perfusate. Intriguingly, the EC$_{50}$ for AITC-evoked Ca^{2+} signals (13.77 µM) was similar to those reported for AITC-evoked Ca^{2+} sparklet in rat brain cECs (4.4 µM) [21] and for AITC-evoked vasodilation in rat cerebral arteries (16.4 µM) [44]. Nevertheless, while AITC-evoked Ca^{2+} signals [21] and vasodilation [44] in rat brain vasculature were sensitive to TRPA1 inhibition, HC-030031 did not affect the Ca^{2+} response to AITC in hCMEC/D3 cells. Intriguingly, TRPA1 protein was expressed in these cells, thereby suggesting that AITC-dependent TRPA1 activation in hCMEC/D3 cells is strongly inhibited by intracellular modulators, such as phosphatidylinositol-4,5-bisphosphate and inorganic polyphosphates [42]. In addition, AITC was still able to evoke robust, albeit slightly lower, Ca^{2+} signals under 0Ca^{2+} conditions in hCMEC/D3 cells, while AITC-evoked Ca^{2+} sparklets were suppressed by removal of extracellular Ca^{2+} and TRPA1 inhibition with HC-030031 in rat brain cECs [21]. Therefore, AITC does not increase the [Ca^{2+}]$_i$ by activating TRPA1 in hCMEC/D3 cells.

Early studies showed that AITC can induce intracellular Ca^{2+} signals that are not supported by TRPA1-mediated Ca^{2+} entry but require cytosolic ROS production in several cancer cell lines [23,24]. Herein, we confirmed that AITC induced rapid cytosolic, but not mitochondrial, ROS production and failed to induce a full increase in [Ca^{2+}]$_i$ in hCMEC/D3 cells pretreated with the widely employed antioxidant NAC. Cytosolic ROS could mobilize the intracellular Ca^{2+} pools that are located in both the ER and the acidic lysosomal compartment by directly gating ER-located InsP$_3$Rs or the lysosomal TRP Mucolipin 1 channel [58,59]. However, AITC still increased the [Ca^{2+}]$_i$ in hCMEC/D3 cells bathed in the absence of extracellular Ca^{2+} and pretreated with either CPA or nigericin that, respectively, deplete the ER and lysosomal Ca^{2+} stores. Mitochondria represent an additional Ca^{2+} reservoir, although it is unlikely to support endothelial Ca^{2+} signalling by mobilizing matrix Ca^{2+} [60]. In accord, depleting mitochondrial Ca^{2+} with FCCP did not affect AITC-evoked intracellular Ca^{2+} mobilization in hCMEC/D3 cells. These unexpected findings led us to hypothesize that AITC did not target an endogenous Ca^{2+} reservoir to increase the [Ca^{2+}]$_i$ under 0Ca^{2+} conditions, but rather inhibited the Ca^{2+}-clearing machinery. This hypothesis was also supported by the evidence that the Ca^{2+} influx component of AITC-evoked Ca^{2+} signals was hampered by the selective blockade of SOCE with La^{3+}, Gd^{3+} and BTP-2. These compounds inhibit Orai1 [48], which represents the pore-forming subunits of Ca^{2+}-selective store-operated channels [48] and mediates SOCE in hCMEC/D3 cells [12]. However, ROS are not able to directly gate extracellular Ca^{2+} entry through Orai1 [58]. Intriguingly, Orai1 channels are closely coupled to PMCA in vascular endothelial cells such that incoming Ca^{2+} is rapidly extruded across the plasma membrane by PMCA-mediated Ca^{2+} removal [48,55]. SOCE is tonically activated in hCMEC/D3 cells [12]. Therefore, we speculated that AITC-evoked ROS production could inhibit PMCA activity, thereby progressively leading to higher [Ca^{2+}]$_i$ and unmasking constitutive SOCE.

Early studies showed that ROS do inhibit PMCA activity and cause a slow and irreversibly sustained increase in [Ca^{2+}]$_i$ similar to that elicited by AITC in hCMEC/D3 cells [41,61]. Therefore, we first evaluated whether AITC slowed the decay rate of CPA-evoked intracellular Ca^{2+} release, which is largely due to PMCA activation in hCMEC/D3 cells, as shown by our preliminary characterization (Figure 5A). Pretreatment with AITC mimicked the effect of $VO3^-$ by preventing the $[Ca^{2+}]_i$ from returning to basal levels after the initial rise in $[Ca^{2+}]_i$ induced by CPA in the absence of extracellular Ca^{2+}. Furthermore, tempering AITC-induced ROS production with NAC did not remarkably affect the recovery of CPA-evoked intracellular Ca^{2+} release. Moreover, acute addition of $VO3^-$ was, per se, able to induce a sustained increase in $[Ca^{2+}]_i$ in hCMEC/D3 cells, while blocking NCX with SEA0400 or preventing mitochondrial Ca^{2+} entry with Ru360 did not alter the basal $[Ca^{2+}]_i$ and did not affect the AITC-evoked Ca^{2+} response. Finally, exogenous H_2O_2 caused a slow Ca^{2+} signal that resembled that induced by either AITC or $VO3^-$, converted the transient increase in $[Ca^{2+}]_i$ induced by CPA into a long-lasting biphasic signal and prevented the subsequent Ca^{2+} response to AITC. These findings are consistent with a model according to which PMCA is primarily involved in maintaining resting $[Ca^{2+}]_i$, such that its inhibition by AITC-induced cytosolic ROS production or $VO3^-$ results in cytosolic Ca^{2+} accumulation. In accord, the pharmacological [62] or genetic [63] blockade of PMCA activity may increase $[Ca^{2+}]_i$ in vascular endothelial cells. Overall, our data indicate that, in hCMEC/D3 cells, AITC inhibits PMCA activity through the production of cytosolic ROS, thereby attenuating Ca^{2+} extrusion across the plasma membrane and resulting in a slow increase in $[Ca^{2+}]_i$. In addition, PMCA is predicted to remove Ca^{2+} that enters into the cytosol through basally active store-operated channels [12], such that constitutive SOCE further enhances cytosolic Ca^{2+} accumulation. This model is supported by evidence that AITC-induced NO release was inhibited by preventing the accompanying increase in $[Ca^{2+}]_i$ with (1) NAC, which prevents ROS accumulation; (2) $VO3^-$, which inhibits PMCA activity; and (3) BTP-2, which inhibits SOCE. In further accordance with this hypothesis, ROS production precedes the increase in $[Ca^{2+}]_i$ that in turn comes first NO release, as testified by the latencies of the corresponding signals. Of note, early studies demonstrated that PMCA1 and PMCA4, which are both expressed in hCMEC/D3 cells [12], negatively regulate eNOS activity in vascular endothelial cells by tethering it to a low Ca^{2+} microdomain [63–65].

5. Conclusions

This investigation demonstrated that AITC stimulates NO release from hCMEC/D3 cells, a widely employed model of human brain cECs, though an increase in $[Ca^{2+}]_i$ that is driven by ROS-dependent inhibition of PMCA activity. Targeting the endothelial ion channel machinery might represent an alternative strategy to restore NO signalling and rescue CBF in brain disorders [18]. Preclinical studies confirmed that intracarotid infusion of AITC caused a remarkable dilation of dural arteries and increase meningeal blood flow [22]. Future studies will have to assess whether AITC administration is also able to increase NO production in the human brain. Interestingly, the production of high levels of ROS plays a crucial role in cerebral ischemia-reperfusion injury by causing endothelial dysfunction and BBB disruption [66–68]. However, it has long been known that low ROS signalling regulates multiple endothelial functions and could be instrumental in rescuing endothelial dysfunction in cerebrovascular and neurological disorders [58,69,70], including atherosclerosis, acute myocardial infarction, stroke and traumatic brain injury. Therefore, the findings reported in the present investigation suggest that dietary supplementation of AITC could be more efficient to prevent ischemic events in the brain by targeting the endothelial ion signalling machinery when ROS production in cerebral microcirculation is still under tight homeostatic control.

Supplementary Materials: The following supporting information can be downloaded at: https://www.mdpi.com/article/10.3390/cells12131732/s1, Figure S1: AITC does not induce bioelectrical signals in hCMEC/D3 cells. Figure S2: 4-HNE evokes TRPA1-mediated Ca^{2+} signals in hCMEC/D3 cells. Figure S3: NAC prevents AITC effect on CPA-evoked Ca^{2+} release in hCMEC/D3 cells. Figure S4: H_2O_2 prevents the decay to the baseline of CPA-evoked Ca^{2+} release.

Author Contributions: Conceptualization, R.B.-R. and F.M.; formal analysis, R.B.-R., V.B., G.P., T.S., U.L. and G.S.; investigation, R.B.-R., V.B., G.P., T.S. and G.S.; data curation, R.B.-R., V.B., G.P., T.S. and G.S.; writing—original draft preparation, F.M.; writing—review and editing, R.B.-R.; visualization, R.B.-R., V.B., G.P., T.S., U.L., G.S. and F.M.; project administration, F.M.; funding acquisition, F.M. All authors have read and agreed to the published version of the manuscript.

Funding: This research has been supported by #NEXTGENERATIONEU (NGEU) and funded by the Ministry of University and Research (MUR), National Recovery and Resilience Plan (NRRP), project MNESYS (PE0000006)—A multiscale integrated approach to the study of the nervous system in health and disease (DN. 1553 11.10.2022) (F.M. and G.S.); Regione Lombardia, regional law n° 9/2020, resolution n° 3776/2020" (F.M.); the EU Horizon 2020 FETOPEN-2018-2020 Program under Grant Agreement N. 828984, LION-HEARTED (F.M.); Italian Ministry of Education, University and Research (MIUR): Dipartimenti di Eccellenza Program (2018–2022)—Dept. of Biology and Biotechnology "L. Spallanzani", University of Pavia (F.M.), and Fondo Ricerca Giovani from the University of Pavia (F.M.). R.B.R. is supported by the National Council of Science and Technology (CONACYT), Identification Number: CVU121216.

Institutional Review Board Statement: Not applicable.

Informed Consent Statement: Not applicable.

Data Availability Statement: Original data are available upon reasonable request to the corresponding author.

Acknowledgments: The authors gratefully acknowledge the Laboratory of Cellular Electrophysiology, Centro Grandi Strumenti of the University of Pavia, for the use of the Port-a-Patch automated system.

Conflicts of Interest: The authors declare no conflict of interest.

References

1. Schaeffer, S.; Iadecola, C. Revisiting the neurovascular unit. *Nat. Neurosci.* **2021**, *24*, 1198–1209. [CrossRef]
2. Longden, T.A.; Zhao, G.; Hariharan, A.; Lederer, W.J. Pericytes and the Control of Blood Flow in Brain and Heart. *Annu. Rev. Physiol.* **2023**, *85*, 137–164. [CrossRef]
3. Longden, T.A.; Dabertrand, F.; Koide, M.; Gonzales, A.L.; Tykocki, N.R.; Brayden, J.E.; Hill-Eubanks, D.; Nelson, M.T. Capillary K(+)-sensing initiates retrograde hyperpolarization to increase local cerebral blood flow. *Nat. Neurosci.* **2017**, *20*, 717–726. [CrossRef]
4. Longden, T.A.; Mughal, A.; Hennig, G.W.; Harraz, O.F.; Shui, B.; Lee, F.K.; Lee, J.C.; Reining, S.; Kotlikoff, M.I.; Konig, G.M.; et al. Local IP3 receptor-mediated Ca^{2+} signals compound to direct blood flow in brain capillaries. *Sci. Adv.* **2021**, *7*, eabh0101. [CrossRef]
5. Thakore, P.; Alvarado, M.G.; Ali, S.; Mughal, A.; Pires, P.W.; Yamasaki, E.; Pritchard, H.A.; Isakson, B.E.; Tran, C.H.T.; Earley, S. Brain endothelial cell TRPA1 channels initiate neurovascular coupling. *eLife* **2021**, *10*, e63040. [CrossRef]
6. Negri, S.; Faris, P.; Soda, T.; Moccia, F. Endothelial signaling at the core of neurovascular coupling: The emerging role of endothelial inward-rectifier K^+ (Kir2.1) channels and N-methyl-d-aspartate receptors in the regulation of cerebral blood flow. *Int. J. Biochem. Cell Biol.* **2021**, *135*, 105983. [CrossRef]
7. Harraz, O.F.; Longden, T.A.; Hill-Eubanks, D.; Nelson, M.T. PIP2 depletion promotes TRPV4 channel activity in mouse brain capillary endothelial cells. *eLife* **2018**, *7*, e38689. [CrossRef]
8. Harraz, O.F.; Longden, T.A.; Dabertrand, F.; Hill-Eubanks, D.; Nelson, M.T. Endothelial GqPCR activity controls capillary electrical signaling and brain blood flow through PIP2 depletion. *Proc. Natl. Acad. Sci. USA* **2018**, *115*, E3569–E3577. [CrossRef]
9. Guerra, G.; Lucariello, A.; Perna, A.; Botta, L.; De Luca, A.; Moccia, F. The Role of Endothelial Ca^{2+} Signaling in Neurovascular Coupling: A View from the Lumen. *Int. J. Mol. Sci.* **2018**, *19*, 938. [CrossRef]
10. Carter, K.J.; Ward, A.T.; Kellawan, J.M.; Eldridge, M.W.; Al-Subu, A.; Walker, B.J.; Lee, J.W.; Wieben, O.; Schrage, W.G. Nitric oxide synthase inhibition in healthy adults reduces regional and total cerebral macrovascular blood flow and microvascular perfusion. *J. Physiol.* **2021**, *599*, 4973–4989. [CrossRef]
11. Hoiland, R.L.; Caldwell, H.G.; Howe, C.A.; Nowak-Fluck, D.; Stacey, B.S.; Bailey, D.M.; Paton, J.F.R.; Green, D.J.; Sekhon, M.S.; Macleod, D.B.; et al. Nitric oxide is fundamental to neurovascular coupling in humans. *J. Physiol.* **2020**, *598*, 4927–4939. [CrossRef]
12. Zuccolo, E.; Laforenza, U.; Negri, S.; Botta, L.; Berra-Romani, R.; Faris, P.; Scarpellino, G.; Forcaia, G.; Pellavio, G.; Sancini, G.; et al. Muscarinic M5 receptors trigger acetylcholine-induced Ca^{2+} signals and nitric oxide release in human brain microvascular endothelial cells. *J. Cell. Physiol.* **2019**, *234*, 4540–4562. [CrossRef]
13. Negri, S.; Faris, P.; Pellavio, G.; Botta, L.; Orgiu, M.; Forcaia, G.; Sancini, G.; Laforenza, U.; Moccia, F. Group 1 metabotropic glutamate receptors trigger glutamate-induced intracellular Ca^{2+} signals and nitric oxide release in human brain microvascular endothelial cells. *Cell. Mol. Life Sci.* **2020**, *77*, 2235–2253. [CrossRef]

14. Berra-Romani, R.; Faris, P.; Pellavio, G.; Orgiu, M.; Negri, S.; Forcaia, G.; Var-Gaz-Guadarrama, V.; Garcia-Carrasco, M.; Botta, L.; Sancini, G.; et al. Histamine induces intracellular Ca^{2+} oscillations and nitric oxide release in endothelial cells from brain microvascular circulation. *J. Cell. Physiol.* **2020**, *235*, 1515–1530. [CrossRef]

15. Negri, S.; Faris, P.; Maniezzi, C.; Pellavio, G.; Spaiardi, P.; Botta, L.; Laforenza, U.; Biella, G.; Moccia, D.F. NMDA receptors elicit flux-independent intracellular Ca^{2+} signals via metabotropic glutamate receptors and flux-dependent nitric oxide release in human brain microvascular endothelial cells. *Cell Calcium* **2021**, *99*, 102454. [CrossRef]

16. Negri, S.; Scolari, F.; Vismara, M.; Brunetti, V.; Faris, P.; Terribile, G.; Sancini, G.; Berra-Romani, R.; Moccia, F. GABA(A) and GABA(B) Receptors Mediate GABA-Induced Intracellular Ca^{2+} Signals in Human Brain Microvascular Endothelial Cells. *Cells* **2022**, *11*, 3860. [CrossRef]

17. Moccia, F.; Negri, S.; Faris, P.; Berra-Romani, R. Targeting the endothelial Ca^{2+} tool kit to rescue endothelial dysfunction in obesity associated-hypertension. *Curr. Med. Chem.* **2019**, *27*, 240–257. [CrossRef]

18. Moccia, F.; Negri, S.; Faris, P.; Angelone, T. Targeting endothelial ion signalling to rescue cerebral blood flow in cerebral disorders. *Vascul. Pharmacol.* **2022**, *145*, 106997. [CrossRef]

19. Thakore, P.; Ali, S.; Earley, S. Regulation of vascular tone by transient receptor potential ankyrin 1 channels. *Curr. Top. Membr.* **2020**, *85*, 119–150. [CrossRef]

20. Alvarado, M.G.; Thakore, P.; Earley, S. Transient Receptor Potential Channel Ankyrin 1: A Unique Regulator of Vascular Function. *Cells* **2021**, *10*, 1167. [CrossRef]

21. Sullivan, M.N.; Gonzales, A.L.; Pires, P.W.; Bruhl, A.; Leo, M.D.; Li, W.; Oulidi, A.; Boop, F.A.; Feng, Y.; Jaggar, J.H.; et al. Localized TRPA1 channel Ca^{2+} signals stimulated by reactive oxygen species promote cerebral artery dilation. *Sci. Signal.* **2015**, *8*, ra2. [CrossRef]

22. Hansted, A.K.; Bhatt, D.K.; Olesen, J.; Jensen, L.J.; Jansen-Olesen, I. Effect of TRPA1 activator allyl isothiocyanate (AITC) on rat dural and pial arteries. *Pharmacol. Rep.* **2019**, *71*, 565–572. [CrossRef]

23. Bo, P.; Lien, J.C.; Chen, Y.Y.; Yu, F.S.; Lu, H.F.; Yu, C.S.; Chou, Y.C.; Yu, C.C.; Chung, J.G. Allyl Isothiocyanate Induces Cell Toxicity by Multiple Pathways in Human Breast Cancer Cells. *Am. J. Chin. Med.* **2016**, *44*, 415–437. [CrossRef]

24. Chiang, J.H.; Tsai, F.J.; Hsu, Y.M.; Yin, M.C.; Chiu, H.Y.; Yang, J.S. Sensitivity of allyl isothiocyanate to induce apoptosis via ER stress and the mitochondrial pathway upon ROS production in colorectal adenocarcinoma cells. *Oncol. Rep.* **2020**, *44*, 1415–1424. [CrossRef]

25. Tusskorn, O.; Senggunprai, L.; Prawan, A.; Kukongviriyapan, U.; Kukongviriyapan, V. Phenethyl isothiocyanate induces calcium mobilization and mitochondrial cell death pathway in cholangiocarcinoma KKU-M214 cells. *BMC Cancer* **2013**, *13*, 571. [CrossRef]

26. Weksler, B.; Romero, I.A.; Couraud, P.O. The hCMEC/D3 cell line as a model of the human blood brain barrier. *Fluids Barriers CNS* **2013**, *10*, 16. [CrossRef]

27. Bintig, W.; Begandt, D.; Schlingmann, B.; Gerhard, L.; Pangalos, M.; Dreyer, L.; Hohnjec, N.; Couraud, P.O.; Romero, I.A.; Weksler, B.B.; et al. Purine receptors and Ca^{2+} signalling in the human blood-brain barrier endothelial cell line hCMEC/D3. *Purinergic Signal.* **2012**, *8*, 71–80. [CrossRef]

28. Bader, A.; Bintig, W.; Begandt, D.; Klett, A.; Siller, I.G.; Gregor, C.; Schaarschmidt, F.; Weksler, B.; Romero, I.; Couraud, P.O.; et al. Adenosine receptors regulate gap junction coupling of the human cerebral microvascular endothelial cells hCMEC/D3 by Ca^{2+} influx through cyclic nucleotide-gated channels. *J. Physiol.* **2017**, *595*, 2497–2517. [CrossRef]

29. Berra-Romani, R.; Faris, P.; Negri, S.; Botta, L.; Genova, T.; Moccia, F. Arachidonic Acid Evokes an Increase in Intracellular Ca^{2+} Concentration and Nitric Oxide Production in Endothelial Cells from Human Brain Microcirculation. *Cells* **2019**, *8*, 689. [CrossRef]

30. Zuccolo, E.; Kheder, D.A.; Lim, D.; Perna, A.; Nezza, F.D.; Botta, L.; Scarpellino, G.; Negri, S.; Martinotti, S.; Soda, T.; et al. Glutamate triggers intracellular Ca^{2+} oscillations and nitric oxide release by inducing NAADP- and InsP3 -dependent Ca^{2+} release in mouse brain endothelial cells. *J. Cell. Physiol.* **2019**, *234*, 3538–3554. [CrossRef]

31. Zuccolo, E.; Laforenza, U.; Ferulli, F.; Pellavio, G.; Scarpellino, G.; Tanzi, M.; Turin, I.; Faris, P.; Lucariello, A.; Maestri, M.; et al. Stim and Orai mediate constitutive Ca^{2+} entry and control endoplasmic reticulum Ca^{2+} refilling in primary cultures of colorectal carcinoma cells. *Oncotarget* **2018**, *9*, 31098–31119. [CrossRef]

32. Lodola, F.; Rosti, V.; Tullii, G.; Desii, A.; Tapella, L.; Catarsi, P.; Lim, D.; Moccia, F.; Antognazza, M.R. Conjugated polymers optically regulate the fate of endothelial colony-forming cells. *Sci. Adv.* **2019**, *5*, eaav4620. [CrossRef]

33. Negri, S.; Faris, P.; Tullii, G.; Vismara, M.; Pellegata, A.F.; Lodola, F.; Guidetti, G.; Rosti, V.; Antognazza, M.R.; Moccia, F. Conjugated polymers mediate intracellular Ca^{2+} signals in circulating endothelial colony forming cells through the reactive oxygen species-dependent activation of Transient Receptor Potential Vanilloid 1 (TRPV1). *Cell Calcium* **2022**, *101*, 102502. [CrossRef]

34. Moccia, F.; Frost, C.; Berra-Romani, R.; Tanzi, F.; Adams, D.J. Expression and function of neuronal nicotinic ACh receptors in rat microvascular endothelial cells. *Am. J. Physiol. Heart Circ. Physiol.* **2004**, *286*, H486–H491. [CrossRef]

35. Moccia, F.; Berra-Romani, R.; Baruffi, S.; Spaggiari, S.; Adams, D.J.; Taglietti, V.; Tanzi, F. Basal nonselective cation permeability in rat cardiac microvascular endothelial cells. *Microvasc. Res.* **2002**, *64*, 187–197.

36. Moccia, F.; Villa, A.; Tanzi, F. Flow-activated Na^+ and K^+ current in cardiac microvascular endothelial cells. *J. Mol. Cell. Cardiol.* **2000**, *32*, 1589–1593. [CrossRef]

37. Sobradillo, D.; Hernandez-Morales, M.; Ubierna, D.; Moyer, M.P.; Nunez, L.; Villalobos, C. A reciprocal shift in transient receptor potential channel 1 (TRPC1) and stromal interaction molecule 2 (STIM2) contributes to Ca^{2+} remodeling and cancer hallmarks in colorectal carcinoma cells. *J. Biol. Chem.* **2014**, *289*, 28765–28782. [CrossRef]

38. Locatelli, F.; Soda, T.; Montagna, I.; Tritto, S.; Botta, L.; Prestori, F.; D'Angelo, E. Calcium Channel-Dependent Induction of Long-Term Synaptic Plasticity at Excitatory Golgi Cell Synapses of Cerebellum. *J. Neurosci.* **2021**, *41*, 3307–3319. [CrossRef]

39. Remigante, A.; Morabito, R.; Spinelli, S.; Trichilo, V.; Loddo, S.; Sarikas, A.; Dossena, S.; Marino, A. d-Galactose Decreases Anion Exchange Capability through Band 3 Protein in Human Erythrocytes. *Antioxidants* **2020**, *9*, 689. [CrossRef]

40. Moccia, F.; Berra-Romani, R.; Baruffi, S.; Spaggiari, S.; Signorelli, S.; Castelli, L.; Magistretti, J.; Taglietti, V.; Tanzi, F. Ca^{2+} uptake by the endoplasmic reticulum Ca^{2+}-ATPase in rat microvascular endothelial cells. *Biochem. J.* **2002**, *364*, 235–244.

41. Kim, M.J.; Choi, K.J.; Yoon, M.N.; Oh, S.H.; Kim, D.K.; Kim, S.H.; Park, H.S. Hydrogen peroxide inhibits Ca^{2+} efflux through plasma membrane Ca^{2+}-ATPase in mouse parotid acinar cells. *Korean J. Physiol. Pharmacol.* **2018**, *22*, 215–223. [CrossRef]

42. Talavera, K.; Startek, J.B.; Alvarez-Collazo, J.; Boonen, B.; Alpizar, Y.A.; Sanchez, A.; Naert, R.; Nilius, B. Mammalian Transient Receptor Potential TRPA1 Channels: From Structure to Disease. *Physiol. Rev.* **2020**, *100*, 725–803. [CrossRef]

43. Moccia, F.; Montagna, D. Transient Receptor Potential Ankyrin 1 (TRPA1) Channel as a Sensor of Oxidative Stress in Cancer Cells. *Cells* **2023**, *12*, 1261. [CrossRef]

44. Earley, S.; Gonzales, A.L.; Crnich, R. Endothelium-dependent cerebral artery dilation mediated by TRPA1 and Ca^{2+}-Activated K^+ channels. *Circ. Res.* **2009**, *104*, 987–994. [CrossRef]

45. Hakim, M.A.; Buchholz, J.N.; Behringer, E.J. Electrical dynamics of isolated cerebral and skeletal muscle endothelial tubes: Differential roles of G-protein-coupled receptors and K^+ channels. *Pharmacol. Res. Perspect.* **2018**, *6*, e00391. [CrossRef]

46. Longden, T.A.; Dunn, K.M.; Draheim, H.J.; Nelson, M.T.; Weston, A.H.; Edwards, G. Intermediate-conductance calcium-activated potassium channels participate in neurovascular coupling. *Br. J. Pharmacol.* **2011**, *164*, 922–933. [CrossRef]

47. Bartekova, M.; Adameova, A.; Gorbe, A.; Ferenczyova, K.; Pechanova, O.; Lazou, A.; Dhalla, N.S.; Ferdinandy, P.; Giricz, Z. Natural and synthetic antioxidants targeting cardiac oxidative stress and redox signaling in cardiometabolic diseases. *Free Radic. Biol. Med.* **2021**, *169*, 446–477. [CrossRef]

48. Moccia, F.; Brunetti, V.; Perna, A.; Guerra, G.; Soda, T.; Berra-Romani, R. The Molecular Heterogeneity of Store-Operated Ca^{2+} Entry in Vascular Endothelial Cells: The Different roles of Orai1 and TRPC1/TRPC4 Channels in the Transition from Ca^{2+}-Selective to Non-Selective Cation Currents. *Int. J. Mol. Sci.* **2023**, *24*, 3259. [CrossRef]

49. Moccia, F.; Negri, S.; Faris, P.; Perna, A.; De Luca, A.; Soda, T.; Romani, R.B.; Guerra, G. Targeting Endolysosomal Two-Pore Channels to Treat Cardiovascular Disorders in the Novel COronaVIrus Disease 2019. *Front. Physiol.* **2021**, *12*, 629119. [CrossRef]

50. Negri, S.; Faris, P.; Moccia, F. Endolysosomal Ca^{2+} signaling in cardiovascular health and disease. *Int. Rev. Cell Mol. Biol.* **2021**, *363*, 203–269. [CrossRef]

51. Berra-Romani, R.; Guzman-Silva, A.; Vargaz-Guadarrama, A.; Flores-Alonso, J.C.; Alonso-Romero, J.; Trevino, S.; Sanchez-Gomez, J.; Coyotl-Santiago, N.; Garcia-Carrasco, M.; Moccia, F. Type 2 Diabetes Alters Intracellular Ca^{2+} Handling in Native Endothelium of Excised Rat Aorta. *Int. J. Mol. Sci.* **2019**, *21*, 250. [CrossRef]

52. Yanda, M.K.; Tomar, V.; Cole, R.; Guggino, W.B.; Cebotaru, L. The Mitochondrial Ca^{2+} import complex is altered in ADPKD. *Cell Calcium* **2022**, *101*, 102501. [CrossRef]

53. Sheikh, A.Q.; Hurley, J.R.; Huang, W.; Taghian, T.; Kogan, A.; Cho, H.; Wang, Y.; Narmoneva, D.A. Diabetes alters intracellular calcium transients in cardiac endothelial cells. *PLoS ONE* **2012**, *7*, e36840. [CrossRef]

54. Klishin, A.; Sedova, M.; Blatter, L.A. Time-dependent modulation of capacitative Ca^{2+} entry signals by plasma membrane Ca^{2+} pump in endothelium. *Am. J. Physiol.* **1998**, *274*, C1117–C1128.

55. Snitsarev, V.A.; Taylor, C.W. Overshooting cytosolic Ca^{2+} signals evoked by capacitative Ca^{2+} entry result from delayed stimulation of a plasma membrane Ca^{2+} pump. *Cell Calcium* **1999**, *25*, 409–417. [CrossRef]

56. Mapelli, L.; Gagliano, G.; Soda, T.; Laforenza, U.; Moccia, F.; D'Angelo, E.U. Granular Layer Neurons Control Cerebellar Neurovascular Coupling Through an NMDA Receptor/NO-Dependent System. *J. Neurosci.* **2017**, *37*, 1340–1351. [CrossRef]

57. Andrei, S.R.; Ghosh, M.; Sinharoy, P.; Damron, D.S. Stimulation of TRPA1 attenuates ischemia-induced cardiomyocyte cell death through an eNOS-mediated mechanism. *Channels* **2019**, *13*, 192–206. [CrossRef]

58. Negri, S.; Faris, P.; Moccia, F. Reactive Oxygen Species and Endothelial Ca^{2+} Signaling: Brothers in Arms or Partners in Crime? *Int. J. Mol. Sci.* **2021**, *22*, 9821. [CrossRef]

59. Zhang, X.; Cheng, X.; Yu, L.; Yang, J.; Calvo, R.; Patnaik, S.; Hu, X.; Gao, Q.; Yang, M.; Lawas, M.; et al. MCOLN1 is a ROS sensor in lysosomes that regulates autophagy. *Nat. Commun.* **2016**, *7*, 12109. [CrossRef]

60. Kluge, M.A.; Fetterman, J.L.; Vita, J.A. Mitochondria and endothelial function. *Circ. Res.* **2013**, *112*, 1171–1188. [CrossRef]

61. Bruce, J.I.; Elliott, A.C. Oxidant-impaired intracellular Ca^{2+} signaling in pancreatic acinar cells: Role of the plasma membrane Ca^{2+}-ATPase. *Am. J. Physiol. Cell Physiol.* **2007**, *293*, C938–C950. [CrossRef]

62. Szewczyk, M.M.; Pande, J.; Akolkar, G.; Grover, A.K. Caloxin 1b3: A novel plasma membrane Ca^{2+}-pump isoform 1 selective inhibitor that increases cytosolic Ca^{2+} in endothelial cells. *Cell Calcium* **2010**, *48*, 352–357. [CrossRef]

63. Long, Y.; Chen, S.W.; Gao, C.L.; He, X.M.; Liang, G.N.; Wu, J.; Jiang, C.X.; Liu, X.; Wang, F.; Chen, F. ATP2B1 Gene Silencing Increases NO Production Under Basal Conditions Through the Ca^{2+}/calmodulin/eNOS Signaling Pathway in Endothelial Cells. *Hypertens. Res.* **2018**, *41*, 246–252. [CrossRef]

64. Zaidi, A. Plasma membrane Ca-ATPases: Targets of oxidative stress in brain aging and neurodegeneration. *World J. Biol. Chem.* **2010**, *1*, 271–280. [CrossRef]

65. Holton, M.; Mohamed, T.M.; Oceandy, D.; Wang, W.; Lamas, S.; Emerson, M.; Neyses, L.; Armesilla, A.L. Endothelial nitric oxide synthase activity is inhibited by the plasma membrane calcium ATPase in human endothelial cells. *Cardiovasc. Res.* **2010**, *87*, 440–448. [CrossRef]

66. Guo, X.; Liu, R.; Jia, M.; Wang, Q.; Wu, J. Ischemia Reperfusion Injury Induced Blood Brain Barrier Dysfunction and the Involved Molecular Mechanism. *Neurochem. Res.* **2023**, *48*, 2320–2334. [CrossRef]

67. Shaw, R.L.; Norton, C.E.; Segal, S.S. Apoptosis in resistance arteries induced by hydrogen peroxide: Greater resilience of endothelium versus smooth muscle. *Am. J. Physiol. Heart Circ. Physiol.* **2021**, *320*, H1625–H1633. [CrossRef]

68. Akki, R.; Siracusa, R.; Morabito, R.; Remigante, A.; Campolo, M.; Errami, M.; La Spada, G.; Cuzzocrea, S.; Marino, A. Neuronal-like differentiated SH-SY5Y cells adaptation to a mild and transient H_2O_2-induced oxidative stress. *Cell Biochem. Funct.* **2018**, *36*, 56–64. [CrossRef]

69. Panieri, E.; Santoro, M.M. ROS signaling and redox biology in endothelial cells. *Cell. Mol. Life Sci.* **2015**, *72*, 3281–3303. [CrossRef]

70. Santoro, M.M. Fashioning blood vessels by ROS signalling and metabolism. *Semin. Cell Dev. Biol.* **2018**, *80*, 35–42. [CrossRef]

Review

Transient Receptor Potential Ankyrin 1 (TRPA1) Channel as a Sensor of Oxidative Stress in Cancer Cells

Francesco Moccia [1,*] and Daniela Montagna [2,3]

[1] Laboratory of General Physiology, Department of Biology and Biotechnology "L. Spallanzani", University of Pavia, 27100 Pavia, Italy

[2] Department of Sciences Clinic-Surgical, Diagnostic and Pediatric, University of Pavia, 27100 Pavia, Italy; d.montagna@unipv.it

[3] Pediatric Clinic, Foundation IRCCS Policlinico San Matteo, 27100 Pavia, Italy

* Correspondence: francesco.moccia@unipv.it; Tel.: +39-0382-987613

Abstract: Moderate levels of reactive oxygen species (ROS), such as hydrogen peroxide (H_2O_2), fuel tumor metastasis and invasion in a variety of cancer types. Conversely, excessive ROS levels can impair tumor growth and metastasis by triggering cancer cell death. In order to cope with the oxidative stress imposed by the tumor microenvironment, malignant cells exploit a sophisticated network of antioxidant defense mechanisms. Targeting the antioxidant capacity of cancer cells or enhancing their sensitivity to ROS-dependent cell death represent a promising strategy for alternative anticancer treatments. Transient Receptor Potential Ankyrin 1 (TRPA1) is a redox-sensitive non-selective cation channel that mediates extracellular Ca^{2+} entry upon an increase in intracellular ROS levels. The ensuing increase in intracellular Ca^{2+} concentration can in turn engage a non-canonical antioxidant defense program or induce mitochondrial Ca^{2+} dysfunction and apoptotic cell death depending on the cancer type. Herein, we sought to describe the opposing effects of ROS-dependent TRPA1 activation on cancer cell fate and propose the pharmacological manipulation of TRPA1 as an alternative therapeutic strategy to enhance cancer cell sensitivity to oxidative stress.

Keywords: cancer; reactive oxygen species; hydrogen peroxide; Transient Receptor Potential Ankyrin 1; Ca^{2+} signaling; nuclear factor erythroid 2-related factor 2; antioxidant defense; apoptosis

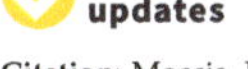

Citation: Moccia, F.; Montagna, D. Transient Receptor Potential Ankyrin 1 (TRPA1) Channel as a Sensor of Oxidative Stress in Cancer Cells. *Cells* **2023**, *12*, 1261. https://doi.org/10.3390/cells12091261

Academic Editors: Alessia Remigante and Rossana Morabito

Received: 20 March 2023
Revised: 20 April 2023
Accepted: 20 April 2023
Published: 26 April 2023

1. Introduction

Reactive oxygen species (ROS) comprise a group of highly reactive oxygen-containing molecules, including the non-radical, hydrogen peroxide (H_2O_2), the free-radicals, hydroxyl radical ($OH^\bullet$), superoxide anion ($O_2^\bullet$), peroxides ($RO^\bullet$), and oxides of nitrogen ($RO^\bullet$) [1–4]. Aberrant redox homeostasis represents a hallmark of cancer cells, since moderate ROS (e.g., 10–50 µM H_2O_2 and $O_2^\bullet$ in the nanomolar range) stimulate cell transformation, hyperproliferation, invasion, metastasis, and angiogenesis (Figure 1) [1–3,5–8]. The rate of basal ROS production in cancer cells is enhanced by multiple mechanisms [2,9], such as metabolic disturbances, adaptation to hypoxia, oncogene activation, and loss of tumor suppressors. However, excessive ROS levels (e.g., >100 µM H_2O_2; $O_2^\bullet$ in the micromolar range; ≈30 µM $OH^\bullet$) could impair tumor development and spread by triggering cell apoptosis, ferroptosis, or senescence [8,10–13]. Therefore, cancer cells cope with oxidative stress by exploiting a sophisticated network of antioxidant defense mechanisms [10,14]. Non-enzymatic small molecules directly scavenge ROS and comprise the endogenously synthesized glutathione (GSH), melatonin, and melanin, as well as the exogenously derived vitamin C, vitamin E, and β-carotene [10]. Noteworthily, GSH expression is up-regulated in a variety of cancer cell types [15], thereby increasing their antioxidant capacity. Enzymatic mechanisms include catalase (CAT), which degrades H_2O_2 to H_2O and oxygen (O_2), peroxiredoxins (PRXs), and glutathione peroxidases (GPXs), which reduce H_2O_2 to O_2, and superoxide dismutases (SODs), which catalyze the conversion of $O_2^\bullet$ to H_2O and O_2

(Figure 1) [10]. Six PRX isoforms have been described not only in the cytosol but also in multiple organelles, including mitochondria, the endoplasmic reticulum (ER), and peroxisomes, whereas eight GPX isoforms scavenge H_2O_2 in the cytosol and mitochondria [10,16]. Reduced thioredoxin (TRX) and reduced glutathione (GSH), respectively, serve as cofactors for PRXs- and GPXs-mediated reduction of H_2O_2 to H_2O. Additionally, GSH is used by glutathione-S-transferases (GSTs) to detoxify reactive compounds generated by oxidative stress [10,16]. Also SODs are spatially distributed in different subcellular compartments to favor $O_2^\bullet$ elimination: Cytoplasmic SOD (SOD-1 or Cu/Zn-SOD), mitochondrial SOD (SOD-2 or Mn-SOD), and extracellular SOD (SOD-3 or EC SOD) [17]. SOD-1 and SOD-2 rapidly dismutate O_2 into H_2O_2, which is less reactive and is reduced to O_2 and H_2O_2 by catalase or converted to H_2O_2 and oxidized glutathione by GPx [17].

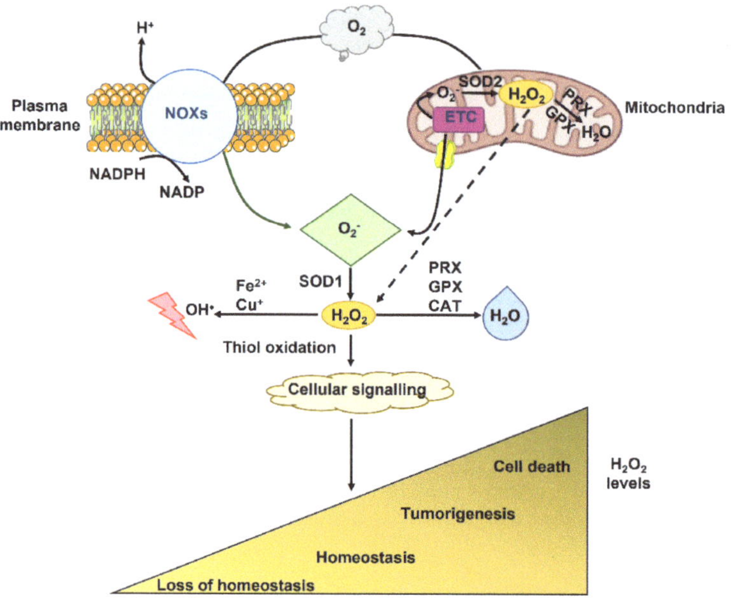

Figure 1. Generation and role of reactive oxygen species (ROS) in cancer cells. The membrane-bound NADPH oxidases (NOXs) and mitochondria are the main ROS generators in mammalian cells. NOXs catalyze the generation of intracellular superoxide anion ($O_2^\bullet$) by operating the transfer of electrons from cytosolic NADPH to molecular oxygen (O_2). Mitochondria produce ROS during cellular respiration: 1–2% of the electrons that are orderly transferred to the terminal electron acceptor, O_2, via the electron transport chain (ETC) leak from the ETC and directly react with O_2 thereby forming $O_2^\bullet$. Mitochondria-derived $O_2^\bullet$ can be released into the intermembrane space, thereby traversing the voltage-dependent anion channel into the cytosol. Herein, NOXs- and mitochondrial-derived $O_2^\bullet$ are converted into hydrogen peroxide (H_2O_2) by cytosolic superoxide dismutase 1 (SOD1). Alternately, mitochondrial-derived $O_2^\bullet$ is released into the mitochondrial matrix, where it is converted into H_2O_2 by the superoxide dismutase 2 (SOD2). H_2O_2 may freely diffuse across the mitochondrial membranes into the cytosol, or it can be detoxified into water (H_2O) in the mitochondrial matrix by glutathione peroxidase (GPX) and peroxiredoxin (PRX). In the cytosol, H_2O_2 can stimulate cellular signaling via thiol oxidation of target proteins; it can be detoxified to H_2O by PRX, GPX, and catalase (CAT); and it can interact with metal cations (Fe^{2+} and Cu^+) to generate hydroxyl radical ($OH^\bullet$), which induce cellular damage by reacting with DNA, proteins, and lipids. A reduction in cytosolic H_2O_2 levels can disrupt cellular signaling and result in a loss of cellular homeostasis. Conversely, excessive H_2O_2 levels can lead to aberrant cellular signaling and favor tumorigenesis. Uncontrolled H_2O_2 levels can result in oxidative stress and cell death.

The transcription factor, nuclear factor-erythroid 2 p45-related factor 2 (NRF2), is the master regulator of redox homeostasis in cancer cells [1,10,14]. NRF2 activity is finely regulated by kelch-like ECH-associated protein 1 (KEAP1) and the Cul3-based E3 ubiquitin ligase, which target NRF2 for proteosomal degradation. High ROS levels cause the oxidation of redox-sensitive cysteine residues in KEAP1, thereby preventing the physical interaction with and subsequent degradation of NRF2. The latter can in turn translocate into the nucleus and drive the expression of numerous antioxidant genes [10,16]. These include ROS-detoxifying enzymes, e.g., GPXs, GSTs, and PRXs, as well as the two subunits comprising the glutamate–cysteine ligase (GCL), i.e., the catalytic subunit (GCLC) and the modifier subunit (GCLM), which catalyzes the rate-limiting step in GSH biosynthesis [1,10,14]. Therefore, an alternative anticancer strategy could be designed by either increasing ROS production or reducing the antioxidant capacity of cancer cells [1,10,14,18].

Transient Receptor Potential Ankyrin 1 (TRPA1), the unique member of the mammalian TRPA sub-family, is a Ca^{2+}-permeable, non-selective cation channel that is able to integrate thermal, mechanical, and chemical signals [19,20]. Among its multiple endogenous agonists, ROS are crucial to activate TRPA1 in many disorders featured by oxidative stress, including neuropathic pain, inflammation, osteoarthritis, migraine, postischemic dysesthesia, diabetes, and respiratory diseases [19,21,22]. TRPA1-dependent extracellular Ca^{2+} entry can support several cancer hallmarks, including hyperproliferation, survival against pro-apoptotic stimuli, and invasive behavior [16,23–25]. In addition, TRPA1-mediated depolarization of peripheral nociceptors is involved in cancer-induced bone pain and cancer-related neuropathic pain [26,27]. A recent series of studies showed that TRPA1 may also serve as a crucial sensor of redox signaling in cancer cells and that extracellular Ca^{2+} entry through TRPA1 can promote either cell survival [28–30] or cell death [31–33] in response to oxidative stress. Herein, we first summarize the current knowledge about the structure and gating mechanisms of TRPA1. Then, we briefly survey the contribution of TRPA1-mediated intracellular Ca^{2+} signals to cancer cell proliferation, migration, and angiogenesis. Finally, we discuss how the redox-sensing capability of TRPA1 could be used by certain cancer cells to engage a non-canonical antioxidant defense program, while ROS-dependent TRPA1 activation leads to intracellular Ca^{2+} overload, mitochondrial dysfunction, and apoptosis in other cancer types. Therefore, TRPA1 channels could represent a novel molecular target for anticancer therapy that could be exploited either to dampen the antioxidant capacity of malignant cells or to exacerbate their sensitivity to ROS signaling.

2. TRPA1: Molecular Structure, Biophysical Properties, and Pharmacological Sensitivity

The mammalian TRP superfamily of non-selective cation channels encompasses 28 members that are subdivided into 6 sub-families based on their sequence homology: Canonical (TRPC1-7), melastatin (TRPM1-8), vanilloid (TRPV1-6), ankyrin (TRPA1), polycystin (TRPP), and mucolipin (TRPML1-3). The TRPP sub-family consists of eight members, but only TRPP2, TRPP3, and TRPP5 function as ion channels [34,35]. TRPA1 is the sole member of the TRPA subfamily and has originally been detected in a subpopulation of Aδ- and C-fiber nociceptive sensory neurons, in which it can serve as a chemical, mechanical, and thermal nocisensor [19,20]. Subsequently, TRPA1 has been found in other cell types such as epithelial cells, fibroblasts, enterochromaffin cells, mast cells, melanocytes, odontoblasts, and β-cells of the Langerhans islets, which may serve as sensory cells and interact with adjoining nociceptors [19,20]. More recently, TRPA1 expression has been confirmed in the central nervous system [36] and in cancer cells [16]. Interestingly, in the dorsal root ganglion, TRPA1 channels are also located in endolysosomes, thereby contributing to mediating intracellular Ca^{2+} release [37].

2.1. The Molecular Structure of TRPA1

The *TRPA1* gene is located in band q21.13 of chromosome 8 in humans and consists of 73.635 bases and 29 exons [38]. The protein channel encoded by the *TRPA1* gene presents an estimated molecular weight of ~127 kDa and long cytosolic NH$_2$- and COOH-terminal tails, which collectively account for ~80% of the total protein mass [38,39]. The functional TRPA1 channel protein results from the assembly of four subunits into a homotetramer through 'domain-swap' interactions [39]. The molecular architecture of the TRPA1 protein has been recently solved at a near-atomic resolution (~4 Å) by using single-particle electron cryo-microscopy [39]. As predicted by the cDNA sequence [38], each subunit consists of six transmembrane (S1–S6) domains and presents an extracellular re-entrant pore loop between S5 and S6 (Figure 2) [39]. The long NH$_2$-terminal of TRPA1 protein houses the most extensive ankyrin repeat domain (ARD) of the TRP superfamily, which comprises 14–16 ankyrin repeats (Figure 2), each consisting of a ~33 amino acids-long α-helix-β-turn-α-helix motif [39]. The ARD is connected to TM1 via the pre-S1 region, containing some cysteine residues (e.g., Cys621, Cys641, and Cys665) that are critical for TRPA1 activation by electrophilic agonists (Figure 2) [39,40]. In addition, the proximal portion of the COOH-terminus contains two residues, i.e., Arg975 and Lys989 (Figure 2), which control the voltage-dependent activation of TRPA1 at highly depolarizing potentials (>+100 mV) [41]. The ion conduction pathway of the TRPA1 channel is featured by two major constrictions, or gates, that resemble those also identified in the central cavity of TRP Vanilloid 1 (TRPV1). The outer gate is contributed to by diagonally opposed Asp915 residues, which are 7 Å apart and control Ca^{2+} permeability (Figure 2). The inner gate is formed by two hydrophobic seals established by Ile957 and Val961, which narrow the funnel to ~4 Å and thereby constrain the permeation of rehydrated cations (Figure 2). The outer pore domain of TRPA1 contains two α-helices, with a string of acidic amino acids (Glu920, Glu924, and Glu930) in the second α-helix that is likely to serve as a negatively charged conduit to repel anions and attract cations (Figure 2) [39].

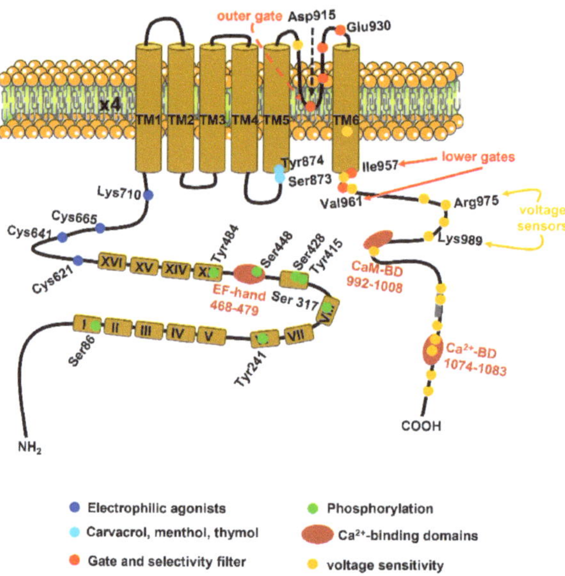

Figure 2. Molecular architecture of the human TRPA1 (hTRPA1) channel protein. Each hTRPA1 monomer comprises six transmembrane (TM) domains, a membrane-reentrant loop lining the channel pore between TM5 and TM6, and cytosolic NH$_2$- and COOH-terminal tails. Brown rounded square shapes indicate the ankyrin repeat domains (ARDs), each indicated by a Roman numeral. ARD9 and ARD10 are not shown. Putative gates selectivity filter (red circles), voltage sensors (orange circles), and Ca^{2+}-binding domains (BD) (dark red circles) are indicated. Single amino acid residues that regulate hTRPA1 channel function, underlie hTRPA1 modulation by various intracellular signaling pathways (e.g., phosphorylation and Ca^{2+}-binding), or underlie agonist- or voltage-dependent gating are shown.

2.2. Biophysical Properties and Gating Mechanisms of TRPA1

TRPA1 is a non-selective cation channel that is permeable to both monovalent (e.g., Na^+ and K^+) and divalent (e.g., Mg^{2+}) cations and may be open in the absence of exogenous stimulation [19,42,43]. In the presence of extracellular Ca^{2+}, TRPA1 presents a single-channel conductance of ~65 pS and ~110 pS for negative and positive membrane potentials, respectively [44]. Non-stimulated TRPA1 channels display a pore diameter of ~11 Å, a monovalent cation permeability sequence of $Rb^+ > K^+ > Cs^+ > Na^+ > Li^+$, high permeability for Ca^{2+} over Na (P_{Ca}/P_{Na} ~6), and a fractional Ca^{2+} current of ~17% [42,45]. Stimulation with electrophilic agonists, such as mustard oil, enhances the P_{Ca}/P_{Na} to ~9 and the fractional Ca^{2+} current to ~23% and changes the monovalent cation permeability sequence to $Ca^{2+} > Ba^{2+} > Mg^{2+} > NH4^+ > Li^+ > Na^+ > K^+ > Rb^+ > Cs^+$ [42,45]. In addition, agonist exposure can change the dimensions of the selectivity filter, thereby resulting in the progressive, but reversible, dilation of the channel pore (by 1–3 Å), which can become permeable to large organic cations, such as N-methyl-D-glucamine and the cationic dye Yo-Pro [19,45–47]. An alternative model proposed that the increase in permeability to large molecules does not reflect a change in channel permeability, but rather an elevation in the intracellular ion concentration [48].

2.3. Ca^{2+}-Dependent Regulation of TRPA1 Activity

Interestingly, TRPA1 activity can be modulated by both extracellular and intracellular Ca^{2+} [36,44]. An increase in $[Ca^{2+}]_i$ could initiate and augment TRPA1-mediated inward currents elicited by several compounds, such as Δ^9-tetrahydrocannabinol (THC) and allyl isothiocyanate (AITC) [49]. Similarly, elevating the extracellular Ca^{2+} concentration also potentiated TRPA1 channel activation and inactivation [50]. These observations led to a model according to which extracellular Ca^{2+} permeates the channel pore and thereafter regulates TRPA1 activity by binding to a site that is located within or very close to the pore [19]. The NH_2 tail contains a putative EF-hand Ca^{2+}-sensing domain that is located on ARD12 and is likely to mediate the Ca^{2+}-dependent activation of TRPA1 via the residues Asp466, Leu474 and Asp477 (Figure 2) [19,36,42,51]. An additional Ca^{2+}-binding site, which is involved in the Ca^{2+}-dependent activation and inactivation of TRPA1, is located at the distal COOH-terminal region and formed by Glu1077 and Asp1080-Asp1082 (Figure 2) [52]. Intriguingly, calmodulin (CaM) may be associated with TRPA1 in the presence of Ca^{2+} via the physical interaction with a CaM-binding domain in the COOH-tail of TRPA1 (Figure 2) [53]. It has been proposed that CaM promotes TRPA1 inactivation following extracellular Ca^{2+} influx [53].

2.4. TRPA1 Activity Is Stimulated by Physical Stimuli: Voltage, Temperature and Membrane Deformation

The gating of TRPA1 channels can be modulated by multiple physical stimuli, including changes in membrane potential, temperature, and plasma membrane tension [19]. TRPA1 can be activated by highly depolarizing voltages (>+100 mV) and presents a half-maximal voltage ($V_{1/2}$) of channel activation ranging between +90 mV and +170 mV [42,44]. As anticipated in Section 2.1, the voltage-dependent gating of TRPA1 depends on four positively charged residues, i.e., Lys969, Arg975, Lys988, and Lys989, which are located within the most proximal helix (H1) of the COOH-terminus (Figure 2) [41]. The voltage sensitivity of the channel is also sensitive to single alanine mutations of multiple residues located in the predicted helix that is centered around Lys1048 and Lys1052 [41]. Membrane depolarization could promote the physical interaction between the distal region and the more proximal voltage sensor, thereby resulting in a conformation change that causes TRPA1 activation [42]. However, the voltage-dependent gating of TRPA1 can be shifted towards more physiological membrane potentials by Ca^{2+} (see Section 2.3), electrophilic and non-electrophilic agonists, and cold or hot temperatures [19,36,42]. In accord, human TRPA1 is intrinsically sensitive to both noxious cold (<12 °C) and noxious (>43 °C) heat [54,55], although its thermosensitive profile may depend on the species [56]. Of note, the heat and cold responsiveness of the human TRPA1 (hTRPA1) is finely tuned by its redox environment [56]. ROS produced in response to dramatic changes in the local temperature could represent the driving force underlying the activation of hTRPA1 by noxious cold or hot temperatures [54]. A recent investigation proposed that the COOH-terminal region of hTRPA1 harbors temperature-sensitive modules that are allosterically coupled to the S5–S6 pore region and the S1–S4 transmembrane domain [55]. The temperature sensitivity of hTRPA1 can be modulated by the NH_2-terminal ARD [57] and is lost under reducing conditions [54]. Finally, TRPA1 channels present mechanosensitive activity that has largely been observed in C fibers exposed to large membrane deformation and in other cell types involved in mechanosensation, such as sensory neurons and Merkel cells [19,58]. Emerging evidence indicates that hTRPA1 is intrinsically mechanosensitive: when reconstituted in artificial lipid bilayers, its single-channel current activity increased in response to an increase in the lipid tension stress [59]. Interestingly, the mechanosensitivity of the human TRPA1 also does not require the NH_2-terminal ARD but is finely tuned by its redox state and is favored by a pro-oxidant environment [59].

2.5. TRPA1 As a Sensor of Redox Signaling

ROS and ROS metabolites, as well as reactive nitrogen species (RNS), are the main endogenous regulators of TRPA1 activity under physiological conditions [17,19,36,42,44]. Therefore, TRPA1 belongs to the class of TRP channels that serve as sensors of the cellular redox state [60] and include TRP Melastatin 2 (TRPM2) [61,62] and TRP Vanilloid 1 (TRPV1) [63,64]. TRPA1 can be directly activated by H_2O_2 [65,66], OH [66], the cyclopentenone prostaglandin 15-deoxy-delta(12,14)-prostaglandin J(2) [15d-PGJ(2)] [65], nitric oxide (NO) [67], and peroxynitrite ($ONOO^-$), [68]. In addition, TRPA1 channels can be stimulated by multiple endogenous aldehydes that are produced in response to lipid peroxidation [4,69], such as 4-hydroxynonenal (4-HNE), 4-oxo-nonenal, and 4-hydroxyhexenal [43,65]. Similarly, TRPA1 activity is enhanced by nitro-oleic acid, which is produced during the nitration of plasma membrane phospholipids [70]. ROS increase the open probability of TRPA1 via covalent modifications of three cysteine residues, i.e., Cys621, Cys641, and Cys665, which are situated within the pre-TM1 region at the NH_2-terminal of the channel protein (Figure 2) [71]. The exceptional reactivity of Cys621 towards electrophiles is facilitated by Lys620, whereas full TRPA1 activation by oxidative stress requires the covalent modification of Cys665 [71]. The combination of cryo-electron microscopy, which solved the full-length structure of hTRPA1 at ~a 4 Å resolution [39], and molecular modeling [72] suggested that the electrophilic reactive Cys621, Cys641, and C665 could form a ligand-binding pocket by coming in close proximity with each other [42]. The pre-TM1 region contains three additional cysteine residues, i.e., Cys619, Cys639, and Cys663, and to a lesser extent Lys708, that are not directly modified by ROS, but react to other electrophilic TRPA1 agonists, such as AITC [73]. The electrophilic activation of TRPA1 channels can also be facilitated by the disruption or formation of disulfide bonds between these, as well as other, cysteine residues at the NH_2-terminal [42,72]. Finally, the TM core of TRPA1 protein presents cysteine and lysine residues, i.e., Cys727, Lys771, and Cys834, that are likely to be exposed to the lipid environment and could, therefore, be reactive to lipophilic electrophiles [36,39].

Quantitative analysis demonstrated that TRPA1 is the most highly redox-sensitive TRP channel and can, therefore, uniquely serve as a sensor of molecular oxygen (O_2) [74]. Under normoxic conditions (~20% O_2), TRPA1 activity is tonically inhibited through hydroxylation of Pro394 in the NH_2-terminal ARD by prolyl hydroxylases (PHDs) [74], which function as the main O_2 sensor of the cell and regulate the stability of hypoxia-inducible transcription factors (HIFs) [75]. The hydroxylation activity of PHDs is decreased upon a reduction in O_2 concentration, thereby relieving the channel from inhibition and leading to TRPA1 activation under hypoxia [74]. In addition, hypoxia can induce the insertion of non-hydroxylated TRPA1 channels into the plasma membrane [74]. On the other hand, hyperoxia can activate TRPA1 through the O_2-dependent oxidation of two cysteine residues at the NH_2-terminal of the channel protein, i.e., Cys633 and Cys656 [74].

Additionally, although seemingly paradoxical, an increase in oxidative stress could occur even during hypoxia. In accord, electron transfer from ubisemiquinone to O_2 at the Q0 site of the mitochondrial complex III dramatically enhances ROS generation in hypoxic cells [76]. Therefore, TRPA1 may also serve as a molecular sensor of the local changes in the O_2 concentration [77].

3. TRPA1-Mediated Ca^{2+} Signals Support Tumorigenesis

Because of the versatility of its gating mechanisms, TRPA1 is uniquely suited to detect the increase in ROS production that supports neoplastic transformation and dissemination [16]. An additional feature of solid malignancies is represented by hypoxia, which is due to their abnormal vascular network, comprising leaky, highly disorganized, and compressed capillary vessels [78,79]. Therefore, the cancer microenvironment has been recognized as the ideal milieu to activate TRPA1 [16,26,60]. In the present section, we will briefly survey the mechanisms whereby extracellular Ca^{2+} entry via TRPA1 channels promotes cancer cell proliferation, survival, migration, and angiogenesis. In Section 4, we will specifically address how redox-sensitive TRPA1 activation could either engage non-canonical antioxidant defense programs or induce apoptosis in cancer cells. In Section 5, we focus on the pharmacology of TRPA1 channels and describe how stimulating or inhibiting TRPA1 activity could represent a novel therapeutic strategy to induce ROS-dependent cancer cell death.

Remodeling of the Ca^{2+} handling machinery, including multiple members of the TRP superfamily, contributes to many cancer hallmarks, such as aberrant proliferation, tissue invasion and metastasis, resistance to pro-apoptotic chemotherapeutics, and sustained angiogenesis [80–88]. Early work showed that the TRPA1 protein was upregulated in several cell lines and in tumor samples of human small-cell lung cancer (SCLC) [33]. TRPA1-mediated extracellular Ca^{2+} entry prevented starvation-induced SCLC cell apoptosis by recruiting extracellular signal-regulated kinases 1/2 (ERK 1/2) in an Src-dependent manner [33]. A parallel investigation demonstrated that TRPA1 expression on the plasma membrane of human lung adenocarcinoma A549 cells can be increased by inflammatory cytokines, such as interleukin (IL)-1α, IL-1β, and tumor necrosis factor α (TNFα) [89]. Similarly, TRPA1 was expressed in prostate cancer stromal cells expanded from different patients, but not in healthy primary cultured prostate epithelial cells [90]. This study showed that the antibacterial agent, triclostan, stimulated TRPA1 to mediate extracellular Ca^{2+} entry, thereby resulting in vascular endothelial growth factor (VEGF) secretion and prostate cancer cell proliferation [90]. In addition, VEGF could target adjacent endothelial cells to induce sprouting angiogenesis and favor prostate cancer vascularization [91]. Interestingly, TRPA1 protein was also largely expressed in prostate tumor-derived endothelial cells (PTECs), while it was absent in its normal counterpart [92]. TRPA1-mediated intracellular Ca^{2+} signals stimulated PTECs to migrate and assemble in capillary-like networks both in vitro and in vivo [92]. Additionally, the TRPA1 protein was expressed and mediated an increase in the intracellular Ca^{2+} concentration ($[Ca^{2+}]_i$) in human prostate cancer-associated fibroblasts (CAFs) [93]. In prostate CAFs, TRPA1 could be activated by the natural polyphenolic antioxidant, resveratrol, and induced the secretion of VEGF and hepatocyte growth factor (HGF), which in turn reduced resveratrol-induced apoptosis in co-cultured human prostate cancer cells [93]. TRPA1 was also detected in human pancreatic ductal adenocarcinoma cells (PDACs), which displayed higher levels of TRPA1 mRNA expression as compared to non-neoplastic cells [24]. TRPA1 activity, both under basal conditions and in the presence of the electrophilic agonist AITC, reduced PDAC cell migration and caused changes in cell cycle progression, i.e., induced a shift from G0/G1 to a sub-G1 phase [24]. TRPA1 could also regulate PDAC cell motility in a flux-independent manner, i.e., without the requirements for extracellular Ca^{2+} entry, as recently suggested for other TRP channels [94,95] and ligand-gated ion channels [96–98]. In agreement with this hypothesis, TRPA1 can physically associate with the fibroblast growth factor receptor 2 (FGFR2) via its NH_2-terminal ARD and thereby stimulate lung adenocarcinoma (LUAD) progression and metastatic spreading in a Ca^{2+}-independent manner [99]. A subsequent report, however, showed that FGFR2 expression is rather low in TRPA1-expressin lung cancer cells and that the TRPA1–FGFR2 interaction is likely to be a rare event in LUAD [29]. Therefore, TRPA1 can contribute to tumorigenesis, although its effect can vary depending on the cancer type, e.g., TRPA1-mediated Ca^{2+} signals stimulate and inhibit migration in PTECs and PDACs, respectively, as discussed above. TRPA1 expression and functional activity have also been reported in human uveal melanoma 92.1 cells [25], human neuroblastoma IMR-32 cells [100], and human oral squamous cell carcinoma (OSCC) samples [23]. Certainly, the validation of TRPA1 as a novel molecular target for anticancer strategies would benefit from a wider knowledge of the impact of TRPA1-mediated Ca^{2+} signals in a more extensive array of cancer types.

4. TRPA1-Mediates ROS-Dependent Intracellular Ca^{2+} Signals in Cancer Cells: Survival vs. Apoptosis

Redox signaling has long been known to regulate cellular fate via distinct spatiotemporal Ca^{2+} signatures [17,64,101,102]. Low-to-moderate ROS levels can induce intracellular Ca^{2+} oscillations that regulate proliferation [103–105], gene expression [103,106], and mitochondrial bioenergetics [107,108], while excessive ROS production results in a continual and persistent rise in $[Ca^{2+}]_i$ that stimulates cell death [109,110]. Based on the evidence that TRPA1 presents a high redox-sensing capability and that intracellular Ca^{2+} signaling finely tunes tumorigenesis, recent investigations sought to unravel whether and how TRPA1 confers cancer cells the ability to cope with or succumb to oxidative stress.

4.1. TRPA1-Mediated Ca^{2+} Influx Promotes Cancer Cell Survival to Oxidative Stress

Takahashi and coworkers recently carried out a systematic investigation to assess the role of extracellular Ca^{2+} entry through TRPA1 in the engagement of an antioxidant defense program in breast and lung cancer cells [29]. This report showed that the TRPA1 transcript and protein were up-regulated in breast and lung tumors as compared to adjacent normal tissue. Furthermore, TRPA1 activation with another electrophilic agonist, i.e., mustard oil, induced a long-lasting increase in $[Ca^{2+}]_i$ that was abolished by the removal of extracellular Ca^{2+} and genetic (via a selective short hairpin RNA) or pharmacological (via AP-18) blockade of TRPA1 [29]. Intriguingly, the same approach demonstrated that TRPA1 mediated H_2O_2-evoked intracellular Ca^{2+} oscillations in breast and lung cancer cell lines [29]. In agreement with the pro-survival role of Ca^{2+} spiking in cancer cells [111,112], extracellular Ca^{2+} entry via TRPA1 was required to promote cell survival in TRPA1-enriched cancer cells challenged with H_2O_2 [29]. In addition, ectopic expression of TRPA1 rescued cell survival and prevented apoptosis in H_2O_2-treated TRPA1-low-expressing cancer cells [29]. By using a more physiological context, the authors unveiled an increase in ROS production in the inner region of breast and lung cancer spheroids, which led to TRPA1-mediated Ca^{2+} entry and resistance to oxidative stress. Importantly, TRPA1 activation did not reduce ROS levels, thereby indicating that TRPA1 does not contribute to scavenging ROS but rather to engaging an antioxidative defense program [29]. Furthermore, TRPA1-mediated Ca^{2+} influx promoted resistance to anoikis [29], i.e., the mode of apoptotic cell death that may occur when cells detach from the extracellular matrix (ECM) and migrate to a distant point to metastasize [113]. ROS generation is crucial to induce anoikis, but TRPA1 activation underlay detachment-induced Ca^{2+} signals and anoikis resistance in the inner region of tumor spheroids without reducing intracellular ROS levels [29]. Additionally, TRPA1-mediated Ca^{2+} entry promoted breast and lung cancer cell resistance to ROS-producing chemotherapeutics, such as carboplatin, doxorubicin, and paclitaxel [29]. In agreement with in vitro findings, this study demonstrated that genetic or pharmacological blockade of TRPA1 retarded breast and lung cancer growth and suppressed chemoresistance in immunocompromised mice and confirmed that tumor cells were also exposed to higher oxidative stress in vivo. Significant levels of 8-hydroxyguanosine (8-OHdG) and 4-HNE, which are common readouts of oxidative stress, were detected in cancer cells [29]. For instance, the 4-HNE concentration can increase up to the low micromolar range in the tumor microenvironment with potential pro-apoptotic effects against cancer cells [8]. Notably, 4-HNE has long been known to stimulate TRPA1-dependent intracellular Ca^{2+} signals [114–116]. The authors then exploited a reverse-phase protein array to unravel the Ca^{2+}-dependent effectors that mediate the antioxidant defense triggered by TRPA1. They reported that TRPA1-mediated extracellular Ca^{2+} entry recruits the Ca^{2+}/calmodulin-dependent proline-rich tyrosine kinase 2 (Pyk2) [117], which in turn engages several pro-survival signaling pathways, such as RAS-ERK, phosphatidylinositol 3-kinase (PI3K)/protein kinase B (AKT), and mammalian target of rapamycin (mTOR), and increases the expression of the anti-apoptotic protein MLC-1 (Figure 3) [29]. Finally, the authors demonstrated that TRPA1 expression in breast and lung cancer cells was regulated by NRF2, which can therefore prevent ROS-induced apoptosis by inducing the expression of both canonical (e.g., antioxidant) and non-canonical (e.g., TRPA1) oxidative stress defense proteins (Figure 3) [16,29]. Interestingly, mutations in *NFE2L2* and *KEAP1 genes*, which, respectively, encode for NRF2 and KEAP1, were associated with higher TRPA1 expression in lung tumors and head-neck squamous carcinoma [29]. Genetic silencing of NRF2 reduced TRPA1 expression in lung cancer cell lines, while it did not affect the expression levels of other TRP isoforms, such as TRPC3 and TRPV1 [29]. This observation is rather interesting since TRPV1 is also sensitive to ROS signaling [64,103,106], while TRPC3 is primarily regulated by diacylglycerol [118]. Therefore, NRF2 is likely to selectively control TRPA1 expression, although NRF2-dependent regulation of other ROS-sensitive TRP isoforms, such as TRPM2 [105] and TRPV4 [119], should also be investigated. Chromatin immunoprecipitation (ChIP)-coupled deep sequencing (ChIP-Seq) identified three putative NRF2-binding sites (Peak 1 to Peak 3) around the TRPA1 gene locus [29]. However, only NRF2 binding to thPeak1 region was able to induce TRPA1 expression [29].

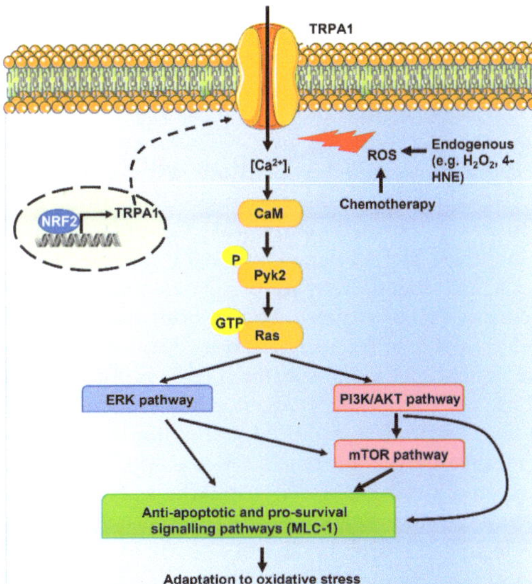

Figure 3. Extracellular Ca^{2+} entry via TRPA1 channels can induce adaptation to oxidative stress in cancer cells. ROS produced by tumor microenvironment (e.g., H_2O_2 and 4-HNE) or in response to chemotherapeutics (e.g., carboplatin, doxorubicin, and paclitaxel) activate TRPA1 on the plasma membrane. Extracellular Ca^{2+} entry, in turn, engages the Ca^{2+}/CaM-dependent Pyk2, which stimulates the monomeric G-protein, RAS, to recruit several anti-apoptotic and pro-survival signaling pathways and thereby induce adaptation to oxidative stress. Furthermore, the redox-sensitive antioxidant transcription factor, NRF2, promotes TRPA1 expression, thereby triggering a positive feedback loop to enhance tolerance to oxidative stress and promote cancer cell survival.

A recent investigation suggested that H_2O_2-induced Ca^{2+} entry via TRPA1 could also engage an antioxidant defense program in melanoma, which presents a rather high oxidative stress in the tumor microenvironment [30]. The authors showed that the number of intratumoral and peritumoral M2 macrophages and the amount of 4-HNE progressively increased with tumor severity in cutaneous melanoma samples, while TRPA1 protein expression remained unchanged [30]. In addition, genetic (via a selective small interfering RNA) and pharmacological (via A967079) blockade of TRPA1 suppressed H_2O_2-evoked intracellular Ca^{2+} signals in the human melanoma cell lines WM266-4 and SK-MEL-28 [30]. Furthermore, TRPA1-mediated Ca^{2+} signals exacerbated H_2O_2-dependent ROS production [30], which could reflect the Ca^{2+}-dependent recruitment of NOX4 [17]. Therefore, TRPA1 activation could amplify oxidative stress in cutaneous melanoma to amplify tumor progression [30]. Future work will have to assess whether TRPA1-induced ROS production delivers a pro-tumorigenic input or somehow tempers the efficacy of the non-canonical defense program engaged by TRPA1-mediated Ca^{2+} signals. Addressing this issue would be instrumental to better predicting the therapeutic outcome of TRPA1 manipulation (see also Section 5.3).

4.2. TRPA1-Mediated Ca^{2+} Influx Promotes ROS-Dependent Apoptosis in Cancer Cells

The findings described in [29] supported the prevailing view that extracellular Ca^{2+} entry via TRPA1 exerts an antioxidant defense effect in breast and lung cancer cells [16]. Nevertheless, parallel investigations provided evidence that the TRPA1-dependent increase in $[Ca^{2+}]_i$ may also support H_2O_2-induced apoptosis in other types of cancer cells. Temozolomide (TMZ) therapy represents the standard of care for the treatment of glioblastoma by inducing lethal DNA damage and subsequent ROS production [120,121]. Unfortunately, the development of TMZ resistance severely hampers its therapeutic efficacy and leads to patients' death [122]. A recent report showed that TMZ induced the expression of O6-methylguanine DNA-methyltransferase (MGMT), a DNA repair enzyme that favors glioblastoma cell resistance, and MnSOD, an antioxidant gene, in the glioblastoma cell lines SHG-44 and U251 [28]. However, previous activation of TRPA1 with Compound 16a (PF-4840154) increased ROS production, enhanced apoptosis, and reduced MGMT/MnSOD expression, thereby reducing TMZ resistance [28]. Similar results were obtained by the ectopic expression of TRPA1 in U521 cells exposed to TMZ. Mechanistic analysis revealed that TRPA1-mediated Ca^{2+} entry boosted ROS production by exacerbating TMZ-dependent damage to mitochondrial dynamics [28]. Consistently, an independent study showed that hypoxia increased TRPA1-dependent membrane currents in another human glioblastoma cell line, i.e., DBTRG, thereby inducing cytosolic Ca^{2+} overload, mitochondrial depolarization, caspase-3 and caspase-9 activation, and apoptosis (Figure 4) [32]. TMZ is also employed to treat relapsed or refractory neuroblastoma [121]. Interestingly, TMZ induced apoptosis in SH-SY5Y neuroblastoma cells via the ROS-dependent activation of TRPA1 followed by mitochondrial dysfunction and caspase activation (Figure 4) [123]. Therefore, these findings indicate that TRPA1 stimulation could represent a promising therapeutic strategy to sensitize certain cancer types to ROS-induced apoptosis. This hypothesis has been supported by two recent investigations showing that AITC induced cytosolic Ca^{2+} overload and reduced viability in OSCC PE/CA-PJ41 cells [23], whereas cinnamaldehyde, another selective electrophilic TRPA1 agonist, induced ROS-dependent apoptosis in colon cancer cells [31].

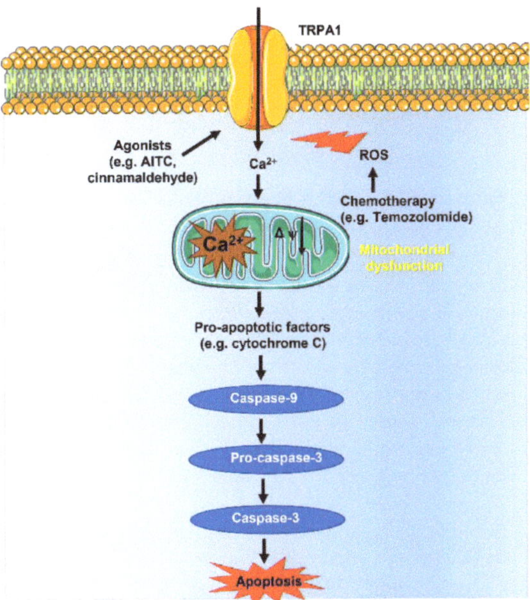

Figure 4. Extracellular Ca^{2+} entry via TRPA1 channels can induce mitochondrial Ca^{2+} overload, caspase-3 activation, and cell death in oxidative stress in cancer cells. ROS produced in response to chemotherapeutic treatments (e.g., Temozolomide) or selective agonists (e.g., AITC and cinnamaldehyde, but see also Table 1) can stimulate TRPA1 channels on the plasma membrane. The following influx of Ca^{2+} can induce mitochondrial Ca^{2+} overload and mitochondrial depolarization as well as the Ca^{2+}-dependent assembly of the mitochondrial permeability transition pore (not shown), thereby releasing pro-apoptotic factors (e.g., cytochrome C and apoptosis-inducing factor, or AIF) into the cytoplasm [124]. Herein, cytochrome C interacts with apoptosis-activating factor 1 (Apaf-1, not shown) to form a supermolecular protein complex that recruits and activates the initiator caspase-9. Caspase-9, in turn, cleaves and activates the executioner, caspase-3 [125].

4.3. Why Does ROS-Sensitive TRPA1-Mediated Ca^{2+} Influx Exert both Anti-Cancer and Pro-Tumorigenic Effects in Cancer Cells?

The highly heterogeneous spatio-temporal profile of Ca^{2+} signals orchestrates the recruitment of downstream Ca^{2+}-dependent effectors [126] and determines whether an increase in $[Ca^{2+}]_i$ induces proliferation [111,127] or senescence [128,129], apoptosis [130–132] or autophagy [84,133], sensitivity [130,131] or resistance [125,134] to anticancer strategies. Therefore, the bimodal pro- and anti-oncogenic effect of ROS-dependent Ca^{2+} influx via TRPA1 in cancer cells is not surprising. Takahashi and coworkers demonstrated that TRPA1 triggers intracellular Ca^{2+} oscillations to promote ROS resistance in lung and breast cancers [29]. Conversely, preliminary evidence suggested that TRPA1 activation led to more protracted (and pervasive) Ca^{2+} elevations in OSCC [23] and glioblastoma [28,32], thereby resulting in cancer cell apoptosis. In accord, repetitive oscillations in $[Ca^{2+}]_i$ are nicely suited to recruit Ca^{2+}-dependent effectors that promote cancer cell proliferation and survival, including Pyk2 [135–138], while avoiding mitochondrial Ca^{2+} overload [127,130,139–141]. It is unclear why TRPA1 activation evoked pro-oncogenic repetitive Ca^{2+} spikes in some, e.g., breast and lung, but not all tumor types that have been examined so far. Extracellular Ca^{2+} entry via other TRP channels, such as TRPM2 [142], TRPV1 [64], and TRPV4 [143], can elicit rhythmic Ca^{2+} release from the ER via Ca^{2+}-induced Ca^{2+} release (CICR) through inositol-1,4,5-trisphosphate ($InsP_3$) receptors ($InsP_3Rs$). Future work should investigate whether TRPA1 channel protein is selectively coupled to ER-located $InsP_3Rs$ in lung and breast cancers rather than in OSCC, glioblastoma, and colorectal carcinoma. Furthermore, TRPA1 has also been detected in acidic lysosomal vesicles [37], which represent an emerging pro-oncogenic Ca^{2+} releasing organelle in cancer cells [84,127] and stimulate $InsP_3$-dependent ER Ca^{2+} spikes via CICR [127,144,145]. Takahashi and coworkers showed that the removal of extracellular Ca^{2+} abolished H_2O_2-evoked, TRPA1-mediated intracellular Ca^{2+} oscillations in lung and breast cancer cells [29]. Nevertheless, the lysosomal expression of TRPA1 in cancer cells surely deserves future investigation.

5. Targeting TRPA1 to Sensitize Cancer Cells to Oxidative Stress

The bimodal effect of the TRPA1-mediated Ca^{2+} influx on the ability of cancer cells to cope or not with oxidative stress could be appropriately exploited for therapeutic purposes. The pharmacological blockade of TRPA1 could provide an effective strategy to reduce the Ca^{2+}-dependent recruitment of antioxidant defense and pro-survival signaling pathways in lung and breast cancers, as recently shown in [146]. Conversely, stimulating TRPA1-mediated Ca^{2+} entry could represent a valuable tool to sensitize ROS-dependent cytosolic Ca^{2+} overload and apoptosis in OSCC, glioblastoma, and colorectal carcinoma. TRPA1 channels are the most broadly tuned chemosensory channels identified so far and are sensitive to a vast panel of small molecular drugs and natural compounds, which can either stimulate or inhibit their activity with a rather high selectivity [19,36,44].

5.1. TRPA1 Activators

TRPA1 channels can be stimulated not only by ROS but also by synthetic drugs, local anesthetics, environmental irritants, and plant-derived pungent compounds [19,36,44,147]. TRPA1 agonists can be broadly categorized as electrophilic compounds, which increase the open probability through covalent modifications, and non-electrophilic activators, which exert a non-covalent modulation of the channel. Electrophilic agonists stimulate TRPA1 activity by targeting the reactive thiol groups of cysteine and lysine at the NH_2-terminal domain that are also able to sense intracellular ROS (see Section 2.5). These compounds include AITC, cinnamaldehyde, allicin, mustard oil, hydrogen sulphide, diallyl sulfide, acrolein, and JT010 [19,36,44,147,148]. Non-electrophilic modulators induce TRPA1-mediated Ca^{2+} entry without inducing covalent modifications of the channel protein and include menthol, thymol, carvacrol, local anesthetics (e.g., lidocaine, tetracaine, and procaine), anesthetic agents (e.g., propofol and etomidate), nonsteroidal anti-inflammatory drugs (e.g., acetaminophen), and several compounds exploited in cosmetics and therapeutics (e.g., alkyl esters of p-hydroxybenzoate or parabens) [19,36,44,147]. The mechanism whereby some non-electrophilic activators stimulate TRPA1 has been elucidated [44]. The non-covalent agonist, GNE551, was recently shown to increase the single-channel open probability of TRPA1 by interacting with a hydrophobic transmembrane binding site (Gln940) [149]. In contrast, the cell-penetrating peptidergic scorpion toxin (WaTx) can prolong the single-channel opening of TRPA1 by associating with an intracellular electrophile ligand-binding domain contributed to by Cys621 and Cys641 [150]. Therefore, a variety of agonists are potentially available to stimulate TRPA1 and enhance ROS-dependent apoptosis in cancer cell types that succumb in response to sustained Ca^{2+} entry via TRPA1. A comprehensive list of electrophilic and non-electrophilic TRPA1 agonists is presented in Table 1.

Table 1. List of hTRPA1 activators.

Agonist	Source	Chemical Nature	EC$_{50}$	Reference
Allyl isothiocyanate (AITC)	Mustard	Electrophilic	64 ± 3 μM	[73]
Cinnamaldehyde	Cinnamon	Electrophilic	400 ± 40 μM	[151]
Allicin	Garlic	Electrophilic	7.5 ± 0.4 μM	[152]
Hydrogen sulphide	Garlic	Electrophilic	1.8 ± 0.08 mM NaHS (mouse TRPA1)	[153]
Diallyl sulfide	Garlic	Electrophilic	254	[154]
Acrolein	Air pollutant	Electrophilic	5 ± 1 μM	[155]

Table 1. *Cont.*

Agonist	Source	Chemical Nature	EC$_{50}$	Reference
JT-010	Synthetic	Electrophilic	0.047 µM	[156]
H$_2$O$_2$	ROS	Electrophilic	290 ± 90 µM	[157]
4-HNE	ROS	Electrophilic	5 µM	[158]
PGJ$_2$	ROS	Electrophilic	5.6 µM (mouse TRPA1)	[65]
4-oxononenal (4-ONE)	ROS	Electrophilic	5.8 µM	[158]
4-hydroxyhexenal (4-HHE)	ROS	Electrophilic	≥4.3 µM	[158]
Menthol	Mint	Non-electrophilic	278 ± 30 µM	[159]
Thymol	Thyme	Non-electrophilic	127 µM	[160]
Carvacrol	Thyme	Non-electrophilic	7 µM	[160]
Lidocaine	Anesthetic	Non-electrophilic	24.000 ± 600 µM	[161]
Propofol	Anesthetic	Non-electrophilic	17 µM	[162]
WaTx	Synthetic	Non-electrophilic	16 µM	[150]

5.2. TRPA1 Blockers

The involvement of TRPA1 in a growing number of disorders (see Section 1) has favored the development of specific TRPA1 antagonists [19,36,44,147], which are listed in Table 2. The first and most widespread TRPA1 antagonist was synthesized by the Hydra Company based on its xanthine structure and was named HC-030031 [147]. Chembridge-5861528 is a derivative of HC-030031 that presents similar potency and specificity, but improved solubility [36,44]. However, the most potent available TRPA1 inhibitors are Compound 10, Compound 31, A-967079, and Glenmark 10, 15, 37 (synthesized by Glenmark), which display half-maximal inhibitory concentration (IC$_{50}$) values within the nanomolar range [19,36,44]. In addition, TRPA1 activity is sensitive to the paracetamol analog (also termed acetaminophen) 6a/b, a novel battery of α-aryl pyrrolidine sulfonamides, and the analgesics tramadol and its metabolite M1 [19]. The *Sambucus ebulus* L. (SEB) fruit extract, which can exert beneficial therapeutic effects against inflammation and ER stress-related disorders [163], has recently been shown to inhibit TRPA1 [31]. However, SEB, as well as early TRPA1 inhibitors such as camphor and ruthenium red, can target other TRP channels and is therefore not selective [31]. On the other hand, the partial agonist AP-18 inhibits TRPA1-mediated Ca^{2+} entry by desensitizing the channel [36,44].

Table 2. List of hTRPA1 inhibitors.

Inhibitor	IC$_{50}$	Reference
HC-03031	5.3–6.2 µM	[164]
Chembridge-5861528	14.3–18.7 µM	[165]
AP-18	3.1 µM	[166]
A-967079	67 nM	[167]
Compound 10	170 nM	[168]
Compound 31	15 nM	[169]

5.3. Could TRPA1 Manipulation Directly Affect Oxidative Stress in the Tumor Microenvironment?

A recent investigation demonstrated that H$_2$O$_2$-dependent TRPA1 activation can amplify oxidative stress in melanoma cell lines [30]. It is, however, still unclear whether TRPA1-induced ROS production delivers a pro-tumorigenic input or somehow tempers the efficacy of the non-canonical defense program engaged by TRPA1-mediated Ca^{2+} signals. Addressing this issue would be instrumental in better predicting the therapeutic outcome of TRPA1 manipulation. In accord, selective TRPA1 stimulation to promote caspase activation and cell death could exacerbate oxidative stress (e.g., in brain tumors), possibly further boosting cancer cell elimination. Nevertheless, if TRPA1-induced ROS production rather facilitates tumor progression, the therapeutic efficacy of this approach could be hampered by a negative-feedback mechanism. Conversely, blocking TRPA1-mediated Ca^{2+} signals to prevent the recruitment of non-canonical antioxidant programs (e.g., in lung and breast cancers) would limit TRPA1-induced ROS production, thereby either increasing (if TRPA1-dependent ROS are pro-tumorigenic) or tempering (if TRPA1-dependent ROS are pro-apoptotic) the therapeutic impact of TRPA1 manipulation.

6. Conclusions

Emerging evidence indicates that the Ca^{2+}-permeable, non-selective cation channel TRPA1 is upregulated in cancer cells. TRPA1 is the most highly redox-sensitive TRP isoform and therefore its ability to mediate Ca^{2+} entry translates the high oxidative stress of the tumor microenvironment in an intracellular Ca^{2+} signal that can dramatically impact cancer cell fate. In breast and lung cancer, ROS-induced TRPA1 activation leads to intracellular Ca^{2+} oscillations that engage antioxidant and pro-survival signaling pathways, resulting in cancer cell tolerance to oxidative stress. TRPA1-mediated Ca^{2+} entry could therefore promote resistance to pro-oxidant therapies based on ROS-producing drugs, such as carboplatin, doxorubicin, and paclitaxel. In other solid malignancies, including glioblastoma and neuroblastoma, ROS-induced TRPA1 activation results in a persistent increase in $[Ca^{2+}]_i$ that causes mitochondrial Ca^{2+} overload and damage, thereby leading to caspase 3 activation and apoptotic cell death. These findings support the notion that the pharmacological manipulation of TRPA1-mediated Ca^{2+} signals could represent an alternative anticancer strategy. For instance, selective TRPA1 agonists (e.g., brain tumors) or blockers (e.g., lung and breast cancers) could be administered as adjuvant drugs of ROS-producing therapeutics, such as carboplatin, doxorubicin, and paclitaxel, to increase cancer cell sensitivity to oxidative stress.

Author Contributions: Conceptualization, F.M.; methodology, F.M. and D.M.; writing—original draft preparation, F.M.; writing—review and editing, F.M. and D.M.; supervision, F.M.; funding acquisition, F.M. and D.M. All authors have read and agreed to the published version of the manuscript.

Funding: This research was funded by the Italian Ministry of Education, University and Research (MIUR): Dipartimenti di Eccellenza Program (2018–2022)-Dept. of Biology and Biotechnology "L. Spallanzani", University of Pavia (F.M.), Fondo Ricerca Giovani from the University of Pavia (F.M.), Program "Ricerca Corrente 08059819" of the Foundation IRCCS Policlinico San Matteo, Pavia (D.M.).

Institutional Review Board Statement: Not applicable.

Informed Consent Statement: Not applicable.

Data Availability Statement: Not applicable.

Acknowledgments: The authors gratefully acknowledge all members of their laboratories for their commitment to cancer research.

Conflicts of Interest: The authors declare no conflict of interest.

References

1. Aboelella, N.S.; Brandle, C.; Kim, T.; Ding, Z.C.; Zhou, G. Oxidative Stress in the Tumor Microenvironment and Its Relevance to Cancer Immunotherapy. *Cancers* **2021**, *13*, 986. [CrossRef]
2. Rezcek, C.R.; Chandel, N.S. The Two Faces of Reactive Oxygen Species in Cancer. *Annu. Rev. Cancer Biol.* **2017**, *1*, 79–89. [CrossRef]
3. Cheung, E.C.; Vousden, K.H. The role of ROS in tumour development and progression. *Nat. Rev. Cancer* **2022**, *22*, 280–297. [CrossRef] [PubMed]
4. Remigante, A.; Morabito, R.; Spinelli, S.; Trichilo, V.; Loddo, S.; Sarikas, A.; Dossena, S.; Marino, A. d-Galactose Decreases Anion Exchange Capability through Band 3 Protein in Human Erythrocytes. *Antioxidants* **2020**, *9*, 689. [CrossRef]
5. Panieri, E.; Santoro, M.M. ROS homeostasis and metabolism: A dangerous liason in cancer cells. *Cell Death Dis.* **2016**, *7*, e2253. [CrossRef]
6. Perillo, B.; Di Donato, M.; Pezone, A.; Di Zazzo, E.; Giovannelli, P.; Galasso, G.; Castoria, G.; Migliaccio, A. ROS in cancer therapy: The bright side of the moon. *Exp. Mol. Med.* **2020**, *52*, 192–203. [CrossRef]
7. Doskey, C.M.; Buranasudja, V.; Wagner, B.A.; Wilkes, J.G.; Du, J.; Cullen, J.J.; Buettner, G.R. Tumor cells have decreased ability to metabolize H(2)O(2): Implications for pharmacological ascorbate in cancer therapy. *Redox Biol.* **2016**, *10*, 274–284. [CrossRef] [PubMed]

8. Burdon, R.H. Superoxide and hydrogen peroxide in relation to mammalian cell proliferation. *Free Radic. Biol. Med.* **1995**, *18*, 775–794. [CrossRef] [PubMed]

9. Sullivan, L.B.; Chandel, N.S. Mitochondrial reactive oxygen species and cancer. *Cancer Metab.* **2014**, *2*, 17. [CrossRef]

10. Hayes, J.D.; Dinkova-Kostova, A.T.; Tew, K.D. Oxidative Stress in Cancer. *Cancer Cell* **2020**, *38*, 167–197. [CrossRef] [PubMed]

11. Redza-Dutordoir, M.; Averill-Bates, D.A. Activation of apoptosis signalling pathways by reactive oxygen species. *Biochim. Biophys. Acta* **2016**, *1863*, 2977–2992. [CrossRef] [PubMed]

12. Ahmad, K.A.; Iskandar, K.B.; Hirpara, J.L.; Clement, M.V.; Pervaiz, S. Hydrogen peroxide-mediated cytosolic acidification is a signal for mitochondrial translocation of Bax during drug-induced apoptosis of tumor cells. *Cancer Res.* **2004**, *64*, 7867–7878. [CrossRef] [PubMed]

13. Wu, L.; Ishigaki, Y.; Zeng, W.; Harimoto, T.; Yin, B.; Chen, Y.; Liao, S.; Liu, Y.; Sun, Y.; Zhang, X.; et al. Generation of hydroxyl radical-activatable ratiometric near-infrared bimodal probes for early monitoring of tumor response to therapy. *Nat. Commun.* **2021**, *12*, 6145. [CrossRef] [PubMed]

14. Gorrini, C.; Harris, I.S.; Mak, T.W. Modulation of oxidative stress as an anticancer strategy. *Nat. Rev. Drug Discov.* **2013**, *12*, 931–947. [CrossRef] [PubMed]

15. Gamcsik, M.P.; Kasibhatla, M.S.; Teeter, S.D.; Colvin, O.M. Glutathione levels in human tumors. *Biomarkers* **2012**, *17*, 671–691. [CrossRef]

16. Reczek, C.R.; Chandel, N.S. ROS Promotes Cancer Cell Survival through Calcium Signaling. *Cancer Cell* **2018**, *33*, 949–951. [CrossRef]

17. Negri, S.; Faris, P.; Moccia, F. Reactive Oxygen Species and Endothelial Ca(2+) Signaling: Brothers in Arms or Partners in Crime? *Int. J. Mol. Sci.* **2021**, *22*, 9821. [CrossRef]

18. Luo, M.; Zhou, L.; Huang, Z.; Li, B.; Nice, E.C.; Xu, J.; Huang, C. Antioxidant Therapy in Cancer: Rationale and Progress. *Antioxidants* **2022**, *11*, 1128. [CrossRef]

19. Talavera, K.; Startek, J.B.; Alvarez-Collazo, J.; Boonen, B.; Alpizar, Y.A.; Sanchez, A.; Naert, R.; Nilius, B. Mammalian Transient Receptor Potential TRPA1 Channels: From Structure to Disease. *Physiol. Rev.* **2020**, *100*, 725–803. [CrossRef]

20. Zygmunt, P.M.; Hogestatt, E.D. Trpa1. *Handb. Exp. Pharmacol.* **2014**, *222*, 583–630. [CrossRef]

21. Singh, R.; Adhya, P.; Sharma, S.S. Redox-sensitive TRP channels: A promising pharmacological target in chemotherapy-induced peripheral neuropathy. *Expert Opin. Ther. Targets* **2021**, *25*, 529–545. [CrossRef] [PubMed]

22. Yamamoto, S.; Shimizu, S. Significance of TRP channels in oxidative stress. *Eur. J. Pharmacol.* **2016**, *793*, 109–111. [CrossRef] [PubMed]

23. Kiss, F.; Kormos, V.; Szoke, E.; Kecskes, A.; Toth, N.; Steib, A.; Szallasi, A.; Scheich, B.; Gaszner, B.; Kun, J.; et al. Functional Transient Receptor Potential Ankyrin 1 and Vanilloid 1 Ion Channels Are Overexpressed in Human Oral Squamous Cell Carcinoma. *Int. J. Mol. Sci.* **2022**, *23*, 1921. [CrossRef] [PubMed]

24. Cojocaru, F.; Selescu, T.; Domocos, D.; Marutescu, L.; Chiritoiu, G.; Chelaru, N.R.; Dima, S.; Mihailescu, D.; Babes, A.; Cucu, D. Functional expression of the transient receptor potential ankyrin type 1 channel in pancreatic adenocarcinoma cells. *Sci. Rep.* **2021**, *11*, 2018. [CrossRef]

25. Mergler, S.; Derckx, R.; Reinach, P.S.; Garreis, F.; Bohm, A.; Schmelzer, L.; Skosyrski, S.; Ramesh, N.; Abdelmessih, S.; Polat, O.K.; et al. Calcium regulation by temperature-sensitive transient receptor potential channels in human uveal melanoma cells. *Cell. Signal.* **2014**, *26*, 56–69. [CrossRef]

26. Duitama, M.; Moreno, Y.; Santander, S.P.; Casas, Z.; Sutachan, J.J.; Torres, Y.P.; Albarracin, S.L. TRP Channels as Molecular Targets to Relieve Cancer Pain. *Biomolecules* **2021**, *12*, 1. [CrossRef]

27. de Almeida, A.S.; Bernardes, L.B.; Trevisan, G. TRP channels in cancer pain. *Eur. J. Pharmacol.* **2021**, *904*, 174185. [CrossRef]

28. Chen, H.; Li, C.; Hu, H.; Zhang, B. Activated TRPA1 plays a therapeutic role in TMZ resistance in glioblastoma by altering mitochondrial dynamics. *BMC Mol. Cell Biol.* **2022**, *23*, 38. [CrossRef]

29. Takahashi, N.; Chen, H.Y.; Harris, I.S.; Stover, D.G.; Selfors, L.M.; Bronson, R.T.; Deraedt, T.; Cichowski, K.; Welm, A.L.; Mori, Y.; et al. Cancer Cells Co-opt the Neuronal Redox-Sensing Channel TRPA1 to Promote Oxidative-Stress Tolerance. *Cancer Cell* **2018**, *33*, 985–1003.e1007. [CrossRef]

30. De Logu, F.; Souza Monteiro de Araujo, D.; Ugolini, F.; Iannone, L.F.; Vannucchi, M.; Portelli, F.; Landini, L.; Titiz, M.; De Giorgi, V.; Geppetti, P.; et al. The TRPA1 Channel Amplifies the Oxidative Stress Signal in Melanoma. *Cells* **2021**, *10*, 3131. [CrossRef]

31. Kaya, M.M.; Kaya, I.; Naziroglu, M. Transient receptor potential channel stimulation induced oxidative stress and apoptosis in the colon of mice with colitis-associated colon cancer: Modulator role of *Sambucus ebulus* L. *Mol. Biol. Rep.* **2022**, *50*, 2207–2220. [CrossRef] [PubMed]

32. Deveci, H.A.; Akyuva, Y.; Nur, G.; Naziroglu, M. Alpha lipoic acid attenuates hypoxia-induced apoptosis, inflammation and mitochondrial oxidative stress via inhibition of TRPA1 channel in human glioblastoma cell line. *Biomed. Pharmacother.* **2019**, *111*, 292–304. [CrossRef] [PubMed]

33. Schaefer, E.A.; Stohr, S.; Meister, M.; Aigner, A.; Gudermann, T.; Buech, T.R. Stimulation of the chemosensory TRPA1 cation channel by volatile toxic substances promotes cell survival of small cell lung cancer cells. *Biochem. Pharmacol.* **2013**, *85*, 426–438. [CrossRef]

34. Gees, M.; Colsoul, B.; Nilius, B. The role of transient receptor potential cation channels in Ca^{2+} signaling. *Cold Spring Harb. Perspect. Biol.* **2010**, *2*, a003962. [CrossRef] [PubMed]

35. Himmel, N.J.; Cox, D.N. Transient receptor potential channels: Current perspectives on evolution, structure, function and nomenclature. *Proc. Biol. Sci.* **2020**, *287*, 20201309. [CrossRef]

36. Alvarado, M.G.; Thakore, P.; Earley, S. Transient Receptor Potential Channel Ankyrin 1: A Unique Regulator of Vascular Function. *Cells* **2021**, *10*, 1167. [CrossRef] [PubMed]

37. Shang, S.; Zhu, F.; Liu, B.; Chai, Z.; Wu, Q.; Hu, M.; Wang, Y.; Huang, R.; Zhang, X.; Wu, X.; et al. Intracellular TRPA1 mediates Ca^{2+} release from lysosomes in dorsal root ganglion neurons. *J. Cell Biol.* **2016**, *215*, 369–381. [CrossRef]

38. Jaquemar, D.; Schenker, T.; Trueb, B. An ankyrin-like protein with transmembrane domains is specifically lost after oncogenic transformation of human fibroblasts. *J. Biol. Chem.* **1999**, *274*, 7325–7333. [CrossRef]

39. Paulsen, C.E.; Armache, J.P.; Gao, Y.; Cheng, Y.; Julius, D. Structure of the TRPA1 ion channel suggests regulatory mechanisms. *Nature* **2015**, *520*, 511–517. [CrossRef]

40. Cordero-Morales, J.F.; Gracheva, E.O.; Julius, D. Cytoplasmic ankyrin repeats of transient receptor potential A1 (TRPA1) dictate sensitivity to thermal and chemical stimuli. *Proc. Natl. Acad. Sci. USA* **2011**, *108*, E1184–E1191. [CrossRef]

41. Samad, A.; Sura, L.; Benedikt, J.; Ettrich, R.; Minofar, B.; Teisinger, J.; Vlachova, V. The C-terminal basic residues contribute to the chemical- and voltage-dependent activation of TRPA1. *Biochem. J.* **2011**, *433*, 197–204. [CrossRef]

42. Meents, J.E.; Ciotu, C.I.; Fischer, M.J.M. TRPA1: A molecular view. *J. Neurophysiol.* **2019**, *121*, 427–443. [CrossRef] [PubMed]

43. Thakore, P.; Alvarado, M.G.; Ali, S.; Mughal, A.; Pires, P.W.; Yamasaki, E.; Pritchard, H.A.; Isakson, B.E.; Tran, C.H.T.; Earley, S. Brain endothelial cell TRPA1 channels initiate neurovascular coupling. *eLife* **2021**, *10*, e63040. [CrossRef] [PubMed]

44. Thakore, P.; Ali, S.; Earley, S. Regulation of vascular tone by transient receptor potential ankyrin 1 channels. *Curr. Top. Membr.* **2020**, *85*, 119–150. [CrossRef] [PubMed]

45. Karashima, Y.; Prenen, J.; Talavera, K.; Janssens, A.; Voets, T.; Nilius, B. Agonist-induced changes in Ca(2+) permeation through the nociceptor cation channel TRPA1. *Biophys. J.* **2010**, *98*, 773–783. [CrossRef]

46. Banke, T.G.; Chaplan, S.R.; Wickenden, A.D. Dynamic changes in the TRPA1 selectivity filter lead to progressive but reversible pore dilation. *Am. J. Physiol. Cell Physiol.* **2010**, *298*, C1457–C1468. [CrossRef]

47. Chen, J.; Kim, D.; Bianchi, B.R.; Cavanaugh, E.J.; Faltynek, C.R.; Kym, P.R.; Reilly, R.M. Pore dilation occurs in TRPA1 but not in TRPM8 channels. *Mol. Pain* **2009**, *5*, 3. [CrossRef]

48. Li, M.; Toombes, G.E.; Silberberg, S.D.; Swartz, K.J. Physical basis of apparent pore dilation of ATP-activated P2X receptor channels. *Nat. Neurosci.* **2015**, *18*, 1577–1583. [CrossRef]

49. Jordt, S.E.; Bautista, D.M.; Chuang, H.H.; McKemy, D.D.; Zygmunt, P.M.; Hogestatt, E.D.; Meng, I.D.; Julius, D. Mustard oils and cannabinoids excite sensory nerve fibres through the TRP channel ANKTM1. *Nature* **2004**, *427*, 260–265. [CrossRef]

50. Wang, Y.Y.; Chang, R.B.; Waters, H.N.; McKemy, D.D.; Liman, E.R. The nociceptor ion channel TRPA1 is potentiated and inactivated by permeating calcium ions. *J. Biol. Chem.* **2008**, *283*, 32691–32703. [CrossRef]

51. Doerner, J.F.; Gisselmann, G.; Hatt, H.; Wetzel, C.H. Transient receptor potential channel A1 is directly gated by calcium ions. *J. Biol. Chem.* **2007**, *282*, 13180–13189. [CrossRef]

52. Sura, L.; Zima, V.; Marsakova, L.; Hynkova, A.; Barvik, I.; Vlachova, V. C-terminal acidic cluster is involved in Ca2+-induced regulation of human transient receptor potential ankyrin 1 channel. *J. Biol. Chem.* **2012**, *287*, 18067–18077. [CrossRef] [PubMed]

53. Hasan, R.; Leeson-Payne, A.T.; Jaggar, J.H.; Zhang, X. Calmodulin is responsible for Ca(2+)-dependent regulation of TRPA1 Channels. *Sci. Rep.* **2017**, *7*, 45098. [CrossRef]

54. Moparthi, L.; Kichko, T.I.; Eberhardt, M.; Hogestatt, E.D.; Kjellbom, P.; Johanson, U.; Reeh, P.W.; Leffler, A.; Filipovic, M.R.; Zygmunt, P.M. Human TRPA1 is a heat sensor displaying intrinsic U-shaped thermosensitivity. *Sci. Rep.* **2016**, *6*, 28763. [CrossRef]

55. Moparthi, L.; Sinica, V.; Moparthi, V.K.; Kreir, M.; Vignane, T.; Filipovic, M.R.; Vlachova, V.; Zygmunt, P.M. The human TRPA1 intrinsic cold and heat sensitivity involves separate channel structures beyond the N-ARD domain. *Nat. Commun.* **2022**, *13*, 6113. [CrossRef]

56. Sinica, V.; Vlachova, V. Transient receptor potential ankyrin 1 channel: An evolutionarily tuned thermosensor. *Physiol. Res.* **2021**, *70*, 363–381. [CrossRef] [PubMed]

57. Moparthi, L.; Survery, S.; Kreir, M.; Simonsen, C.; Kjellbom, P.; Hogestatt, E.D.; Johanson, U.; Zygmunt, P.M. Human TRPA1 is intrinsically cold- and chemosensitive with and without its N-terminal ankyrin repeat domain. *Proc. Natl. Acad. Sci. USA* **2014**, *111*, 16901–16906. [CrossRef] [PubMed]

58. Reeh, P.W.; Fischer, M.J.M. Nobel somatosensations and pain. *Pflug. Arch.* **2022**, *474*, 405–420. [CrossRef]

59. Moparthi, L.; Zygmunt, P.M. Human TRPA1 is an inherently mechanosensitive bilayer-gated ion channel. *Cell Calcium* **2020**, *91*, 102255. [CrossRef]

60. Sakaguchi, R.; Mori, Y. Transient receptor potential (TRP) channels: Biosensors for redox environmental stimuli and cellular status. *Free Radic. Biol. Med.* **2020**, *146*, 36–44. [CrossRef]

61. Ferrera, L.; Barbieri, R.; Picco, C.; Zuccolini, P.; Remigante, A.; Bertelli, S.; Fumagalli, M.R.; Zifarelli, G.; La Porta, C.A.M.; Gavazzo, P.; et al. TRPM2 Oxidation Activates Two Distinct Potassium Channels in Melanoma Cells through Intracellular Calcium Increase. *Int. J. Mol. Sci.* **2021**, *22*, 8359. [CrossRef]

62. Remigante, A.; Spinelli, S.; Marino, A.; Pusch, M.; Morabito, R.; Dossena, S. Oxidative Stress and Immune Response in Melanoma: Ion Channels as Targets of Therapy. *Int. J. Mol. Sci.* **2023**, *24*, 887. [CrossRef]

63. Ogawa, N.; Kurokawa, T.; Fujiwara, K.; Polat, O.K.; Badr, H.; Takahashi, N.; Mori, Y. Functional and Structural Divergence in Human TRPV1 Channel Subunits by Oxidative Cysteine Modification. *J. Biol. Chem.* **2016**, *291*, 4197–4210. [CrossRef] [PubMed]

64. Negri, S.; Faris, P.; Tullii, G.; Vismara, M.; Pellegata, A.F.; Lodola, F.; Guidetti, G.; Rosti, V.; Antognazza, M.R.; Moccia, F. Conjugated polymers mediate intracellular Ca(2+) signals in circulating endothelial colony forming cells through the reactive oxygen species-dependent activation of Transient Receptor Potential Vanilloid 1 (TRPV1). *Cell Calcium* **2022**, *101*, 102502. [CrossRef]

65. Andersson, D.A.; Gentry, C.; Moss, S.; Bevan, S. Transient receptor potential A1 is a sensory receptor for multiple products of oxidative stress. *J. Neurosci.* **2008**, *28*, 2485–2494. [CrossRef]

66. Bessac, B.F.; Sivula, M.; von Hehn, C.A.; Escalera, J.; Cohn, L.; Jordt, S.E. TRPA1 is a major oxidant sensor in murine airway sensory neurons. *J. Clin. Investig.* **2008**, *118*, 1899–1910. [CrossRef]

67. Miyamoto, T.; Dubin, A.E.; Petrus, M.J.; Patapoutian, A. TRPV1 and TRPA1 mediate peripheral nitric oxide-induced nociception in mice. *PLoS ONE* **2009**, *4*, e7596. [CrossRef] [PubMed]

68. Andersson, D.A.; Filipovic, M.R.; Gentry, C.; Eberhardt, M.; Vastani, N.; Leffler, A.; Reeh, P.; Bevan, S. Streptozotocin Stimulates the Ion Channel TRPA1 Directly: Involvement of peroxynitrite. *J. Biol. Chem.* **2015**, *290*, 15185–15196. [CrossRef] [PubMed]

69. Crupi, R.; Morabito, R.; Remigante, A.; Gugliandolo, E.; Britti, D.; Cuzzocrea, S.; Marino, A. Susceptibility of erythrocytes from different sources to xenobiotics-induced lysis. *Comp. Biochem. Physiol. C Toxicol. Pharmacol.* **2019**, *221*, 68–72. [CrossRef]

70. Taylor-Clark, T.E.; Ghatta, S.; Bettner, W.; Undem, B.J. Nitrooleic acid, an endogenous product of nitrative stress, activates nociceptive sensory nerves via the direct activation of TRPA1. *Mol. Pharmacol.* **2009**, *75*, 820–829. [CrossRef]

71. Bahia, P.K.; Parks, T.A.; Stanford, K.R.; Mitchell, D.A.; Varma, S.; Stevens, S.M., Jr.; Taylor-Clark, T.E. The exceptionally high reactivity of Cys 621 is critical for electrophilic activation of the sensory nerve ion channel TRPA1. *J. Gen. Physiol.* **2016**, *147*, 451–465. [CrossRef]

72. Zayats, V.; Samad, A.; Minofar, B.; Roelofs, K.E.; Stockner, T.; Ettrich, R. Regulation of the transient receptor potential channel TRPA1 by its N-terminal ankyrin repeat domain. *J. Mol. Model.* **2013**, *19*, 4689–4700. [CrossRef] [PubMed]

73. Hinman, A.; Chuang, H.H.; Bautista, D.M.; Julius, D. TRP channel activation by reversible covalent modification. *Proc. Natl. Acad. Sci. USA* **2006**, *103*, 19564–19568. [CrossRef] [PubMed]

74. Takahashi, N.; Kuwaki, T.; Kiyonaka, S.; Numata, T.; Kozai, D.; Mizuno, Y.; Yamamoto, S.; Naito, S.; Knevels, E.; Carmeliet, P.; et al. TRPA1 underlies a sensing mechanism for O_2. *Nat. Chem. Biol.* **2011**, *7*, 701–711. [CrossRef]

75. Nguyen, T.L.; Duran, R.V. Prolyl hydroxylase domain enzymes and their role in cell signaling and cancer metabolism. *Int. J. Biochem. Cell Biol.* **2016**, *80*, 71–80. [CrossRef]

76. Chandel, N.S.; McClintock, D.S.; Feliciano, C.E.; Wood, T.M.; Melendez, J.A.; Rodriguez, A.M.; Schumacker, P.T. Reactive oxygen species generated at mitochondrial complex III stabilize hypoxia-inducible factor-1alpha during hypoxia: A mechanism of O_2 sensing. *J. Biol. Chem.* **2000**, *275*, 25130–25138. [CrossRef]

77. Mori, Y.; Takahashi, N.; Kurokawa, T.; Kiyonaka, S. TRP channels in oxygen physiology: Distinctive functional properties and roles of TRPA1 in O(2) sensing. *Proc. Jpn. Acad. Ser. B Phys. Biol. Sci.* **2017**, *93*, 464–482. [CrossRef] [PubMed]

78. Eelen, G.; Treps, L.; Li, X.; Carmeliet, P. Basic and Therapeutic Aspects of Angiogenesis Updated. *Circ. Res.* **2020**, *127*, 310–329. [CrossRef]

79. Armani, G.; Pozzi, E.; Pagani, A.; Porta, C.; Rizzo, M.; Cicognini, D.; Rovati, B.; Moccia, F.; Pedrazzoli, P.; Ferraris, E. The heterogeneity of cancer endothelium: The relevance of angiogenesis and endothelial progenitor cells in cancer microenvironment. *Microvasc. Res.* **2021**, *138*, 104189. [CrossRef]

80. Monteith, G.R.; Prevarskaya, N.; Roberts-Thomson, S.J. The calcium-cancer signalling nexus. *Nat. Rev. Cancer* **2017**, *17*, 367–380. [CrossRef] [PubMed]

81. Prevarskaya, N.; Skryma, R.; Shuba, Y. Ion Channels in Cancer: Are Cancer Hallmarks Oncochannelopathies? *Physiol. Rev.* **2018**, *98*, 559–621. [CrossRef] [PubMed]

82. Scarpellino, G.; Genova, T.; Avanzato, D.; Bernardini, M.; Bianco, S.; Petrillo, S.; Tolosano, E.; de Almeida Vieira, J.R.; Bussolati, B.; Fiorio Pla, A.; et al. Purinergic Calcium Signals in Tumor-Derived Endothelium. *Cancers* **2019**, *11*, 766. [CrossRef]

83. Scarpellino, G.; Munaron, L.; Cantelmo, A.R.; Fiorio Pla, A. Calcium-Permeable Channels in Tumor Vascularization: Peculiar Sensors of Microenvironmental Chemical and Physical Cues. *Rev. Physiol. Biochem. Pharmacol.* **2020**, *182*, 111–137. [CrossRef]

84. Faris, P.; Shekha, M.; Montagna, D.; Guerra, G.; Moccia, F. Endolysosomal Ca(2+) Signalling and Cancer Hallmarks: Two-Pore Channels on the Move, TRPML1 Lags Behind! *Cancers* **2018**, *11*, 27. [CrossRef]

85. Zhong, T.; Zhang, W.; Guo, H.; Pan, X.; Chen, X.; He, Q.; Yang, B.; Ding, L. The regulatory and modulatory roles of TRP family channels in malignant tumors and relevant therapeutic strategies. *Acta Pharm. Sin. B* **2022**, *12*, 1761–1780. [CrossRef] [PubMed]

86. Lodola, F.; Laforenza, U.; Cattaneo, F.; Ruffinatti, F.A.; Poletto, V.; Massa, M.; Tancredi, R.; Zuccolo, E.; Khdar, A.D.; Riccardi, A.; et al. VEGF-induced intracellular Ca^{2+} oscillations are down-regulated and do not stimulate angiogenesis in breast cancer-derived endothelial colony forming cells. *Oncotarget* **2017**, *8*, 95223–95246. [CrossRef] [PubMed]

87. Fliniaux, I.; Germain, E.; Farfariello, V.; Prevarskaya, N. TRPs and Ca(2+) in cell death and survival. *Cell Calcium* **2018**, *69*, 4–18. [CrossRef]

88. Perna, A.; Sellitto, C.; Komici, K.; Hay, E.; Rocca, A.; De Blasiis, P.; Lucariello, A.; Moccia, F.; Guerra, G. Transient Receptor Potential (TRP) Channels in Tumor Vascularization. *Int. J. Mol. Sci.* **2022**, *23*, 4253. [CrossRef] [PubMed]

89. Takahashi, K.; Ohta, T. Membrane translocation of transient receptor potential ankyrin 1 induced by inflammatory cytokines in lung cancer cells. *Biochem. Biophys. Res. Commun.* **2017**, *490*, 587–593. [CrossRef]

90. Derouiche, S.; Mariot, P.; Warnier, M.; Vancauwenberghe, E.; Bidaux, G.; Gosset, P.; Mauroy, B.; Bonnal, J.L.; Slomianny, C.; Delcourt, P.; et al. Activation of TRPA1 Channel by Antibacterial Agent Triclosan Induces VEGF Secretion in Human Prostate Cancer Stromal Cells. *Cancer Prev. Res. (Phila.)* **2017**, *10*, 177–187. [CrossRef]

91. Moccia, F.; Poletto, V. May the remodeling of the Ca(2)(+) toolkit in endothelial progenitor cells derived from cancer patients suggest alternative targets for anti-angiogenic treatment? *Biochim. Biophys. Acta* **2015**, *1853*, 1958–1973. [CrossRef] [PubMed]

92. Bernardini, M.; Brossa, A.; Chinigo, G.; Grolez, G.P.; Trimaglio, G.; Allart, L.; Hulot, A.; Marot, G.; Genova, T.; Joshi, A.; et al. Transient Receptor Potential Channel Expression Signatures in Tumor-Derived Endothelial Cells: Functional Roles in Prostate Cancer Angiogenesis. *Cancers* **2019**, *11*, 956. [CrossRef] [PubMed]

93. Vancauwenberghe, E.; Noyer, L.; Derouiche, S.; Lemonnier, L.; Gosset, P.; Sadofsky, L.R.; Mariot, P.; Warnier, M.; Bokhobza, A.; Slomianny, C.; et al. Activation of mutated TRPA1 ion channel by resveratrol in human prostate cancer associated fibroblasts (CAF). *Mol. Carcinog.* **2017**, *56*, 1851–1867. [CrossRef]

94. Genova, T.; Grolez, G.P.; Camillo, C.; Bernardini, M.; Bokhobza, A.; Richard, E.; Scianna, M.; Lemonnier, L.; Valdembri, D.; Munaron, L.; et al. TRPM8 inhibits endothelial cell migration via a non-channel function by trapping the small GTPase Rap1. *J. Cell Biol.* **2017**, *216*, 2107–2130. [CrossRef]

95. Grolez, G.P.; Chinigo, G.; Barras, A.; Hammadi, M.; Noyer, L.; Kondratska, K.; Bulk, E.; Oullier, T.; Marionneau-Lambot, S.; Le Mee, M.; et al. TRPM8 as an Anti-Tumoral Target in Prostate Cancer Growth and Metastasis Dissemination. *Int. J. Mol. Sci.* **2022**, *23*, 6672. [CrossRef] [PubMed]

96. Soda, T.; Brunetti, V.; Berra-Romani, R.; Moccia, F. The Emerging Role of N-Methyl-D-Aspartate (NMDA) Receptors in the Cardiovascular System: Physiological Implications, Pathological Consequences, and Therapeutic Perspectives. *Int. J. Mol. Sci.* **2023**, *24*, 3914. [CrossRef] [PubMed]

97. Negri, S.; Scolari, F.; Vismara, M.; Brunetti, V.; Faris, P.; Terribile, G.; Sancini, G.; Berra-Romani, R.; Moccia, F. GABA(A) and GABA(B) Receptors Mediate GABA-Induced Intracellular Ca(2+) Signals in Human Brain Microvascular Endothelial Cells. *Cells* **2022**, *11*, 3860. [CrossRef]

98. Negri, S.; Faris, P.; Maniezzi, C.; Pellavio, G.; Spaiardi, P.; Botta, L.; Laforenza, U.; Biella, G.; Moccia, D.F. NMDA receptors elicit flux-independent intracellular Ca(2+) signals via metabotropic glutamate receptors and flux-dependent nitric oxide release in human brain microvascular endothelial cells. *Cell Calcium* **2021**, *99*, 102454. [CrossRef]

99. Berrout, J.; Kyriakopoulou, E.; Moparthi, L.; Hogea, A.S.; Berrout, L.; Ivan, C.; Lorger, M.; Boyle, J.; Peers, C.; Muench, S.; et al. TRPA1-FGFR2 binding event is a regulatory oncogenic driver modulated by miRNA-142-3p. *Nat. Commun.* **2017**, *8*, 947. [CrossRef]

100. Louhivuori, L.M.; Bart, G.; Larsson, K.P.; Louhivuori, V.; Nasman, J.; Nordstrom, T.; Koivisto, A.P.; Akerman, K.E. Differentiation dependent expression of TRPA1 and TRPM8 channels in IMR-32 human neuroblastoma cells. *J. Cell. Physiol.* **2009**, *221*, 67–74. [CrossRef]

101. Bansaghi, S.; Golenar, T.; Madesh, M.; Csordas, G.; RamachandraRao, S.; Sharma, K.; Yule, D.I.; Joseph, S.K.; Hajnoczky, G. Isoform- and species-specific control of inositol 1,4,5-trisphosphate (IP3) receptors by reactive oxygen species. *J. Biol. Chem.* **2014**, *289*, 8170–8181. [CrossRef]

102. Joseph, S.K.; Booth, D.M.; Young, M.P.; Hajnoczky, G. Redox regulation of ER and mitochondrial Ca(2+) signaling in cell survival and death. *Cell Calcium* **2019**, *79*, 89–97. [CrossRef] [PubMed]

103. Lodola, F.; Rosti, V.; Tullii, G.; Desii, A.; Tapella, L.; Catarsi, P.; Lim, D.; Moccia, F.; Antognazza, M.R. Conjugated polymers optically regulate the fate of endothelial colony-forming cells. *Sci. Adv.* **2019**, *5*, eaav4620. [CrossRef] [PubMed]

104. Negri, S.; Faris, P.; Rosti, V.; Antognazza, M.R.; Lodola, F.; Moccia, F. Endothelial TRPV1 as an Emerging Molecular Target to Promote Therapeutic Angiogenesis. *Cells* **2020**, *9*, 1341. [CrossRef] [PubMed]

105. Martinotti, S.; Laforenza, U.; Patrone, M.; Moccia, F.; Ranzato, E. Honey-Mediated Wound Healing: H(2)O(2) Entry through AQP3 Determines Extracellular Ca(2+) Influx. *Int. J. Mol. Sci.* **2019**, *20*, 764. [CrossRef]

106. Moccia, F.; Negri, S.; Faris, P.; Ronchi, C.; Lodola, F. Optical excitation of organic semiconductors as a highly selective strategy to induce vascular regeneration and tissue repair. *Vascul. Pharmacol.* **2022**, *144*, 106998. [CrossRef]

107. Booth, D.M.; Varnai, P.; Joseph, S.K.; Hajnoczky, G. Oxidative bursts of single mitochondria mediate retrograde signaling toward the ER. *Mol. Cell* **2021**, *81*, 3866–3876. [CrossRef]

108. Booth, D.M.; Enyedi, B.; Geiszt, M.; Varnai, P.; Hajnoczky, G. Redox Nanodomains Are Induced by and Control Calcium Signaling at the ER-Mitochondrial Interface. *Mol. Cell* **2016**, *63*, 240–248. [CrossRef]

109. Wang, S.; Lv, J.; Pang, Y.; Hu, S.; Lin, Y.; Li, M. Ion channel-targeting near-infrared photothermal switch with synergistic effect for specific cancer therapy. *J. Mater. Chem. B* **2022**, *10*, 748–756. [CrossRef]

110. Nardin, C.; Peres, C.; Mazzarda, F.; Ziraldo, G.; Salvatore, A.M.; Mammano, F. Photosensitizer Activation Drives Apoptosis by Interorganellar Ca(2+) Transfer and Superoxide Production in Bystander Cancer Cells. *Cells* **2019**, *8*, 1175. [CrossRef]

111. Hausmann, D.; Hoffmann, D.C.; Venkataramani, V.; Jung, E.; Horschitz, S.; Tetzlaff, S.K.; Jabali, A.; Hai, L.; Kessler, T.; Azorin, D.D.; et al. Autonomous rhythmic activity in glioma networks drives brain tumour growth. *Nature* **2023**, *613*, 179–186. [CrossRef]

112. Rosa, N.; Ivanova, H.; Wagner, L.E., 2nd; Kale, J.; La Rovere, R.; Welkenhuyzen, K.; Louros, N.; Karamanou, S.; Shabardina, V.; Lemmens, I.; et al. Bcl-xL acts as an inhibitor of IP(3)R channels, thereby antagonizing Ca(2+)-driven apoptosis. *Cell Death Differ.* **2022**, *29*, 788–805. [CrossRef] [PubMed]

113. Adeshakin, F.O.; Adeshakin, A.O.; Afolabi, L.O.; Yan, D.; Zhang, G.; Wan, X. Mechanisms for Modulating Anoikis Resistance in Cancer and the Relevance of Metabolic Reprogramming. *Front. Oncol.* **2021**, *11*, 626577. [CrossRef] [PubMed]
114. Sullivan, M.N.; Gonzales, A.L.; Pires, P.W.; Bruhl, A.; Leo, M.D.; Li, W.; Oulidi, A.; Boop, F.A.; Feng, Y.; Jaggar, J.H.; et al. Localized TRPA1 channel Ca^{2+} signals stimulated by reactive oxygen species promote cerebral artery dilation. *Sci. Signal.* **2015**, *8*, ra2. [CrossRef]
115. Oehler, B.; Kloka, J.; Mohammadi, M.; Ben-Kraiem, A.; Rittner, H.L. D-4F, an ApoA-I mimetic peptide ameliorating TRPA1-mediated nocifensive behaviour in a model of neurogenic inflammation. *Mol. Pain* **2020**, *16*, 1744806920903848. [CrossRef] [PubMed]
116. Negri, S.; Faris, P.; Soda, T.; Moccia, F. Endothelial signaling at the core of neurovascular coupling: The emerging role of endothelial inward-rectifier K(+) (Kir2.1) channels and N-methyl-d-aspartate receptors in the regulation of cerebral blood flow. *Int. J. Biochem. Cell Biol.* **2021**, *135*, 105983. [CrossRef]
117. Moccia, F.; Zuccolo, E.; Poletto, V.; Turin, I.; Guerra, G.; Pedrazzoli, P.; Rosti, V.; Porta, C.; Montagna, D. Targeting Stim and Orai Proteins as an Alternative Approach in Anticancer Therapy. *Curr. Med. Chem.* **2016**, *23*, 3450–3480. [CrossRef]
118. Moccia, F.; Lucariello, A.; Guerra, G. TRPC3-mediated Ca(2+) signals as a promising strategy to boost therapeutic angiogenesis in failing hearts: The role of autologous endothelial colony forming cells. *J. Cell. Physiol.* **2018**, *233*, 3901–3917. [CrossRef]
119. Suresh, K.; Servinsky, L.; Reyes, J.; Baksh, S.; Undem, C.; Caterina, M.; Pearse, D.B.; Shimoda, L.A. Hydrogen peroxide-induced calcium influx in lung microvascular endothelial cells involves TRPV4. *Am. J. Physiol. Lung Cell. Mol. Physiol.* **2015**, *309*, L1467–L1477. [CrossRef]
120. Pandey, A.; Tripathi, S.C.; Mai, J.; Hanash, S.M.; Shen, H.; Mitra, S.; Rostomily, R.C. Combinatorial Effect of PLK1 Inhibition with Temozolomide and Radiation in Glioblastoma. *Cancers* **2021**, *13*, 5114. [CrossRef]
121. Thomas, A.; Tanaka, M.; Trepel, J.; Reinhold, W.C.; Rajapakse, V.N.; Pommier, Y. Temozolomide in the Era of Precision Medicine. *Cancer Res.* **2017**, *77*, 823–826. [CrossRef]
122. Singh, N.; Miner, A.; Hennis, L.; Mittal, S. Mechanisms of temozolomide resistance in glioblastoma—A comprehensive review. *Cancer Drug Resist.* **2021**, *4*, 17–43. [CrossRef]
123. Ozkal, B.; Ovey, I.S. Selenium enhances TRPA1 channel-mediated activity of temozolomide in SH-SY5Y neuroblastoma cells. *Childs Nerv. Syst.* **2020**, *36*, 1283–1292. [CrossRef]
124. Bonora, M.; Giorgi, C.; Pinton, P. Molecular mechanisms and consequences of mitochondrial permeability transition. *Nat. Rev. Mol. Cell Biol.* **2022**, *23*, 266–285. [CrossRef]
125. Patergnani, S.; Danese, A.; Bouhamida, E.; Aguiari, G.; Previati, M.; Pinton, P.; Giorgi, C. Various Aspects of Calcium Signaling in the Regulation of Apoptosis, Autophagy, Cell Proliferation, and Cancer. *Int. J. Mol. Sci.* **2020**, *21*, 8323. [CrossRef]
126. Bootman, M.D.; Bultynck, G. Fundamentals of Cellular Calcium Signaling: A Primer. *Cold Spring Harb. Perspect. Biol.* **2020**, *12*, a038802. [CrossRef] [PubMed]
127. Faris, P.; Pellavio, G.; Ferulli, F.; Di Nezza, F.; Shekha, M.; Lim, D.; Maestri, M.; Guerra, G.; Ambrosone, L.; Pedrazzoli, P.; et al. Nicotinic Acid Adenine Dinucleotide Phosphate (NAADP) Induces Intracellular Ca(2+) Release through the Two-Pore Channel TPC1 in Metastatic Colorectal Cancer Cells. *Cancers* **2019**, *11*, 542. [CrossRef] [PubMed]
128. Farfariello, V.; Gordienko, D.V.; Mesilmany, L.; Touil, Y.; Germain, E.; Fliniaux, I.; Desruelles, E.; Gkika, D.; Roudbaraki, M.; Shapovalov, G.; et al. TRPC3 shapes the ER-mitochondria Ca(2+) transfer characterizing tumour-promoting senescence. *Nat. Commun.* **2022**, *13*, 956. [CrossRef]
129. Ziegler, D.V.; Vindrieux, D.; Goehrig, D.; Jaber, S.; Collin, G.; Griveau, A.; Wiel, C.; Bendridi, N.; Djebali, S.; Farfariello, V.; et al. Calcium channel ITPR2 and mitochondria-ER contacts promote cellular senescence and aging. *Nat. Commun.* **2021**, *12*, 720. [CrossRef] [PubMed]
130. Kerkhofs, M.; Bittremieux, M.; Morciano, G.; Giorgi, C.; Pinton, P.; Parys, J.B.; Bultynck, G. Emerging molecular mechanisms in chemotherapy: Ca(2+) signaling at the mitochondria-associated endoplasmic reticulum membranes. *Cell Death Dis.* **2018**, *9*, 334. [CrossRef]
131. Morciano, G.; Marchi, S.; Morganti, C.; Sbano, L.; Bittremieux, M.; Kerkhofs, M.; Corricelli, M.; Danese, A.; Karkucinska-Wieckowska, A.; Wieckowski, M.R.; et al. Role of Mitochondria-Associated ER Membranes in Calcium Regulation in Cancer-Specific Settings. *Neoplasia* **2018**, *20*, 510–523. [CrossRef] [PubMed]
132. Astesana, V.; Faris, P.; Ferrari, B.; Siciliani, S.; Lim, D.; Biggiogera, M.; De Pascali, S.A.; Fanizzi, F.P.; Roda, E.; Moccia, F.; et al. [Pt(O,O'-acac)(gamma-acac)(DMS)]: Alternative Strategies to Overcome Cisplatin-Induced Side Effects and Resistance in T98G Glioma Cells. *Cell. Mol. Neurobiol.* **2020**, *41*, 563–587. [CrossRef] [PubMed]
133. Dubois, C.; Kondratskyi, A.; Bidaux, G.; Noyer, L.; Vancauwenberghe, E.; Farfariello, V.; Toillon, R.A.; Roudbaraki, M.; Tierny, D.; Bonnal, J.L.; et al. Co-targeting Mitochondrial Ca(2+) Homeostasis and Autophagy Enhances Cancer Cells' Chemosensitivity. *iScience* **2020**, *23*, 101263. [CrossRef] [PubMed]
134. Tandl, D.; Sponagel, T.; Alansary, D.; Fuck, S.; Smit, T.; Hehlgans, S.; Jakob, B.; Fournier, C.; Niemeyer, B.A.; Rodel, F.; et al. X-ray irradiation triggers immune response in human T-lymphocytes via store-operated Ca^{2+} entry and NFAT activation. *J. Gen. Physiol.* **2022**, *154*, e202112865. [CrossRef]
135. Lu, F.; Sun, J.; Zheng, Q.; Li, J.; Hu, Y.; Yu, P.; He, H.; Zhao, Y.; Wang, X.; Yang, S.; et al. Imaging elemental events of store-operated Ca(2+) entry in invading cancer cells with plasmalemmal targeted sensors. *J. Cell Sci.* **2019**, *132*, 224923. [CrossRef]

136. Lee, D.; Hong, J.H. Activated PyK2 and Its Associated Molecules Transduce Cellular Signaling from the Cancerous Milieu for Cancer Metastasis. *Int. J. Mol. Sci.* **2022**, *23*, 15475. [CrossRef]
137. Chen, Y.F.; Chiu, W.T.; Chen, Y.T.; Lin, P.Y.; Huang, H.J.; Chou, C.Y.; Chang, H.C.; Tang, M.J.; Shen, M.R. Calcium store sensor stromal-interaction molecule 1-dependent signaling plays an important role in cervical cancer growth, migration, and angiogenesis. *Proc. Natl. Acad. Sci. USA* **2011**, *108*, 15225–15230. [CrossRef]
138. Vismara, M.; Negri, S.; Scolari, F.; Brunetti, V.; Trivigno, S.M.G.; Faris, P.; Galgano, L.; Soda, T.; Berra-Romani, R.; Canobbio, I.; et al. Platelet-Derived Extracellular Vesicles Stimulate Migration through Partial Remodelling of the Ca(2+) Handling Machinery in MDA-MB-231 Breast Cancer Cells. *Cells* **2022**, *11*, 3120. [CrossRef]
139. Rhythmic Ca^{2+} Communication Promotes Glioma Cell Proliferation. *Cancer Discov.* **2023**, *13*, 259. [CrossRef]
140. Alharbi, A.; Zhang, Y.; Parrington, J. Deciphering the Role of Ca(2+) Signalling in Cancer Metastasis: From the Bench to the Bedside. *Cancers* **2021**, *13*, 179. [CrossRef]
141. Sharma, A.; Ramena, G.T.; Elble, R.C. Advances in Intracellular Calcium Signaling Reveal Untapped Targets for Cancer Therapy. *Biomedicines* **2021**, *9*, 1077. [CrossRef] [PubMed]
142. Tao, R.; Sun, H.Y.; Lau, C.P.; Tse, H.F.; Lee, H.C.; Li, G.R. Cyclic ADP ribose is a novel regulator of intracellular Ca2+ oscillations in human bone marrow mesenchymal stem cells. *J. Cell. Mol. Med.* **2011**, *15*, 2684–2696. [CrossRef] [PubMed]
143. Balducci, V.; Faris, P.; Balbi, C.; Costa, A.; Negri, S.; Rosti, V.; Bollini, S.; Moccia, F. The human amniotic fluid stem cell secretome triggers intracellular Ca(2+) oscillations, NF-kappaB nuclear translocation and tube formation in human endothelial colony-forming cells. *J. Cell. Mol. Med.* **2021**, *25*, 8074–8086. [CrossRef] [PubMed]
144. Galione, A.; Davis, L.C.; Martucci, L.L.; Morgan, A.J. NAADP-Mediated Ca(2+) Signalling. *Handb. Exp. Pharmacol.* **2022**, 1–32. [CrossRef]
145. Zuccolo, E.; Kheder, D.A.; Lim, D.; Perna, A.; Nezza, F.D.; Botta, L.; Scarpellino, G.; Negri, S.; Martinotti, S.; Soda, T.; et al. Glutamate triggers intracellular Ca(2+) oscillations and nitric oxide release by inducing NAADP- and InsP3 -dependent Ca(2+) release in mouse brain endothelial cells. *J. Cell. Physiol.* **2019**, *234*, 3538–3554. [CrossRef]
146. Li, B.; Ren, S.; Gao, D.; Li, N.; Wu, M.; Yuan, H.; Zhou, M.; Xing, C. Photothermal Conjugated Polymer Nanoparticles for Suppressing Breast Tumor Growth by Regulating TRPA1 Ion Channels. *Adv. Healthc. Mater.* **2022**, *11*, e2102506. [CrossRef] [PubMed]
147. Mahajan, N.; Khare, P.; Kondepudi, K.K.; Bishnoi, M. TRPA1: Pharmacology, natural activators and role in obesity prevention. *Eur. J. Pharmacol.* **2021**, *912*, 174553. [CrossRef]
148. Munaron, L.; Avanzato, D.; Moccia, F.; Mancardi, D. Hydrogen sulfide as a regulator of calcium channels. *Cell Calcium* **2013**, *53*, 77–84. [CrossRef]
149. Liu, C.; Reese, R.; Vu, S.; Rouge, L.; Shields, S.D.; Kakiuchi-Kiyota, S.; Chen, H.; Johnson, K.; Shi, Y.P.; Chernov-Rogan, T.; et al. A Non-covalent Ligand Reveals Biased Agonism of the TRPA1 Ion Channel. *Neuron* **2021**, *109*, 273–284.e274. [CrossRef]
150. Lin King, J.V.; Emrick, J.J.; Kelly, M.J.S.; Herzig, V.; King, G.F.; Medzihradszky, K.F.; Julius, D. A Cell-Penetrating Scorpion Toxin Enables Mode-Specific Modulation of TRPA1 and Pain. *Cell* **2019**, *178*, 1362–1374.e1316. [CrossRef]
151. Hu, H.; Tian, J.; Zhu, Y.; Wang, C.; Xiao, R.; Herz, J.M.; Wood, J.D.; Zhu, M.X. Activation of TRPA1 channels by fenamate nonsteroidal anti-inflammatory drugs. *Pflug. Arch.* **2010**, *459*, 579–592. [CrossRef] [PubMed]
152. Bautista, D.M.; Movahed, P.; Hinman, A.; Axelsson, H.E.; Sterner, O.; Hogestatt, E.D.; Julius, D.; Jordt, S.E.; Zygmunt, P.M. Pungent products from garlic activate the sensory ion channel TRPA1. *Proc. Natl. Acad. Sci. USA* **2005**, *102*, 12248–12252. [CrossRef] [PubMed]
153. Andersson, D.A.; Gentry, C.; Bevan, S. TRPA1 has a key role in the somatic pro-nociceptive actions of hydrogen sulfide. *PLoS ONE* **2012**, *7*, e46917. [CrossRef] [PubMed]
154. Koizumi, K.; Iwasaki, Y.; Narukawa, M.; Iitsuka, Y.; Fukao, T.; Seki, T.; Ariga, T.; Watanabe, T. Diallyl sulfides in garlic activate both TRPA1 and TRPV1. *Biochem. Biophys. Res. Commun.* **2009**, *382*, 545–548. [CrossRef]
155. Bautista, D.M.; Jordt, S.E.; Nikai, T.; Tsuruda, P.R.; Read, A.J.; Poblete, J.; Yamoah, E.N.; Basbaum, A.I.; Julius, D. TRPA1 mediates the inflammatory actions of environmental irritants and proalgesic agents. *Cell* **2006**, *124*, 1269–1282. [CrossRef] [PubMed]
156. Takaya, J.; Mio, K.; Shiraishi, T.; Kurokawa, T.; Otsuka, S.; Mori, Y.; Uesugi, M. A Potent and Site-Selective Agonist of TRPA1. *J. Am. Chem. Soc.* **2015**, *137*, 15859–15864. [CrossRef]
157. Bessac, B.F.; Jordt, S.E. Breathtaking TRP channels: TRPA1 and TRPV1 in airway chemosensation and reflex control. *Physiology* **2008**, *23*, 360–370. [CrossRef] [PubMed]
158. Taylor-Clark, T.E.; McAlexander, M.A.; Nassenstein, C.; Sheardown, S.A.; Wilson, S.; Thornton, J.; Carr, M.J.; Undem, B.J. Relative contributions of TRPA1 and TRPV1 channels in the activation of vagal bronchopulmonary C-fibres by the endogenous autacoid 4-oxononenal. *J. Physiol.* **2008**, *586*, 3447–3459. [CrossRef]
159. Bianchi, B.R.; Zhang, X.F.; Reilly, R.M.; Kym, P.R.; Yao, B.B.; Chen, J. Species comparison and pharmacological characterization of human, monkey, rat, and mouse TRPA1 channels. *J. Pharmacol. Exp. Ther.* **2012**, *341*, 360–368. [CrossRef]
160. Lee, S.P.; Buber, M.T.; Yang, Q.; Cerne, R.; Cortes, R.Y.; Sprous, D.G.; Bryant, R.W. Thymol and related alkyl phenols activate the hTRPA1 channel. *Br. J. Pharmacol.* **2008**, *153*, 1739–1749. [CrossRef] [PubMed]
161. Leffler, A.; Lattrell, A.; Kronewald, S.; Niedermirtl, F.; Nau, C. Activation of TRPA1 by membrane permeable local anesthetics. *Mol. Pain* **2011**, *7*, 62. [CrossRef] [PubMed]

162. Nishimoto, R.; Kashio, M.; Tominaga, M. Propofol-induced pain sensation involves multiple mechanisms in sensory neurons. *Pflug. Arch.* **2015**, *467*, 2011–2020. [CrossRef] [PubMed]
163. Tasinov, O.; Dincheva, I.; Badjakov, I.; Kiselova-Kaneva, Y.; Galunska, B.; Nogueiras, R.; Ivanova, D. Phytochemical Composition, Anti-Inflammatory and ER Stress-Reducing Potential of *Sambucus ebulus* L. Fruit Extract. *Plants* **2021**, *10*, 2446. [CrossRef] [PubMed]
164. McNamara, C.R.; Mandel-Brehm, J.; Bautista, D.M.; Siemens, J.; Deranian, K.L.; Zhao, M.; Hayward, N.J.; Chong, J.A.; Julius, D.; Moran, M.M.; et al. TRPA1 mediates formalin-induced pain. *Proc. Natl. Acad. Sci. USA* **2007**, *104*, 13525–13530. [CrossRef]
165. Wei, H.; Chapman, H.; Saarnilehto, M.; Kuokkanen, K.; Koivisto, A.; Pertovaara, A. Roles of cutaneous versus spinal TRPA1 channels in mechanical hypersensitivity in the diabetic or mustard oil-treated non-diabetic rat. *Neuropharmacology* **2010**, *58*, 578–584. [CrossRef]
166. Defalco, J.; Steiger, D.; Gustafson, A.; Emerling, D.E.; Kelly, M.G.; Duncton, M.A. Oxime derivatives related to AP18: Agonists and antagonists of the TRPA1 receptor. *Bioorg. Med. Chem. Lett.* **2010**, *20*, 276–279. [CrossRef]
167. McGaraughty, S.; Chu, K.L.; Perner, R.J.; Didomenico, S.; Kort, M.E.; Kym, P.R. TRPA1 modulation of spontaneous and mechanically evoked firing of spinal neurons in uninjured, osteoarthritic, and inflamed rats. *Mol. Pain* **2010**, *6*, 14. [CrossRef]
168. Copeland, K.W.; Boezio, A.A.; Cheung, E.; Lee, J.; Olivieri, P.; Schenkel, L.B.; Wan, Q.; Wang, W.; Wells, M.C.; Youngblood, B.; et al. Development of novel azabenzofuran TRPA1 antagonists as in vivo tools. *Bioorg. Med. Chem. Lett.* **2014**, *24*, 3464–3468. [CrossRef]
169. Rooney, L.; Vidal, A.; D'Souza, A.M.; Devereux, N.; Masick, B.; Boissel, V.; West, R.; Head, V.; Stringer, R.; Lao, J.; et al. Discovery, optimization, and biological evaluation of 5-(2-(trifluoromethyl)phenyl)indazoles as a novel class of transient receptor potential A1 (TRPA1) antagonists. *J. Med. Chem.* **2014**, *57*, 5129–5140. [CrossRef]

Article

Palmitate-Induced Cardiac Lipotoxicity Is Relieved by the Redox-Active Motif of SELENOT through Improving Mitochondrial Function and Regulating Metabolic State

Carmine Rocca [1,*,†], Anna De Bartolo [1,2,†], Rita Guzzi [3,4], Maria Caterina Crocco [3,5], Vittoria Rago [6], Naomi Romeo [1], Ida Perrotta [7], Ernestina Marianna De Francesco [8], Maria Grazia Muoio [8], Maria Concetta Granieri [1], Teresa Pasqua [9], Rosa Mazza [1], Loubna Boukhzar [2], Benjamin Lefranc [2,10], Jérôme Leprince [2,10], Maria Eugenia Gallo Cantafio [11], Teresa Soda [9], Nicola Amodio [11], Youssef Anouar [2,10] and Tommaso Angelone [1,12,*]

1   Cellular and Molecular Cardiovascular Pathophysiology Laboratory, Department of Biology, Ecology and Earth Sciences (DiBEST), University of Calabria, 87036 Rende, Italy
2   UNIROUEN, Inserm U1239, Neuroendocrine, Endocrine and Germinal Differentiation and Communication (NorDiC), Rouen Normandie University, 76000 Mont-Saint-Aignan, France
3   Department of Physics, Molecular Biophysics Laboratory, University of Calabria, 87036 Rende, Italy
4   CNR-NANOTEC, Department of Physics, University of Calabria, 87036 Rende, Italy
5   STAR Research Infrastructure, University of Calabria, Via Tito Flavio, 87036 Rende, Italy
6   Department of Pharmacy, Health and Nutritional Sciences, University of Calabria, 87036 Rende, Italy
7   Centre for Microscopy and Microanalysis (CM2), Department of Biology, Biology, Ecology and Earth Sciences (DiBEST), University of Calabria, 87036 Rende, Italy
8   Endocrinology, Department of Clinical and Experimental Medicine, University of Catania, Garibaldi-Nesima Hospital, 95124 Catania, Italy
9   Department of Health Science, University Magna Graecia of Catanzaro, 88100 Catanzaro, Italy
10   UNIROUEN, UMS-UAR HERACLES, PRIMACEN, Cell Imaging Platform of Normandy, Institute for Research and Innovation in Biomedicine (IRIB), 76183 Rouen, France
11   Department of Experimental and Clinical Medicine, Magna Graecia University, 88100 Catanzaro, Italy
12   National Institute of Cardiovascular Research (INRC), 40126 Bologna, Italy
*   Correspondence: carmine.rocca@unical.it (C.R.); tommaso.angelone@unical.it (T.A.)
†   These authors contributed equally to this work.

Citation: Rocca, C.; De Bartolo, A.; Guzzi, R.; Crocco, M.C.; Rago, V.; Romeo, N.; Perrotta, I.; De Francesco, E.M.; Muoio, M.G.; Granieri, M.C.; et al. Palmitate-Induced Cardiac Lipotoxicity Is Relieved by the Redox-Active Motif of SELENOT through Improving Mitochondrial Function and Regulating Metabolic State. *Cells* **2023**, *12*, 1042. https://doi.org/10.3390/cells12071042

Academic Editors: Alessia Remigante and Rossana Morabito

Received: 24 February 2023
Revised: 22 March 2023
Accepted: 27 March 2023
Published: 29 March 2023

Abstract: Cardiac lipotoxicity is an important contributor to cardiovascular complications during obesity. Given the fundamental role of the endoplasmic reticulum (ER)-resident Selenoprotein T (SELENOT) for cardiomyocyte differentiation and protection and for the regulation of glucose metabolism, we took advantage of a small peptide (PSELT), derived from the SELENOT redox-active motif, to uncover the mechanisms through which PSELT could protect cardiomyocytes against lipotoxicity. To this aim, we modeled cardiac lipotoxicity by exposing H9c2 cardiomyocytes to palmitate (PA). The results showed that PSELT counteracted PA-induced cell death, lactate dehydrogenase release, and the accumulation of intracellular lipid droplets, while an inert form of the peptide (I-PSELT) lacking selenocysteine was not active against PA-induced cardiomyocyte death. Mechanistically, PSELT counteracted PA-induced cytosolic and mitochondrial oxidative stress and rescued SELENOT expression that was downregulated by PA through FAT/CD36 (cluster of differentiation 36/fatty acid translocase), the main transporter of fatty acids in the heart. Immunofluorescence analysis indicated that PSELT also relieved the PA-dependent increase in CD36 expression, while in SELENOT-deficient cardiomyocytes, PA exacerbated cell death, which was not mitigated by exogenous PSELT. On the other hand, PSELT improved mitochondrial respiration during PA treatment and regulated mitochondrial biogenesis and dynamics, preventing the PA-provoked decrease in PGC1-α and increase in DRP-1 and OPA-1. These findings were corroborated by transmission electron microscopy (TEM), revealing that PSELT improved the cardiomyocyte and mitochondrial ultrastructures and restored the ER network. Spectroscopic characterization indicated that PSELT significantly attenuated infrared spectral-related macromolecular changes (i.e., content of lipids, proteins, nucleic acids, and carbohydrates) and also prevented the decrease in membrane fluidity induced by PA. Our findings further

---

delineate the biological significance of SELENOT in cardiomyocytes and indicate the potential of its mimetic PSELT as a protective agent for counteracting cardiac lipotoxicity.

Keywords: antioxidants; cardiomyocyte; lipotoxicity; peptides; selenoproteins

1. Introduction

Obesity represents a major public health problem that seriously increases the risk of developing cardiovascular diseases (CVD) and directly contributes to incident cardiovascular risk factors [1]. In obese patients, the increased left ventricular mass and myocardial changes correlate with adiposity, which is responsible for cardiac steatosis by promoting the ectopic deposition of triglycerides in the heart [2]. Following lipid accumulation within the cells of the cardiovascular system, several complex mechanisms drive myocardial dysfunction, leading to heart failure (HF). In this process, known as cardiac lipotoxicity, saturated long chain fatty acids (FAs)–such as palmitic acid (PA)–play a major role due to their ability to alter cellular structures, triggering oxidative stress, endoplasmic reticulum (ER) stress, defective insulin signaling, mitochondrial dysfunction, inflammation, and myofibrillar dysfunction, culminating in cell death [3,4]. Despite the high metabolic flexibility of the heart in terms of substrate utilization, as well as its capacity to adjust the rate of FA uptake to match myocardial demand to obtain energy [5], the augmented circulating levels of free FAs during obesity dramatically increase the myocardial uptake of lipids that may be stored as triglycerides. However, since cardiomyocytes possess a limited storage capacity, the excess free FAs are also shunted into non-oxidative pathways, leading to myocardial lipotoxic injury [6]. Accordingly, elevated serum levels of PA, which represents the major circulating saturated FA, have been proposed as a potential factor contributing to lipotoxic cardiomyopathy development [7]. During chronic lipid overload conditions, excessive cytosolic FAs can increase mitochondrial uncoupling, which in turn may generate reactive oxygen species (ROS) and activate stress-sensitive pathways, leading to oxidative stress [6]. Additionally, ER protein misfolding secondary to the excess of lipids also promotes ER and nuclear oxidative stress [8]. Therefore, endogenous antioxidant enzymes are crucial to detoxify lipid hydroperoxides and other reactive species to maintain the normal cellular machinery, in particular during lipid overload conditions. In this context, selenoproteins are considered among the most potent antioxidant defense systems, with crucial implication in several pathophysiological contexts, including CVD [9,10]. Selenoprotein T (SELENOT) is an ER-resident thioredoxin-like enzyme and the only ER-located selenoprotein whose gene knockout in mice is lethal early during embryogenesis [11]. As a member of the selenoprotein family, SELENOT contains a selenocysteine (Sec, U) residue in its CVSU amino acid redox motif that is fundamental for its biological functions, including the regulation of ER homeostasis, N-glycosylation, and intracellular Ca^{2+} mobilization and neuroendocrine secretion. We previously showed in transgenic cell and animal models that a reduced expression of SELENOT is associated with oxidative and nitrosative stress, unfolded protein response (UPR) activation, and the depletion of Ca^{2+} stores leading to altered hormone secretion [12,13]. Considerable interest has also arisen in the recent years regarding the biological significance of SELENOT in glucose homeostasis and cardiac pathophysiology. SELENOT is strongly expressed in human and mouse pancreatic β- and δ-cells, and conditional pancreatic β-cell SELENOT-knockout mice exhibited impaired glucose tolerance with a deficit in insulin production/secretion [14]. On the other hand, SELENOT is highly expressed during the early hyperplastic growth of cardiomyocytes, suggesting its involvement in cardiac development during embryogenesis, and can also protect cardiomyocytes following myocardial ischemia/reperfusion (MI/R) insult [15,16].

The biological impact of SELENOT in cardiac and metabolic pathophysiology raised an intriguing hypothesis concerning its role in protecting cardiomyocytes exposed to a dysmetabolic condition generated by lipid overload. More precisely, considering that the

CVSU motif of SELENOT carries a selenosulfide oxidoreductase that interacts with other cellular components via redox reactions and is thus essential for SELENOT biological action, in the present study, we designed a small SELENOT mimetic peptide (called PSELT), encompassing the redox motif of the full protein, and tested its ability to protect H9c2 cardiomyocytes against PA-induced lipotoxicity.

2. Materials and Methods

2.1. Peptides and Drugs

The SELENOT-derived peptide 43–52 (corresponding to the sequence H-Phe-Gln-Ile-Cys-Val-Ser-Sec-Gly-Tyr-Arg-OH) designated as PSELT and its inactive form (Ser 46,49, designated as the inert-PSELT or I-PSELT) were chemically synthesized as described previously [17]. Dulbecco's Modified Eagle Medium F-12 (DMEM/F-12), Dulbecco's phosphate-buffered saline (DPBS), penicillin/streptomycin, 0.25% Trypsin-EDTA (1×), and fetal bovine serum (FBS), 4′,6-diamidino-2-phenylindole (DAPI) were purchased from Thermo Fisher Scientific (Waltham, MA, USA). Sodium palmitate (PA), 3-(4,5-Dimethylthiazol-)2,5-diphenyl Tetrazolium Bromide (MTT), β-nicotinamide adenine dinucleotide (NADH), and tween-20 were purchased from Sigma Aldrich (Saint Louis, MO, USA). Sulfo-Nsuccinimidyl oleate (SSO) was purchased from Cayman Chemical (Ann Arbor, MI, USA). Dimethyl sulfoxide (DMSO), bovine serum albumin (BSA), and nonfat dried milk were purchased from PanReac AppliChem (Glenview, IL, USA). Before each experiment, all solutions were freshly prepared. Absolute ethanol, hydrochloric acid, and methanol were purchased from Carlo Erba Reagents (Cornaredo, Milan, Italy).

2.2. Cell Culture

H9c2 cardiomyoblast cells were obtained from the American Type Culture Collection (ATCC) (Manassas, VA, USA) (Cat# CRL-1446), cultured in Dulbecco's Modified Eagle Medium/Nutrient Mixture F-12 (DMEM/F-12, Gibco, Thermo Fisher Scientific, Waltham, MA, USA) supplemented with 10% fetal bovine serum (FBS, Gibco), 1% penicillin/streptomycin (Thermo Fisher Scientific), and incubated in humidified atmosphere (5% CO_2) at 37 °C. Upon reaching 80% cell confluence, H9c2 cardiomyocytes were digested at a 1:2 ratio using 0.25% Trypsin-EDTA (1×) (Gibco) according to the manufacturer's instructions (ATCC). For experiments, cells were seeded in complete medium and incubated for 48 h at 37 °C, 5% CO_2, as previously reported [16,18–20].

2.3. Cell Viability Assay

MTT assay was used to determine the effect of PSELT or I-PSELT on cell viability following PA exposure. H9c2 cells (5000 per well) were seeded in a 96-well plate and then treated with palmitate (PA) (from 100 to 500 μmol/L) for 24 h or vehicle (BSA) as control. PA was purchased from Sigma and dissolved to make a 10 mM stock solution with 10% fatty acid-free BSA [21]. Once the concentration for PA-induced cell death was established, H9c2 cardiomyocytes were exposed to PA (100 μmol/L) and PSELT (from 5 to 100 nmol/L) or its inactive form, indicated as inert-PSELT (I-PSELT), from 5 to 100 nmol/L, for 24 h. After treatments, 100 μL of 2 mg/mL MTT solution (Sigma Aldrich) was added to each well after removal of the culture medium, and then cells were incubated for 4 h at 37 °C, 5% CO_2. Finally, MTT solution was replaced by DMSO, and formazan crystals were dissolved. The absorbance was measured at 570 nm using a microplate reader (Multiskan™ SkyHigh, Thermo Fisher Scientific Inc.). The means of the absorbance values of six wells in each experimental group were expressed as the percentage cell viability relative to the control cells. The experiment was repeated three independent times [16,18,19].

2.4. Lactate Dehydrogenase (LDH) Assessment

The entity of damage induced by PA in H9c2 cardiomyocytes was assessed by analyzing the levels of LDH released in the culture medium following the method of McQueen (1972) [22], and as previously indicated by Rocca et al. (2019) [23]. To analyze the levels of

LDH released in the culture medium, H9c2 cells (100,000 per mL) were seeded in a 24-well plate and treated with vehicle (control), PA (100 µmol/L), PA + PSELT or PSELT (5 nmol/L) for 24 h. At the end of the treatments, 100 µL per well of the culture medium was used for LDH activity determination. The enzyme activity was evaluated spectrophotometrically using Multiskan™ SkyHigh (Thermo Fisher Scientific), following the method of McQueen (1972) [22]. The reaction velocity was determined by a decrease in absorbance at 340 nm resulting from the oxidation of NADH (indicative of LDH activity) that was expressed in IU/L [24].

2.5. Oil Red O Staining

Intracellular lipid accumulation was measured by Oil Red O staining. H9c2 cells were seeded in a 6-well plate, treated with vehicle, PA (100 µmol/L), PA + PSELT or PSELT (5 nmol/L) and incubated in a humidified atmosphere at 37 °C for 24 h. After the treatments, H9c2 cells were washed three times with DPBS and incubated with the Oil Red O kit according to the manufacturer's instructions (#04–220923, Bio Optica, Milan, Italy). H9c2 cardiomyocytes were incubated with reagent A for 20 min, then washed and incubated with reagent B for 30 s followed by the addition of distilled water for 3 min. Nuclei were counterstained with hematoxylin provided by the kit [25–27]. After staining, cells were visualized under an Olympus BX41 microscope, and the images were taken with CSV1.14 software, using a CAM XC-30 for image acquisition.

2.6. Detection of Intracellular Reactive Oxygen Species (ROS) and Mitochondrial Superoxide Generation by MitoSOX

Intracellular reactive oxygen species (ROS) generation was measured using the cell-permeable probes CM-H$_2$DCFDA (5-(and-6)-chloromethyl-2′,7′-dichlorodihydrofluorescein diacetate acetyl ester) (Thermo Fisher Scientific) and 2′,7′-dichlorodihydrofluorescein diacetate (H$_2$DCFDA) (Thermo Fisher Scientific), while mitochondrial superoxide generation was evaluated by MitoSOX™ Red (Thermo Fisher #M36008, Waltham, MA, USA), a cell-permeable reagent that selectively targets mitochondria in live cells.

For the detection of intracellular ROS, H9c2 cells (100,000 per well) were seeded on coverslips in 6-well plates and exposed to PA and PSELT, alone or in co-treatment for 24 h. At the end of the experimental protocol, H9c2 cardiomyocytes were incubated with 10 µmol/L CM-H$_2$DCFDA at 37 °C for 30 min in the dark, and then the cells were carefully rinsed with DPBS and visualized under a fluorescence microscope (Olympus; 20× objective) [28,29]. For measuring the production of total ROS by flow cytometry, H9c2 cells were treated as above, collected, washed with DPBS, and stained with H2DCFDA. ROS were measured by flow cytometry (BD Fortessa X-20) according to the producer's guidelines. The data were analyzed with FlowJo 10.8.1 software.

For mitochondrial superoxide generation, H9c2 cardiomyocytes were seeded on a coverslip in a 6-multiwell plate at a density of 100,000 cells per well and after 48 h, incubated with PA 100 (µmol/L) and PSELT (5 nmol/L) alone or in co-treatment for 24 h. At the end of the treatment, MitoSOX reagent was first dissolved in dimethylsulfoxide (5 mmol/L) and then diluted to 5 µmol/L in DMEM/F-12 phenol-free and serum-free medium [29]. A suitable amount of the solution was added to the cells followed by incubation for 10 min at 37 °C protected from light. Cells were then washed twice with DPBS. Fluorescence was detected using an Olympus fluorescence microscope (20× objective) and quantified using ImageJ 1.6 software (National Institutes of Health, Bethesda, MD, USA).

2.7. Immunofluorescence Analysis for CD36 Evaluation

H9c2 cardiomyocytes were seeded on chamber slides (50,000 cells per chamber), incubated for 48 h at 37 °C, 5% CO$_2$, and then treated with vehicle, PA (100 µmol/L), PA+ PSELT or PSELT (5 nmol/L) for 24 h. At the end of the treatment, H9c2 cells were washed three times with DPBS and fixed for 10 min with ice-cold methanol. Then, fixed cardiomyocytes were rinsed with cold-DPBS two times, and the permeabilization step was performed using 0.1%

Triton X-100 in DPBS for 30 min at RT. Permeabilized H9c2 cells were then washed with DPBS following by blocking with 1% BSA in DPBS for 30 min at RT [16]. For immunofluorescence staining, H9c2 cells were incubated with a primary antibody against CD36 (diluted 1:200) for 2 h at room temperature and then stained with donkey anti-rabbit secondary antibody, Alexa Fluor 555 (diluted 1:1200), for 1 h at room temperature following the manufacturer's instruction. The cells were then washed twice with DPBS and stained with DAPI for nuclei counterstaining. Images were obtained using an Olympus fluorescence BX41 microscope and acquired with CSV1.14 software, using a CAM XC-30 for acquisition. The fluorescence quantification was carried out using ImageJ 1.6 software.

2.8. Short Interfering RNA (siRNA) Transfection for SELENOT Silencing

SELENOT gene silencing in H9c2 cardiomyocytes was performed as previously described by Rocca et al. (2022) [16]. Briefly, H9c2 cardiomyocytes (5000 per well) were seeded in 96-well plates and incubated for 48 h at 37 °C, 5% CO2. SELENOT siRNA (100 nmol/L) was transfected into H9c2 cardiomyocytes in serum-free medium using the Lipofectamine 2000 transfection reagent following the manufacturer's instructions (Invitrogen, Thermo Fisher Scientific, Waltham, MA, USA). Negative control si-RNA (si-NC) was used to detect non-specific effects of siRNA delivery and to compare siRNA-treated samples. Both si-NC and siRNA for SELENOT were purchased from Santa Cruz Biotechnology. H9c2 cardiomyocytes were transfected in serum-free medium for 6 h, after which the medium was replaced with full-medium and cells were incubated for 36 h at 37 °C, 5% CO$_2$. H9c2 cells were treated with PA (100 µmol/L) or co-treated with PA and increasing concentrations of PSELT (from 5 to 100 nmol/L) for 24 h. At the end of the treatments, the viability of the H9c2 cells was evaluated by MTT assay. The cell viability was reported as the percentage cell survival relative to the si-NC transfected cells in six wells for each experimental group [16]. The experiment was repeated three independent times.

2.9. Assessment of Mitochondrial Respiratory Function Using the Seahorse XF Analyzer

For the evaluation of mitochondrial respiration, real-time oxygen consumption rates (OCR) were determined using the Seahorse Extracellular Flux (XFe-96) (Seahorse Bioscience, Agilent Technologies, Inc), as previously described [30]. H9c2 cardiomyocytes were seeded at a density of 10,000 cells per well in a XF96 Seahorse microplate, incubated for 24 h in complete medium at 37 °C and 5% CO2, and then exposed to vehicle, PA (100 µmol/L), and PSELT (5 nmol/L), alone and in combination for 24 h. At the end of the treatments, media were replaced with pre-warmed Seahorse XF assay medium (Agilent Technologies, Inc.) supplemented with 10 mmol/L glucose and 1 mmol/L pyruvate and adjusted to pH 7.4. Cells were maintained in 175 µL of XF assay medium per well at 37 °C in a non-CO2 incubator for 1 h. During the incubation, 10 µmol/L oligomycin, 9 µmol/L FCCP, 10 µmol/L Rotenone, and 10 µmol/L antimycin A were loaded in XF assay medium into the injection ports in the XFe-96 sensor cartridge for OCR measurement. Data were analyzed by XFe-96 software, and measurements were normalized by the protein content, which was determined by Sulphorhodamine B assay, as previously described [30].

2.10. Western Blot

After the H9c2 cells were treated with vehicle, PA (100 µmol/L), PA + PSELT or PSELT (5 nmol/L) for 24 h, the cardiomyocytes were washed with DPBS and the total proteins were extracted using RIPA lysis buffer (Sigma Aldrich, St. Louis, MI, USA) supplemented with protease inhibitors [16]. Cell lysates were transferred in microcentrifuge tubes, incubated on ice for 30 min with intermittent mixing, and centrifuged at 12,000× *g* for 15 min at 4 °C. The supernatant was collected, and the protein concentration was determined by Bradford reagent using bovine serum albumin (BSA) as a standard. Equal amounts of protein (50 µg for all antigens and 30 µg for SELENOT) were loaded on 12% SDS-PAGE gel for N-terminus peroxisome proliferator-activated receptor-gamma coactivator-1 alpha (PGC-1α) and Superoxide dismutase 2 (SOD-2) and SELENOT; on

10% SDS-PAGE gel for Catalase (CAT); and on 8% SDS-PAGE gel for dynamin-related protein 1 (DRP-1) and optic atrophy 1 (OPA1). Gels were subjected to electrophoresis and transferred to polyvinyl difluoride membranes (GE Healthcare, Chicago, IL, USA). Membranes were blocked in 5% non-fat dried milk at room temperature for 1 h, washed three times with tris-buffered saline containing 0.1% Tween 20 (TBST), incubated overnight at 4 °C with primary specific antibodies for the antigens above mentioned, and diluted 1:1000 (for SELENOT and OPA-1), 1:500 (for PGC-1α, SOD-2, CAT) in TBST and 5% BSA, and 1:1000 for DRP-1 in TBST and non-fat dried milk 1%. β-actin antibody was used as a loading control. Following incubation with primary antibodies, the membranes were washed three times with TBST and then incubated with secondary antibodies peroxidase-conjugated at room temperature for 1 h (anti-mouse diluted 1:2000 and anti-rabbit diluted 1:3000) (Sigma Aldrich) in TBST containing 5% non-fat dried milk. Immunodetection was carried out using a chemiluminescence kit (Santa Cruz Biotechnology, Dallas, TX, USA) or Clarity Western ECL Substrate (Bio-rad, Hercules, CA, USA) when necessary. Densitometric analyses were performed using ImageJ 1.6 software (National Institutes of Health, Bethesda, MD, USA) as previously indicated [16,18,23].

2.11. Attenuated Total Reflectance Fourier-Transform Infrared (ATR-FTIR) Spectroscopic Measurements

For these analyses, H9c2 cells exposed to vehicle, PA (100 µmol/L), PA + PSELT or PSELT 5 nmol/L for 24 h were collected using trypsin, centrifuged ($1500\times g$, 5 min), and then 1,000,000 cells were resuspended in 300 µL of the complete medium. The infrared spectra of live H9c2 cells were collected in the attenuated total reflectance (ATR) mode at 37 °C by using a Tensor II FTIR spectrometer (Bruker Optics, Ettlingen, Germany) equipped with a thermostated BioATR II sample holder and a mercury-cadmium-telluride detector. Then, 20 µL of the cell suspension was deposited on the ATR silicon crystal and left to equilibrate for 2 min before measurement. Spectra were recorded for 180 min (each 10 min) using the kinetic option in Opus acquisition software in order to maximize the absorption signal due to cell sedimentation on the crystal. Each spectrum was an average of 120 scans at 4 cm^{-1} spectral resolution. The background spectrum was recorded with the cell culture medium under the same experimental setup. At least three replicates for each cell type (grown independently) were measured. Spectral processing and analysis: the kinetic spectra of each measurement were baseline corrected with a rubber band function and averaged starting from the saturation stage using Opus 7 software. This average absorbance spectrum was further averaged with the replicas of the same cell type and normalized for the area under the amide II band (1597–1481 cm^{-1}). The normalized mean spectra of each cell type were used in the calculation of the difference spectra. The statistical comparison between the normalized mean spectra of untreated and differently treated H9c2 cells was performed by using Student's *t*-test (two-tailed, nonparametric Wilcoxon, 99% CI) at each wavenumber in Prism 5. The spectral regions where a significant difference in the absorption occurred (*p*-value < 0.0001) were considered in the discussion. The lipid/protein ratio was calculated from the area under the 3050–2800 cm^{-1} region (lipid content) and the 1700–1600 cm^{-1} region (protein content). The bandwidth of the CH2 symmetric band was measured at 75% of the height of the peak maximum from the baseline-corrected spectra.

2.12. Transmission Electron Microscopy (TEM) Analysis

H9c2 cardiomyocytes were seeded in 100 mm plates and upon reaching 80% confluence, they were treated with vehicle, PA (100 µmol/L), PA + PSELT (5 nmol/L) or PSELT (5 nmol/L). Cells were harvested using trypsin, centrifuged ($1500\times g$, 5 min), and then fixed with 3% glutaraldehyde in 0.1 M phosphate buffer overnight at 4 °C. Post-fixation proceeded in buffered osmium tetroxide, followed by dehydration in a graded acetone series (30%, 50%, 70%, 90%, and 100%) and embedding in Epon812. Ultrathin sections (60–90 nm in thickness) were cut with a diamond knife, mounted on copper grids (G300 Cu), and imaged using a Jeol JEM 1400-Plus electron microscope operating at 80 kV. For each condition, at least 100 cells from randomly chosen fields were observed.

2.13. Statistical Analysis

Data, shown as means $\pm$ SEM, were analyzed by one-way ANOVA followed by Dunnett's multiple comparison test and Newman–Keuls multiple comparison test (for post-ANOVA comparisons), and unpaired *t*-test when appropriate. Values with (*) $p < 0.05$, (**) $p < 0.01$, (***) $p < 0.001$, and (****) $p < 0.0001$ were considered statistically significant. The statistical analysis was conducted using Prism 5 (GraphPad Software, La Jolla, CA, USA).

3. Results

3.1. PSELT Mitigates PA-Induced Cytotoxicity and Lipid Accumulation in H9c2 Cardiomyocytes

To determine the concentration range at which PA induces toxic action on cardiomyocytes in term of cell viability, H9c2 cells were treated with increasing concentrations of PA (100–500 µmol/L) for 24 h. MTT assay showed that, compared to control cells, PA dose-dependently decreased cardiomyocyte viability starting from 100 µmol/L (Figure 1A). After this first cardiotoxic dose of PA was established, H9c2 cells were exposed to PA (100 µmol/L) and co-treated with increasing concentrations of PSELT (5–100 nmol/L) for 24 h. The results indicated that PA induced a significant decrease in cell viability compared to the control cells, while PSELT was able to significantly mitigate PA-dependent cell death at each tested concentration, starting from 5 nmol/L (Figure 1B). Thus, the first effective concentration of PSELT (5 nmol/L) was considered for the subsequent analyses. Conversely, in the concentration range of 5–100 nmol/L, the inert counterpart of PSELT (inert-PSELT) was ineffective in counteracting PA-induced cell viability decrease (Figure 1C).

The effect of PSELT against PA-induced cytotoxicity was also evaluated by measuring, in the H9c2 culture medium, the enzymatic activity of LDH, whose release indicates damage of the cell membrane. As shown in Figure 1D, PA treatment significantly increased LDH release with respect to the control group, while H9c2 cells exposed to PA + PSELT exhibited a significant decrease in LDH activity compared to cells treated with PA alone. In the cells exposed to PSELT alone, no significant change in LDH levels was detected compared to control cells (Figure 1D).

Oil Red O staining and relative spectrophotometric quantification were performed to measure the accumulation of intracellular lipids in H9c2 cells. This analysis revealed a significant increase in the intracellular lipid droplets in cardiomyocytes after PA exposure compared to control cells. However, in H9c2 cardiomyocytes treated with PA + PSELT, intracellular lipids were significantly reduced compared to PA alone (Figure 1E).

3.2. PSELT Protects H9c2 Cells against PA-Induced Oxidative Stress

To evaluate whether PSELT could counteract PA-induced oxidative stress, we evaluated intracellular ROS generation by using the specific fluorescent probe CM-H_2DCFDA, mitochondrial superoxide generation by MitoSOX-Red staining, and the expression levels of endogenous antioxidant enzymes. Figure 2A indicates that PA significantly increased ROS generation, as evidenced by the enhanced fluorescence intensity of the probe observed in PA-treated cells compared to control cells. PSELT significantly decreased this PA-induced fluorescence intensity, and PSELT alone did not generate any significant intracellular ROS compared to the control group. ROS determination by the H_2DCFDA assay and measured by flow cytometry in H9c2 cells exposed to PA with or without PSELT reflected the same trend (Supplementary Figure S1).

Mitosox-Red staining was then conducted to detect mitochondrial superoxide generation in the H9c2 cardiomyocytes. As revealed by increased red fluorescence-stained cells in Figure 2B, PA treatment resulted in a significant increase in $O_2{}^-$ levels compared to those of control cells. PSELT treatment significantly reduced $O_2{}^-$ generation in the presence of PA compared to PA alone (Figure 2B). The cell oxidative status was also evaluated by assessing the expression levels of the endogenous antioxidant enzymes SOD-2 and CAT; Western blot and densitometric analyses of the H9c2 cell extracts showed that SOD-2 and CAT expression significantly increased in PA-treated cells with respect to the control group, whereas in the PA + PSELT group, the expression of both enzymes was significantly reduced compared to the group with PA alone (Figure 2C).

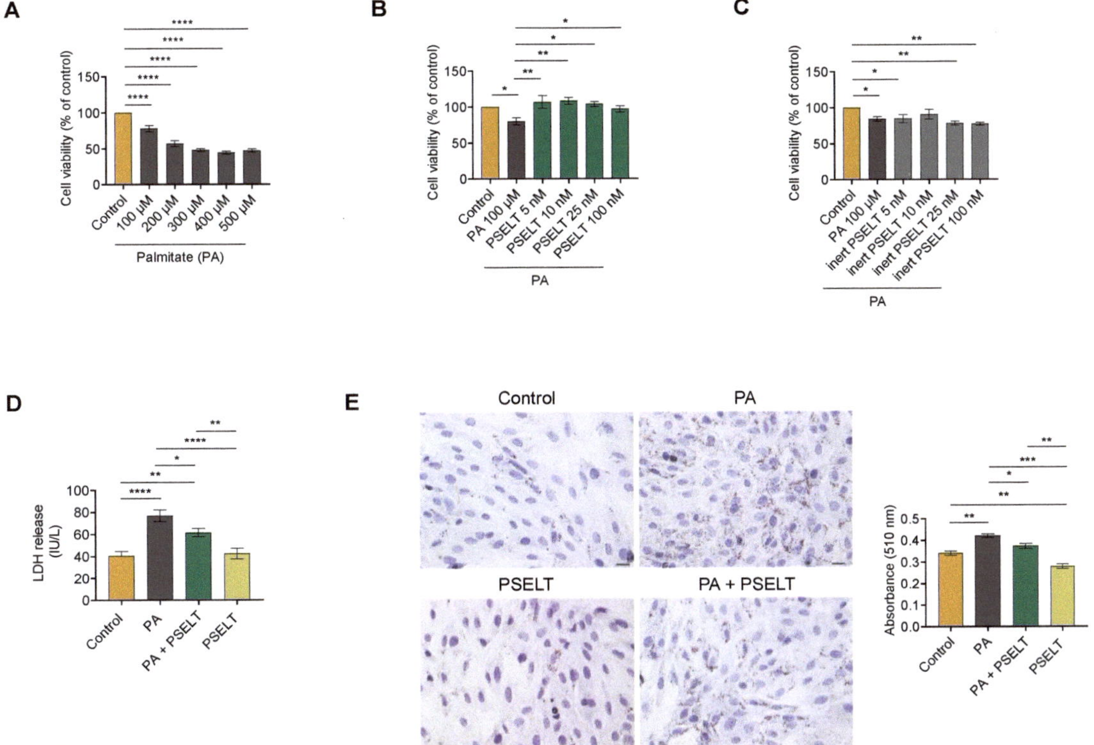

Figure 1. Effects of palmitate (PA) with or without PSELT on cell viability, cytotoxicity, and lipid accumulation in H9c2 cardiomyocytes. H9c2 cells were exposed to vehicle (Control) or increasing concentrations of (**A**) PA (100–500 μmol/L) or (**B**) PA 100 μmol/L + PSELT (5–100 nmol/L) or (**C**) PA 100 μmol/L + inert PSELT (I-PSELT) (5–100 nmol/L) for 24 h. The viability of H9c2 cells was determined using MTT assay and was expressed as the percentage of control cells. Results are reported as the mean ± SEM (n = 6 per group). Significant differences were detected by one-way ANOVA followed by Dunnett's test, $p < 0.05$ (*); $p < 0.01$ (**); and $p < 0.0001$ (****) vs. the Control group. (**D**) Lactate dehydrogenase (LDH) release in the culture medium of H9c2 cardiomyocytes treated with vehicle, PA (100 μmol/L) and PSELT (5 nmol/L) alone or in co-treatment for 24 h. The LDH activity is expressed as IU/L. Data are shown as the mean ± SEM of six separate experiments. Significant differences were detected by one-way ANOVA and Newman-Keuls multiple comparison test, $p < 0.05$ (*); $p < 0.01$ (**); and $p < 0.0001$ (****). (**E**) Representative images of Oil Red O staining for lipid droplet assessment and relative quantification. H9c2 cardiomyocytes treated with vehicle (Control), PA 100 μmol/L, PA + PSELT 5 nmol/L or PSELT for 24 h. Scale bar: 25 μm. Quantification of the stained lipid droplets was performed by measuring the absorbance at 510 nm. Values are the mean ± SEM of three different experiments. $p < 0.05$ (*); $p < 0.01$ (**); $p < 0.001$ (***); and $p < 0.0001$ (****).

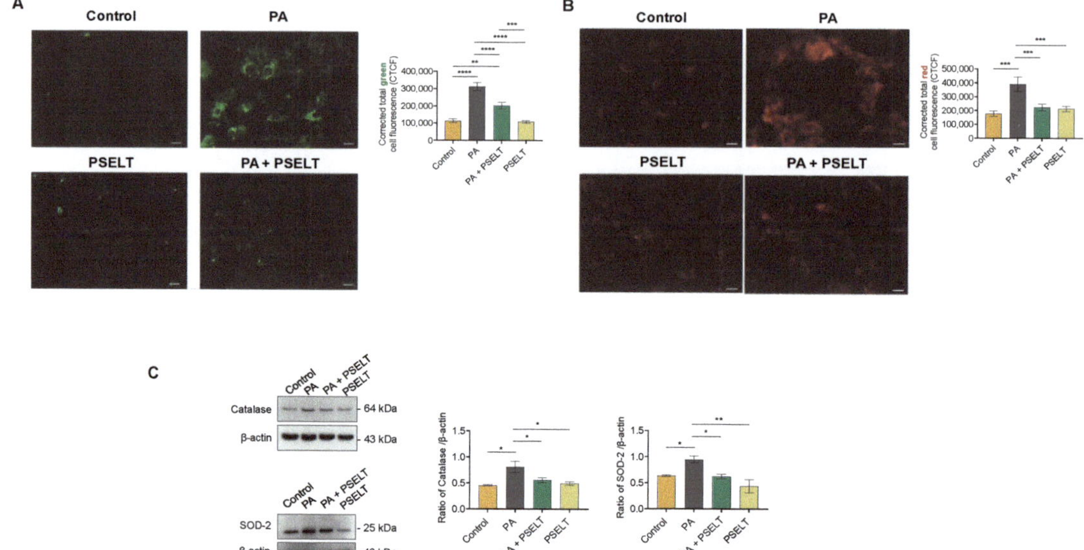

Figure 2. Effects of PSELT on PA-induced oxidative stress in H9c2 cardiomyocytes. (**A**) Representative images of a fluorescent probe (CM-H$_2$ DCFDA) for ROS detection with relative fluorescence quantification. Data are expressed as the mean ± SEM (*n* = 3 different experiments). Scale bar: 25 μm. Significant differences in corrected total cell fluorescence (CTCF) were detected by one-way ANOVA and Newman-Keuls multiple comparison test $p < 0.01$ (**); $p < 0.001$ (***); and $p < 0.0001$ (****). (**B**) Superoxide (O$_2{}^-$) levels assessed by the MitoSOX fluorescent probe and quantification of the fluorescence intensity in H9c2 cardiomyocytes treated with vehicle (Control), PA 100 μmol/L, PA + PSELT 5 nmol/L or PSELT for 24 h. Scale bar: 25 μm. Data are expressed as the mean ± SEM (*n* = 3 different experiments). Significant differences in CTCF were detected by one-way ANOVA and Newman-Keuls multiple comparison test, $p < 0.001$ (***). Western blot analysis of (**C**) Catalase and SOD-2 in H9c2 cardiomyocytes treated with vehicle (Control), PA 100 μmol/L, PA + PSELT 5 nmol/L or PSELT for 24 h. Histograms represent the ratio of the densitometric analysis of protein:loading control. Statistical differences were analyzed by Newman-Keuls multiple comparison test (one-way ANOVA) $p < 0.05$ (*); $p < 0.01$ (**).

3.3. PSELT Rescues the PA-Induced Reduction of Endogenous SELENOT Expression in H9c2 Cells, and Endogenous SELENOT Is Fundamental for PSELT-Induced Cell Protection against the PA Effect

To evaluate whether PA could affect the expression of the endogenous SELENOT, we performed Western blot and relative densitometric analysis of SELENOT in H9c2 cells exposed to PA for 12, 18, and 24 h. As shown in Figure 3A, PA treatment reduced SELENOT levels in a time-dependent manner compared to the control, particularly at 18 and 24 h where the reduction of SELENOT expression was statistically significant. Then, we evaluated whether PA could decrease endogenous SELENOT expression through FAT/CD36 (cluster of differentiation 36/fatty acid translocase), the main transporter of fatty acids in the heart. To this aim, H9c2 cardiomyocytes were first treated for 1 h with 1 μmol/L of sulfo-N-succinimidyl oleate (SSO) [31]–an irreversible inhibitor of CD36 able to block CD36-mediated FA uptake–followed by exposure to PA for 24 h. As shown in Figure 3B, PA significantly reduced SELENOT expression compared to the control, while in the cells exposed to SSO + PA, SELENOT expression was preserved. To confirm that SSO was effective in preventing PA-induced cytotoxicity, we performed an MTT assay showing the ability of SSO to significantly mitigate PA-induced cell death in H9c2 cardiomyocytes compared to PA alone (Figure 3C). In order to evaluate the influence of exogenous PSELT on endogenous SELENOT expression during PA treatment, we carried out a Western blot analysis of H9c2 cells exposed to PA for 24 h in the presence or absence of PSELT. Figure 3D confirmed the ability of PA to reduce SELENOT expression compared to the control and indicated that PSELT significantly rescued the PA-induced reduction of SELENOT.

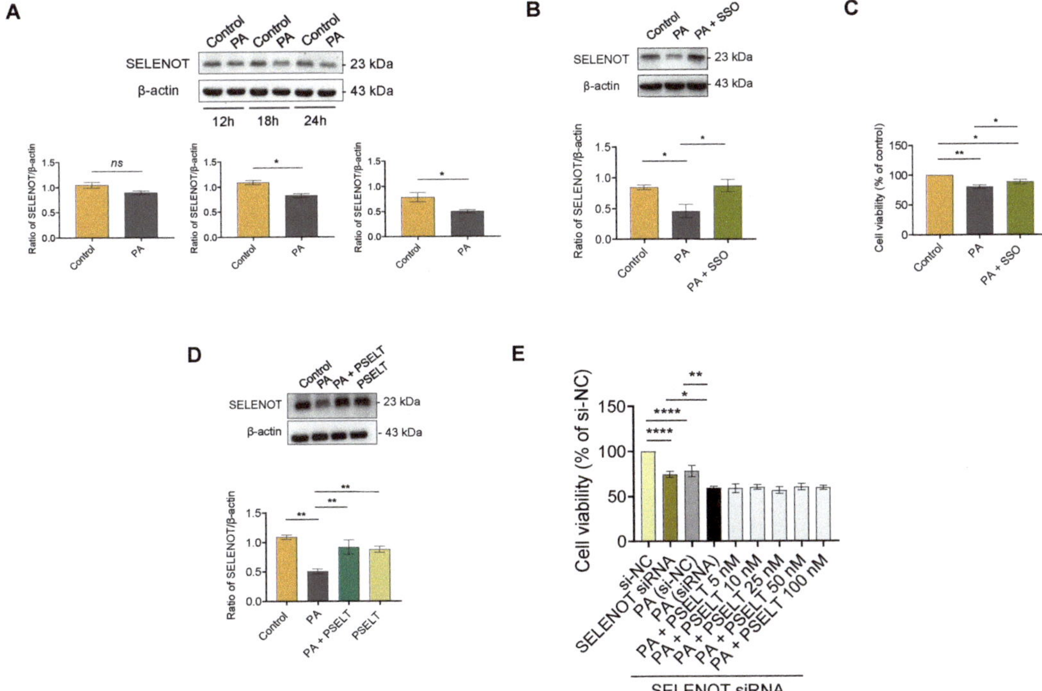

Figure 3. Effect of PA and PSELT on endogenous SELENOT expression and the role of SELENOT in PSELT-induced cell protection against PA in H9c2 cardiomyocytes. (**A**) Effect of PA on endogenous SELENOT expression at different times. Western blot analysis of SELENOT in H9c2 cardiomyocytes treated with vehicle (Control) or PA 100 μmol/L for 12 h, 18 h, and 24 h. Histograms represent the ratio of the densitometric analysis of protein:loading control. Data are expressed as the mean ± SEM, and significant differences were detected by *t*-tests; $p < 0.05$ (*). (**B**) Representative Western blot analysis of SELENOT in H9c2 cardiomyocytes treated with vehicle (Control), PA 100 μmol/L, or sulfosuccinimidyl oleate (SSO) 1 μmol/L for 1 h, an irreversible inhibitor of CD36 + PA, for an additional 24 h. Data are expressed as the mean ± SEM ($n = 3$ independent experiments). (**C**) MTT assay in H9c2 cardiomyocytes exposed to vehicle (Control), PA 100 μmol/L, or SSO 1 μmol/L for 1 h (an irreversible inhibitor of CD36) + PA for an additional 24 h. Cell viability was expressed as the percentage of control cells exposed only to vehicle (indicated as Control). (**D**) Western blot analysis of SELENOT in H9c2 cardiomyocytes treated with vehicle (Control), PA 100 μmol/L, PA + PSELT 5 nmol/L or PSELT for 24 h. Data are expressed as the mean ± SEM ($n = 3$ independent experiments). Histograms represent the ratio of the densitometric analysis of protein:loading control. (**E**) Effect of si-SELENOT gene silencing on cell viability in H9c2 cells treated with PA 100 μmol/L alone and in the presence of increasing concentrations of PSELT (5–100 nmol/L) for 24 h. Cell viability was evaluated using MTT assay and was expressed as the percentage of control cells transfected only with the si-RNA-negative control (indicated as si-NC). Results are presented as the mean ± SEM ($n = 6$ per group). Significant differences were detected by one-way ANOVA followed by Newman–Keuls multiple comparison test. $p < 0.05$ (*); $p < 0.01$ (**); and $p < 0.0001$ (****).

To establish the direct role of endogenous SELENOT in the cytoprotection mediated by exogenous PSELT during PA exposure, we tested the effect of SELENOT knockdown by using a SELENOT siRNA. We first observed that the transfection of the siRNA SELENOT (si-SELENOT) significantly reduced H9c2 cell viability compared to the control [i.e., cells transfected with control siRNA (si-NC)] (Figure 3E). As shown above, this analysis also showed that PA reduced cell viability in si-NC transfected cells compared to control cells (si-NC). Further, SELENOT knockdown worsened PA-induced cytotoxicity since the extent of the PA-induced cell death was higher in SELENOT-knockdown cells compared to the control cells exposed to PA [PA (si-NC) group] (Figure 3E). Additionally, none of the PSELT concentrations (5–100 nmol/L) in si-SELENOT-transfected cells exposed to PA were able to significantly mitigate PA-induced cell death compared with cells silenced for SELENOT and exposed only to PA (Figure 3E).

3.4. PSELT Reduces the PA-Dependent Upregulation of CD36 in H9c2 Cardiomyocytes

To evaluate whether PSELT could affect the expression levels of the fatty acid transporter CD36, we performed an immunofluorescence analysis on H9c2 cells exposed to PA and PSELT, alone and in co-treatment, for 24 h. As shown in Figure 4, PA significantly increased CD36 expression compared with control cells, as revealed by the enhanced fluorescence intensity in the PA group. Conversely, PSELT was able to significantly mitigate CD36 expression in H9c2 cardiomyocytes exposed to PA with respect to cells treated with PA alone, as revealed by a lower fluorescence intensity in the PA + PSELT group. Moreover, a slight but significant increase in fluorescence intensity was also detected in PSELT-treated H9c2 cells compared to control cells.

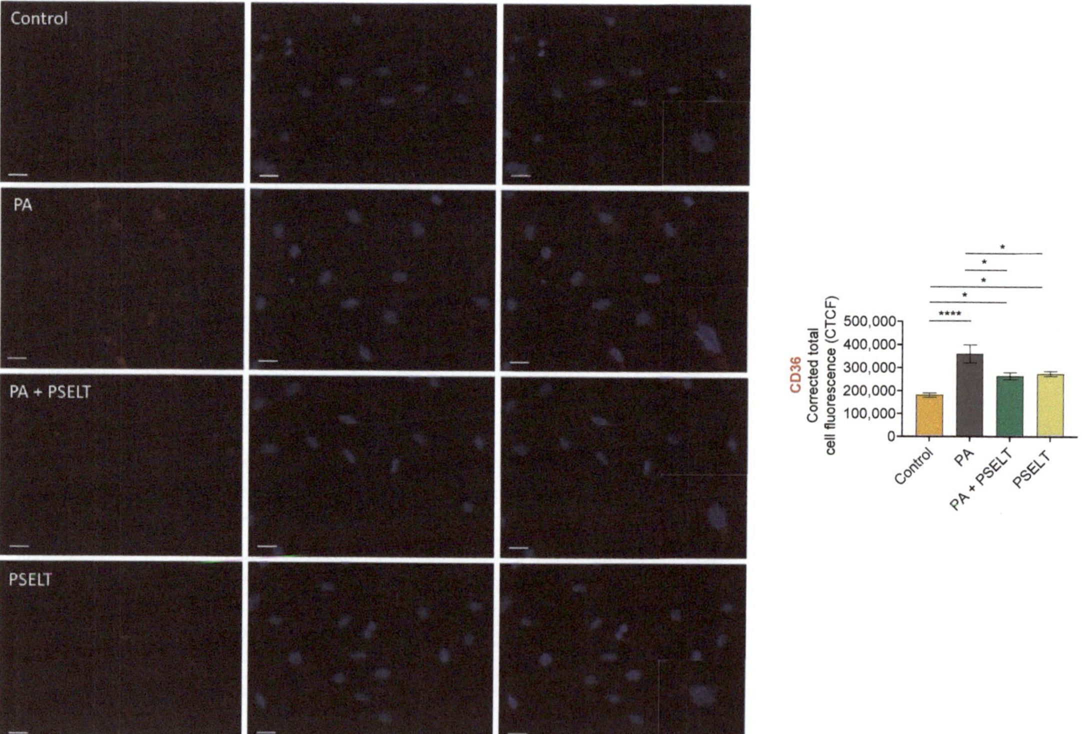

Figure 4. Effects of PSELT on CD36 expression in PA-treated H9c2 cardiomyocytes. Representative images of cardiomyocytes treated with vehicle or PA (100 μmol/L) and PSELT (5 nmol/L) alone or in co-treatment for 24 h. CD36 expression was assessed by using an anti-CD36 primary antibody (Thermo

Fisher scientific) and a donkey anti-rabbit secondary antibody—Alexa Fluor™ 555 (red) (Invitrogen). Nuclei were counterstained with DAPI (blue). Scale bars: 25 µm. Data are expressed as the mean ± SEM (n = 3 different experiments). Significant differences in CTCF were detected by one-way ANOVA and Newman-Keuls multiple comparison test, $p < 0.05$ (*) and $p < 0.0001$ (****).

3.5. PSELT Mitigates the Detrimental Effects of PA on Mitochondrial Function, Biogenesis, and Dynamics

To evaluate the effects of PSELT on mitochondrial function during PA treatment, we performed metabolic flux analysis using the Seahorse XFe96. In H9c2 cardiomyocytes, a dramatic reduction in the oxygen consumption rate (OCR) was observed after treatment with PA, whereas PSELT mitigated this effect (Figure 5A and Supplementary Figure S2).

We next investigated whether the protective action of PSELT could also be linked to mitochondrial biogenesis and dynamics; to this aim, we carried out Western blot analyses aimed at assessing specific markers involved in these molecular processes. Figure 5B shows that PA significantly reduced the levels of PGC1-α, a master regulator of mitochondrial biogenesis, compared to control cells. However, PSELT treatment significantly increased these levels in the presence of PA compared to cells exposed only to PA (Figure 5B). Moreover, we assessed whether PSELT could influence the expression levels of DRP-1, a key regulator of mitochondrial fission, and OPA-1, a gatekeeper of stress-sensitive mitochondrial fusion. Our results indicate that PA triggered a significant increase in both DRP-1 and OPA-1 expression compared to the control group, which was significantly reversed in the presence of PSELT (Figure 5C,D).

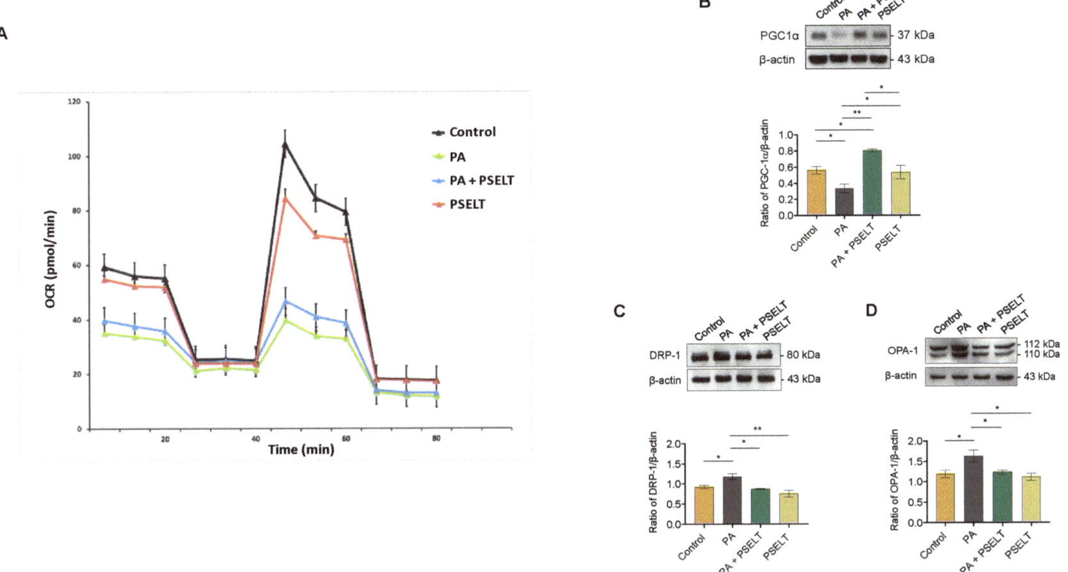

Figure 5. Effects of PSELT on mitochondrial respiration, biogenesis, and dynamics in H9c2 cells exposed to PA. (**A**) Evaluation of the mitochondrial respiratory capacity in H9c2 cardiomyocytes by Seahorse XF Analyzer after 24 h treatment with PA or PSELT alone or in combination and quantification of the oxygen consumption rates (OCR). Western blot analysis of (**B**) PGC-1α, (**C**) DRP-1, and (**D**) OPA-1 in H9c2 cardiomyocytes treated with vehicle (Control), PA 100 µmol/L, PA + PSELT 5 nmol/L or PSELT for 24 h. Histograms represent the ratio of the densitometric analysis of protein:loading control. Data are expressed as the mean ± SEM (n = 3 independent experiments). Significant differences were detected by one-way ANOVA and Newman-Keuls multiple comparison test. $p < 0.05$ (*); $p < 0.01$ (**).

3.6. PSELT Mitigates PA-Dependent Ultrastructural Alterations in H9c2 Cardiomyocytes

To evaluate the protective action of PSELT against PA in H9c2 cardiomyocytes at the ultrastructural level, we carried out TEM analyses. Figure 6A,B shows that the cytoplasm of control cells was rich in ER, Golgi bodies, and mitochondria (see higher magnification images). Dilated cisternae of ER were found to be a typical component of control cells, while mitochondria possessed their typical cristae morphology and structure. After 24 h of PA treatment, the ER network reorganized into stacked and concentrically whorled membranes. The cristae disoriented in most of the mitochondria that also changed shape to oval or round, became swollen, and were often embedded in the ER (Figure 6C,D). Treatment with PSELT restored an apparent normal ultrastructure with no evidence of swelling or injury in cytoplasm and organelles. Most of the mitochondria showed regularly spaced lamellar cristae; the mitochondrial matrix also possessed the typical homogeneous staining pattern of modest electron density, see Figure 6E,F. PSELT treatment alone did not induce any significant effect on the structural and subcellular organization of the cells (Figure 6G,H).

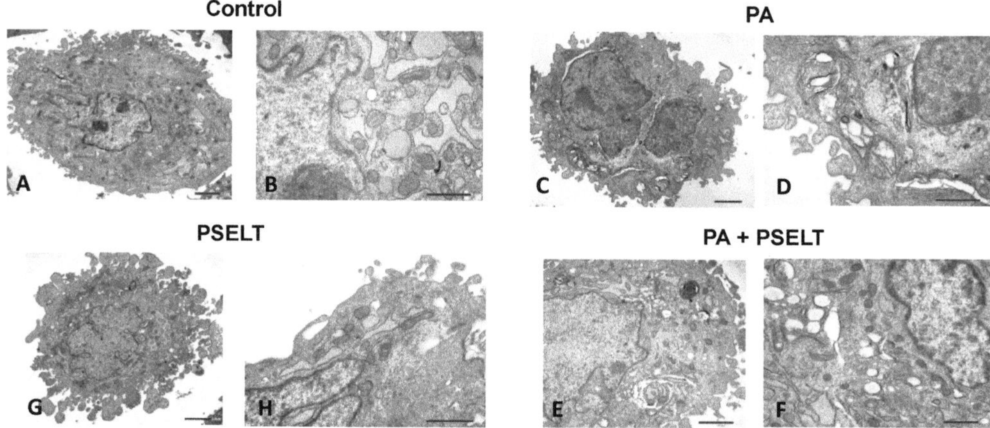

Figure 6. Representative images of transmission electron microscopy (TEM) regarding the effect of PA and PSELT, alone and in combination, on H9c2 cell ultrastructure. Transmission electron micrographs of control cells with intact architecture of cellular organelles, normal mitochondria with distinct cristea, and dilated ER profiles (**A,B**). PA treatment led to reorganization of the ER into concentrically whorled membranes that in some instances wrapped around degenerated mitochondria with swollen cristae (**C,D**). The beneficial effect of PSELT treatment was revealed by improved cell ultrastructure. ER membranes showing the ultrastructure of their naïve counterparts; mitochondria had highly organized cristae with regular thickness and electron density (**E,F**). PSELT treatment alone did not affect the regular arrangement of the intracellular structures and organelles (**G,H**).

3.7. PSELT Mitigates Specific FTIR Spectral Alterations Induced by PA in H9c2 Cardiomyocytes

Figure 7A shows a representative mean absorbance FTIR spectrum of H9c2 cells in aqueous solution in the spectral regions 3050–2800 and 1800–900 cm^{-1}, which contain the signature of key biomolecules. The absorption in the first region originated mainly from the CH_2 and CH_3 stretching vibration (both symmetric and asymmetric) of the lipid acyl chain and reflects the molecular properties of cellular membranes. The second spectral region (fingerprint region) exhibited more complex spectral features where the contribution of the different biomolecules partially overlapped. The main peaks were due to lipids (1750–1715 cm^{-1} C = O stretching, and CH_2 bending at 1470 cm^{-1}), proteins (amide I, 1646 cm^{-1}, amide II, 1546 cm^{-1}, and amide III at 1300 cm^{-1}), nucleic acids (asymmetric and symmetric phosphate stretching at 1125 and 1085 cm^{-1}, respectively), and carbohydrate and glycogen (1155 and 1030 cm^{-1}) [32,33]. For a more detailed description of the absorption peaks numbered in Figure 7A, see Table 1. Typically, variations of the IR bands may include peak shifts, linewidth changes, and variation in the peak intensity/area that can be correlated with specific modifications of the functional groups of the relevant molecules within the sample [34]. In particular, peak shifts and bandwidth provide structural and dynamical information whereas band intensity is related to the concentration of the corresponding functional groups of the biomolecules (according to the Beer–Lambert law).

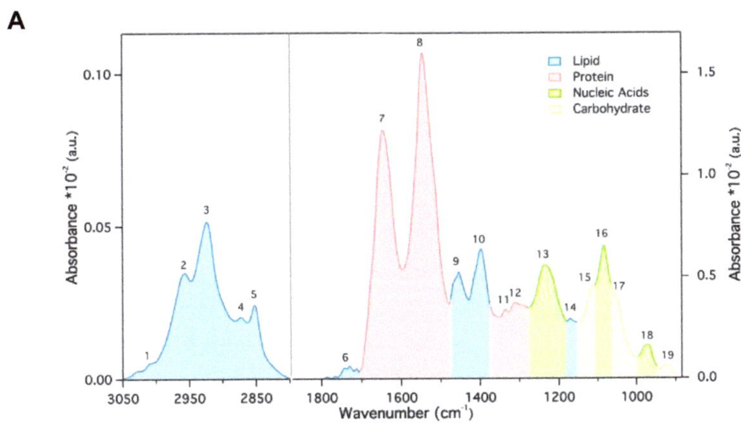

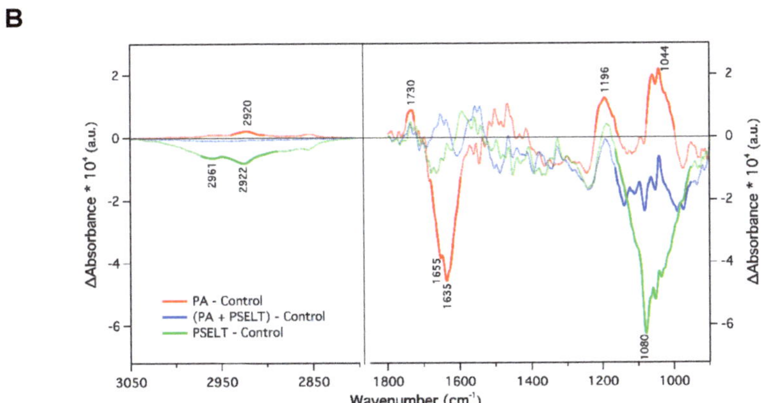

Figure 7. Effects of PSELT on FTIR spectral-related macromolecular changes of PA-treated H9c2 cardiomyocytes. (**A**) Representative mean ATR-FTIR spectrum of untreated H9c2 cells recorded at 37 °C in solution. The lipid region (3050–2800 cm^{-1}) and the fingerprint region (1800–900 cm^{-1}) are shown with the absorption of the different biomolecules. The labelled peaks indicate the absorbance of specific functional groups (**B**) Infrared difference spectra between differently treated cells H9c2 minus the mean spectrum obtained for untreated control H9c2 cells. Pre-processing spectral analysis included baseline correction and normalization to the area under the amide II band. The statistical significance ($\alpha = 0.01$) was determined by using Student's *t*-test at each wavenumber, and the spectral regions with *p*-values < 0.0001 are shown as thicker colored lines. (a.u. indicated arbitrary units).

Table 1. Band Assignments of the Main Peak Absorption of Cells in the 3050–900 cm^{-1} IR Spectral Region [34,35].

Peak Number	Wavenumber (cm^{-1})	Assignment of Functional Groups
1	3010	Olefinic = CH stretching vibration (unsaturated lipids)
2	2957	CH$_3$ asymmetric stretching (lipids)
3	2923	CH$_2$ asymmetric stretching (lipids)
4	2872	CH$_3$ symmetric stretching (proteins and lipids)
5	2852	CH$_2$ symmetric stretching (mainly lipids)
6	1750–1715	C=O stretching (ester functional groups in lipids)
7	1646	Amide I (protein C=O stretching)
8	1546	Amide II (protein NH bending, CN stretching)
9	1456	CH$_2$ bending (mainly lipids)
10	1399	COO$^-$ symmetric stretching (fatty acids)
11	1337	CH$_3$ symmetric bending (lipids)
12	1313	CH$_2$ wagging (lipids)
13	1236	PO$_2^-$ asymmetric stretching, fully hydrogen-bonded (mainly nucleic acids)
14	1173	CO–O–C asymmetric stretching (ester bonds in cholesteryl esters)
15	1116	Ribose ring vibrations (RNA)
16	1086	PO$_2^-$ symmetric stretching (nucleic acids and phospholipids)
17	1050	C–O stretching (polysaccharides, glycogen)
18	971	C–N$^\pm$–C stretching (nucleic acids)
19	924	Ribose ring vibrations (RNA)

Based on a comparison of the mean absorbance spectra of the four different cell samples analyzed, no variation of the band position was observed (Supplementary Figure S3), indicating that there were no significant structural changes. However, changes in the peak intensity were observed.

To better visualize the effects of the different treatments on the molecular composition of H9c2 cardiomyocytes, we calculated the difference between the mean absorbance spectra of the treated cells minus the mean spectrum of the untreated cells representing our control sample (Figure 7B). The regions with the highest statistical significance ($p < 0.0001$) in absorbance were highlighted with a thicker line. Positive/negative peaks in the plot indicated higher/lower concentration of the molecular components within the cell.

When we compared H9c2 cells treated with PA with control cells (red line), significant differences were observed in the amide I region, which reflects a decrease in protein concentration. The peaks positions at 1655 and 1635 cm^{-1} are associated with α-helix and β-sheet proteins [32]. Positive peaks were found around 1196 and 1046 cm^{-1}. These regions are assigned due to the vibrations of specific groups of nucleic acids and carbohydrates and suggest changes in the content of such molecules. Positive peaks are also found for the lipid component at about 2920 and 1730 cm^{-1}. These variations were also reflected in the lipid/protein ratio for the two samples, which increased from 0.200 ± 0.079 (control) to 0.337 ± 0.043 for the PA sample. Moreover, the bandwidth of the CH$_2$ symmetric band at 2852 cm^{-1} decreased from 16.747 ± 0.003 cm^{-1} (control) to 16.030 ± 0.401 cm^{-1} (PA). Such a reduction suggests a decrease in cell membrane fluidity under PA treatment. When the cells were co-treated with PSELT (blue line), the difference in absorption almost vanished below 1200 cm^{-1} and only a reduction in the components related to nucleic acids and carbohydrates was observed.

Finally, the effect of PSELT alone on the cells (green line) involved a reduction of the lipid content (see the negative peak at 2922 cm^{-1}) and also of the bands related to nucleic acids and carbohydrates. Regarding the effects induced by PA in the region between 1165 and 955 cm^{-1}, PSELT co-treatment mitigated the PA-dependent increase (blue line).

4. Discussion

Cardiac lipotoxicity can induce cell dysfunction and cell death, thus increasing both atherosclerotic coronary heart disease and HF, which represent important contributors to cardiovascular complications among obese individuals [6]. Although diverse studies have been conducted in this field, it is mandatory to improve our knowledge on the molecular mechanisms that drive cardiac lipotoxicity in order to characterize specific approaches to minimize obesity-related cardiac complications. In the present study, we employed H9c2 cardiomyocytes exposed to PA, a widely used in vitro model for recapitulating the cardiac harmful effects of consuming a high-fat diet, and provided novel evidence on the beneficial action of the antioxidant SELENOT mimetic peptide (PSELT) against cardiomyocyte damage induced by lipotoxicity.

4.1. PSELT Exerts Protective Effects against PA-Induced Cytotoxicity and Lipotoxicity through Its Redox Site Containing the Sec Residue

Several reports indicate that PA causes lipotoxicity and cell dysfunction in many cell types, including cardiomyocytes [36–40]. Therefore, we exposed H9c2 cells, which are widely used as a cell line to model cardiomyocytes in vitro due to their biochemical, morphological, and electrical/hormonal properties [41,42], to PA in order to model hyperlipidaemia in vitro and to focus on the mechanism affecting the myocardium (i.e., lipotoxicity), which is implicated in the pathogenesis of HF in obesity. In line with previous studies, our results showed first that PA caused cardiomyocyte death in a dose-dependent manner, indicating that the in vitro model of cytotoxicity was successfully established. In accordance, the use of PA represents the most commonly method to induce cardiac lipotoxicity in in vitro systems [43,44].

Our previous in vivo and in vitro data demonstrated that SELENOT exerts a crucial role in preserving redox and ER homeostasis and is essential for cardiomyocyte differentiation and protection, as well as for glucose metabolism through its ability to regulate insulin production/secretion [12–16]. Therefore, we hypothesized that SELENOT could play a role in protecting cardiomyocytes exposed to a dysmetabolic condition generated by lipid overload. To address this issue, we took advantage of a selective SELENOT-derived small peptide (PSELT) able to mimic the activity of the full-length protein through its CVSU redox motif, as previously reported by our groups in different pathophysiological contexts [15–17]. In this regard, therapeutic peptides are emerging as very promising tools due to their ability to selectively target specific molecules and pathways, which may circumvent some limitations of the conventional therapeutics such as those related to the use of a full-length protein in its recombinant form [45].

Our results indicated that PSELT counteracts lipotoxic cardiomyocyte death through its redox active site, as an analogous control peptide lacking the Sec residue in the catalytic site (i.e., inert PSELT) was ineffective, indicating that the cytoprotective action of PSELT is attributable to the Sec residue in the CVSU motif. The protective effects of PSELT against PA-induced lipotoxicity were also confirmed by its ability to reduce the release of LDH in the culture medium (i.e., an important indicator of cytotoxicity and membrane damage) and to counteract intracellular lipid droplets [46].

4.2. PSELT Counteracts PA-Induced Oxidative Stress and the Reduction of Endogenous SELENOT

It is widely known that oxidative stress plays a key role in the onset and progression of several multifactorial diseases, including obesity-related cardiovascular disorders [47–49]. Therefore, selective pharmacological interventions aimed at inhibiting ROS overproduction could represent suitable strategies to mitigate lipotoxic cardiomyocyte death and cardiac dysfunction [50]. Here, we report that PSELT mitigated intracellular ROS production and the mitochondrial superoxide generation provoked by PA and decreased the PA-dependent activation of key endogenous antioxidant enzymes, such as SOD and catalase. It has been reported that an increase in ROS production induces lipid peroxidation, stimulating the activation of antioxidant defenses and inducing redox homeostasis imbalance [51,52]. For instance, the decrease in glutathione (GSH) reductase and glutathione peroxidase (GPX) activities in the presence of PA may activate the antioxidant enzymes SOD and CAT as the result of an adaptive response employed by the cell to counteract the lipotoxic stress condition [53]. The increase in SOD-2 levels found in our study during PA treatment could be due to the augmented production of mitochondrial superoxide, which in turn may contribute to ROS generation. The increase in CAT during PA could be linked to excessive hydrogen peroxide production due to the increase in SOD-2 levels. Therefore, our results suggest that PSELT may induce direct antioxidant activity and a consequent equilibrium of the redox status by rebalancing the levels of ROS-metabolizing enzymes. It is known that a close relationship between increased levels of cytosolic/mitochondrial Ca^{2+} and excessive ROS generation promoting cardiomyocyte and endothelial dysfunction exists [54,55]. It is also widely accepted that PA triggers oxidative stress, leading to disrupted redox-dependent regulatory mechanisms of ER homeostasis and ER stress resulting in Ca^{2+} dysregulation, which may actively participate in apoptotic cell death [56]. Although we did not investigate the cytosolic and mitochondrial Ca^{2+} overload and its potential role in PA-cardiac lipotoxicity, literature evidence has reported Ca^{2+} overload in cardiomyoctes during PA exposure, even if at higher doses compared to those used in our study [21,57]. Considering that SELENOT crucially modulates ER thiol redox balance and contributes to Ca^{2+} signaling by modulating Ca^{2+} flux into and from the ER lumen and by a redox mechanism involving thiol groups in calcium channels and pumps [12,58], it is possible that PSELT and/or SELENOT can be functionally involved in Ca^{2+} regulation for mediating their beneficial action during lipid overload conditions.

The ability of PSELT to mitigate the alteration of cardiomyocyte redox status following PA prompted us to investigate whether PA could also affect the expression of SELENOT (i.e., another key endogenous antioxidant enzyme) and to evaluate the functional role of endogenous SELENOT in the exogenous PSELT-mediated cytoprotection. Our results showed first that PA time-dependently decreased SELENOT expression, indicating that the lipid overload-induced cardiomyocyte injury may depend, at least in part, on the decrease in a selenoprotein faithfully involved in ER homeostasis and exerting a crucial protective role in cardiomyocytes, which also acts as a redox-sensing protein [12,15,16]. We then determined whether PA-dependent SELENOT downregulation could be mediated by FAT/CD36, which plays a pivotal role in the uptake of long-chain fatty acids under both physiological and pathological conditions and is responsible for more than 70% of fatty acid uptake/oxidation in the heart [59,60]. To this aim, we employed SSO, an irreversible inhibitor of FAT/CD36 blocking the uptake of fatty acids, and found that pre-treatment with this compound not only prevented SELENOT downregulation in cells exposed to PA but also inhibited lipotoxic cell death, indicating that PA can affect endogenous SELENOT expression through CD36, indicating this selenoprotein as a novel molecular actor in CD36 signaling for which a crucial role in lipid-overloaded hearts has been reported in several studies [[61] and references therein]. Intriguingly, PSELT rescued SELENOT protein expression after PA treatment, indicating important physiological crosstalk between the exogenous peptide and the endogenous protein. To further corroborate this hypothesis, we assessed the effect of PSELT during PA treatment in SELENOT-silenced cardiomyocytes. The results showed that H9c2 cell viability was compromised by SELENOT deficiency per se, a phenomenon that was worsened in SELENOT-silenced cells exposed to PA, confirming the essential role of this selenoprotein in cardiomyocyte survival and function [15,16]. However, PSELT addition did not mitigate lipotoxic damage in SELENOT-silenced cardiomyocytes, suggesting that SELENOT is required for PSELT action and that the combined action of the protein and the peptide is fundamental to protect cardiomyocytes from the lipotoxic insult. The ability of exogenous PSELT to prevent PA-induced SELENOT downregulation could be of particular interest, as SELENOT not only plays a crucial role in cardiomyocyte differentiation and protection but also represents an essential protein for life, as revealed by the fact that it is the only ER selenoprotein whose gene disruption leads to early embryonic lethality, it is the most highly conserved selenoprotein throughout evolution, and it represents one of the highest-priority selenoproteins [62,63].

Based on the ability of CD36 to affect SELENOT expression and to potentially use this mechanism, among others, for mediating cardiac lipotoxicity, we wondered whether PSELT could influence CD36 expression during PA exposure. Our immunofluorescence analysis revealed that the peptide mitigated the PA-induced upregulation of CD36, indicating that PSELT can protect cardiomyocytes and prevent SELENOT decrease during PA treatment by also reducing the uptake of PA through this transporter. Although we currently do not have the possibility to unequivocally demonstrate that PSELT may influence the biological activity of CD36 through a direct binding to this transporter since a specific antibody against the portion of SELENOT encompassing the PSELT sequence does not exist, our immunofluorescence analysis, which also indicated that PSELT was able to stimulate CD36 under basal conditions (without PA), suggested that the regulation of CD36 expression through direct binding may exist. On the other hand, we cannot exclude that PSELT, acting intracellularly [16,17], may regulate SELENOT expression during stressful conditions induced by PA, making the PSELT/SELENOT crosstalk particularly complicated.

4.3. PSELT Improves Mitochondrial Ultrastructure and Function in Terms of Respiration, Biogenesis, and Dynamics in PA-Treated Cardiomyocytes

There is mounting evidence that maintaining mitochondrial integrity and function is crucial for cardiac cells [64,65]. Here, we determined the effect of PA with or without PSELT on mitochondrial respiration in H9c2 cardiomyocytes. Our data showing a decrease in the oxygen consumption rate in PA-treated cells are consistent with the ability of PA to act as a partial inhibitor of the electron transport chain and to induce oxidative imbalance in mitochondria [21,66]. Interestingly, PSELT improved mitochondrial respiration and mitigated PA-induced mitochondrial respiratory dysfunction.

Mitochondrial alterations can be also related to their biogenesis and dynamic deficiency, which plays a critical role in lipotoxicity and lipid overload-induced metabolic disorders [67]. To further delineate which of these mechanisms are involved in PSELT-dependent mitochondrial protection during PA deleterious effects, we first analyzed the expression levels of PGC1-α, a transcriptional coactivator that plays a key regulatory role in mitochondrial biogenesis. Consistent with several studies [68,69], our results showed that PA markedly reduced PGC1-α expression, an effect that was completely reversed by PSELT, suggesting a potential action of the peptide in promoting mitochondrial biogenesis. Consolidated data have also indicated the presence of cross-regulatory circuits coordinating mitophagy, mitochondrial dynamics, and mitochondrial biogenesis aimed at maintaining the quantity and quality of the mitochondrial network [70]. Therefore, we also studied the potential action of PSELT in mitochondrial dynamics by assessing the expression levels of DRP1, an essential regulator of mitochondrial fission and OPA-1, a critical mediator of the fusion process. In our study, PA induced an increase in both markers, indicating a significant imbalance of fission/fusion that will inevitably lead to mitochondrial dysfunction. The increased mitochondrial fusion observed during PA exposure could represent an attempt of the cell to address and mitigate lipotoxic stimulus, while the increased fission could be required for generating new mitochondria on the one hand, and for removing injured mitochondria through apoptosis on the other hand, to control mitochondrial quality [71]. Although different stimuli, including the excess of lipids, can differentially regulate mitochondrial dynamics, activating or inhibiting mitochondrial fission/fusion depending on the context [72], our data showed that PSELT may restore the levels of both DRP-1 and OPA-1 to normal conditions, thus counteracting PA-induced mitochondrial alterations by re-establishing mitochondrial fission/fusion balance.

Important evidence also highlighted the active participation of ER in mitochondrial division establishing a new model linking mitochondrial dynamics and cell death [71,73]. Moreover, saturated fatty acids have been reported to be lipotoxic in cardiomyocytes due to their generation of ER stress [3,74]. Thus, in this study we evaluated ultrastructural changes during PA and PSELT treatments, focusing on mitochondrial morphology and the ER network. Our TEM analyses showed impaired mitochondrial morphology characterized by the swelling of mitochondria and injured mitochondrial cristae as well as mitochondrial embedding within the ER and the presence of alterations in the ER network in cardiomyocytes exposed to PA. These findings are in line with several studies reporting ultrastructural changes secondary to PA exposure that are mainly linked to ROS overproduction and oxidative stress [47,75–77]. Interestingly, PSELT not only improved PA-induced cristae remodeling but also restored the conventional mitochondrial ultrastructure and promoted the ER network, reinforcing the hypothesis that the peptide can protect cardiomyocytes against PA by ameliorating the mitochondrial structure and function and ER network. Notably, we previously demonstrated that SELENOT localizes in the cardiac ER and plays a crucial role in the regulation of ER proteostasis, acting as a "guardian" of ER homeostasis through its redox center [12,15]. Therefore, the ability of PSELT to cross the plasma membrane and potentially localize/target ER should be regarded with particular interest in the context of key functional cooperation between exogenous PSELT and endogenous SELENOT in ER homeostasis under both physiological and lipid overload conditions.

4.4. PSELT Attenuates FTIR Spectral-Related Macromolecular Changes Induced by PA in H9c2 Cardiomyocytes

The beneficial action of PSELT against PA prompted us to deepen our understanding of the effect of the peptide on the macromolecular content by a spectroscopic characterization. FTIR spectroscopy analysis revealed positive peaks in the lipid component of PA-treated cells, confirming the ability of saturated fatty acids to induce lipid overload and cardiac lipotoxicity [78–80]. We also found that PA negatively affected the content of the α-helix and β-sheet structures of proteins; this effect is likely related to the oxidative burst and ER stress generated by PA that may activate UPR, impairing protein synthesis. This hypothesis is consistent with recent findings according to which PA causes a decline in protein synthesis in skeletal muscle by inducting ER stress [46]. We then found that PA also promoted a substantial increase in the content of nucleic acids, which could reflect a first-line response adopted by cardiomyocytes against cellular damage, aimed at reconstructing new mitochondria and microsomes. This hypothesis is in line with previous studies indicating the capability of myocardial tissue to enhance the RNA/DNA concentration for the early renewal of injured cell material following specific insults, such as myocardial infarction [81].

Intriguingly, PSELT markedly reduced PA-induced lipid accumulation, further confirming the direct action of the peptide in reducing the accumulation of lipid droplets, and mitigated the PA-dependent reduction in protein synthesis and increase in nucleic acids. These findings agree with our recent studies showing that SELENOT, as a novel subunit of the oligosaccharyl transferase (OST) complex, is crucial for the regulation of ER proteostasis, hormone *N*-glycosylation, folding, and secretion [11,13]. Additionally, TEM analyses showing the ability of PSELT to improve the ER ultrastructure and network, together with our recent work indicating that PSELT exerts cardioprotection by relieving ER stress in a rat model of MI/R [16], corroborate the idea that this peptide may preserve the ER structure and function in cardiomyocytes and thus counteract PA-dependent protein synthesis decline.

Another aspect of our spectroscopic findings refers to a possible decrease in membrane fluidity observed in H9c2 cells treated with PA, which could be related to a decrease in the content of cholesterol, unsaturated and saturated lipids, and proteins, as well as to the action of free radicals that can promote molecular oxidation through different mechanisms [82]. Interestingly, PSELT mitigated the loss in cardiomyocyte membrane fluidity, corroborating its ability to restore the lipid profile and to exert antioxidant defense. Indeed, consolidated evidence reported the ability of oxygen free radicals to alter the cell membrane fluidity. Particularly, different oxyradical generating systems can affect the function of cardiac membranes by decreasing the phospholipid N-methylation activity, which is known to determine the membrane fluidity [83]. This effect may represent the result of lipid peroxidation that occurs when a free radical oxidizes an unsaturated lipid chain, leading to the formation of a hydroperoxidized lipid and an alkyl radical, thus altering membrane integrity, fluidity, and function [84,85]. Together, these data further delineate the biological significance of SELENOT in controlling the homoeostasis of cardiomyocytes during oxidative stress triggered by various insults including lipotoxicity.

5. Conclusions

In conclusion, this study reports the ability of the SELENOT mimetic PSELT to inhibit PA-provoked detrimental effects in H9c2 cardiomyocytes by counteracting cell death, lipid accumulation, redox alteration, CD36 upregulation, and mitochondrial and ER dysfunction (Figure 8). Our results also indicate that PSELT can preserve PA-dependent alterations in the cellular macromolecular content and membrane fluidity. Interestingly, these findings suggest that exogenous PSELT requires endogenous SELENOT–a crucial protein for cardiomyocyte differentiation and protection–to exert its beneficial action during lipotoxicity, further delineating the biological significance of SELENOT in cardiomyocytes and highlighting important physiological crosstalk between the exogenous peptide and the endogenous protein (Figure 8).

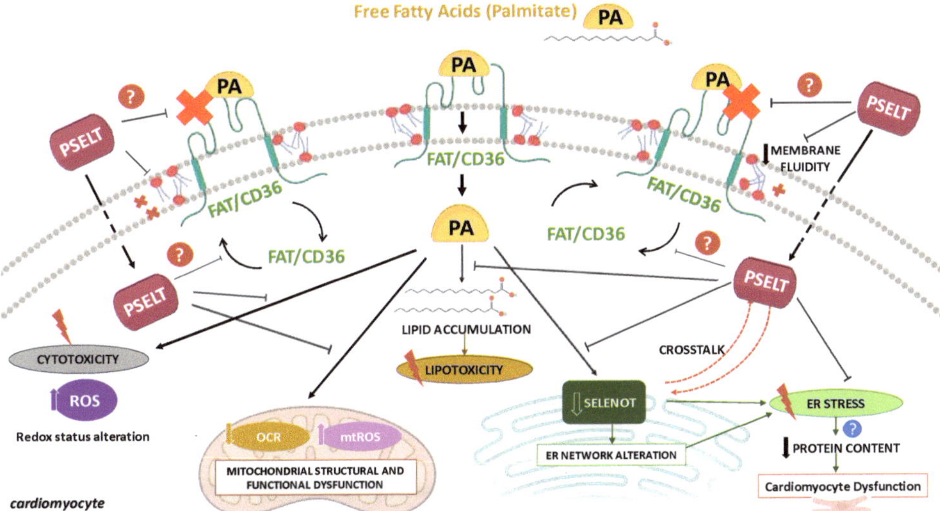

Figure 8. Schematic molecular mechanisms underlying the protective effect of the redox-active motif of SELENOT (PSELT) against palmitate-induced lipotoxicity in cardiomyocytes. ER: endoplasmic reticulum; FAT/CD36: cluster of differentiation 36/fatty acid translocase; OCR: oxygen consumption rate; PA: palmitate; mtROS: mitochondrial reactive oxygen species; ROS: reactive oxygen species; SELENOT: selenoprotein T. For further explanation, see text.

Supplementary Materials: The following supporting information can be downloaded at https://www.mdpi.com/article/10.3390/cells12071042/s1. Figure S1: PSELT mitigates PA-induced intracellular ROS generation. Intracellular ROS levels assessed by H_2DCFDA molecular probe in H9c2 cells treated with vehicle (Control), palmitate (PA), PA and PSELT (PA + PSELT), or PSELT only (PSELT). (A) Representative DCFH fluorescence intensity evaluated by flow cytometry (B) DCFH fluorescent intensity analysis. Values are the mean ± SD of three different experiments. $p < 0.01$ (**) and $p < 0.0001$ (****); Figure S2: PSELT mitigates the detrimental effect of palmitate (PA) on mitochondrial respiration. A significant reduction in basal respiration was observed in cardiomyocytes stimulated with PA, whereas this effect was mitigated by PSELT. Data shown are the mean ± SEM of 2 independent experiments performed in triplicate. $* p < 0.05$; $*** p < 0.001$ indicates significance, all relative to the control cells; Figure S3: ATR-FTIR mean spectra of control H9c2 cells (Control) and differently treated cells, specifically with palmitate (PA), sodium palmitate and peptide (PA + PSELT), or peptide only (PSELT). The cells were dispersed in their culture medium at a concentration of $10^{-6}/300$ μL and maintained at 37 °C in the BioATR II sample holder. The spectral resolution is 4 cm^{-1}. The spectra were normalized to the area under the amide II band (1597–1481 cm^{-1}).

Author Contributions: Conceptualization, C.R., A.D.B. and T.A.; methodology, C.R., A.D.B., R.G., M.C.C., V.R., N.R., I.P., E.M.D.F., M.G.M., M.C.G., T.P., R.M., L.B., B.L., J.L., M.E.G.C., T.S. and N.A.; validation, C.R., A.D.B., R.G., M.C.C., V.R., N.R., I.P., E.M.D.F., M.G.M., M.C.G., T.P., R.M., L.B., B.L., J.L., M.E.G.C., T.S., N.A., Y.A. and T.A.; formal analysis, C.R., A.D.B., R.G., M.C.C., V.R., N.R., I.P., E.M.D.F., M.G.M., M.C.G., T.P., R.M., L.B., B.L., J.L., M.E.G.C., T.S., N.A., Y.A. and T.A.; investigation, C.R., A.D.B., R.G., M.C.C., V.R., I.P., E.M.D.F., N.A., Y.A. and T.A.; writing—original draft preparation, C.R., A.D.B., R.G., I.P. and T.A.; writing—review and editing, C.R., Y.A. and T.A.; supervision, C.R. and T.A.; funding acquisition, C.R., E.M.D.F., N.A. and T.A. All authors have read and agreed to the published version of the manuscript.

Funding: This research was funded by the Ministry of Education, Universities and Research (MIUR) of Italy [project Proof of concept- grant number POC01_00049–"Il dominio catalitico 43–52 derivato dalla selenoproteina T (PSELT) come nuovo farmaco nella protezione dall'infarto del miocardio"]; the University of Calabria "ex-60%", The Italian Foundation for Cancer Research (AIRC) under Start-Up 2018–ID. 21651–P.I. Ernestina Marianna De Francesco (EMDF), and The Italian Foundation for Cancer Research (AIRC) under IG24449-P.I. Nicola Amodio (NA).

Institutional Review Board Statement: Not applicable.

Informed Consent Statement: Not applicable.

Data Availability Statement: The data presented in this study are available in the article.

Acknowledgments: Carmine Rocca (C.R.) acknowledges University of Calabria (D.M. 168 del 28 febbraio 2018, D.M. 83 del 14.05.2020, D.M. 856 del 16.11.2020) for financial support for the RTDb position. The authors also thank the National Institute of Cardiovascular Research (INRC), Bologna, Italy, for the financial support received in the context of the research project "The challenge of dealing with heart failure with preserved ejection fraction: multiple phenotypes with a common pathophysiological substratum? Donazione Anna Maria Ruvinetti", which was partially used for this work.

Conflicts of Interest: The authors declare no conflict of interest.

Abbreviations

CAT	catalase
CM-H$_2$DCFDA	(5-(and-6)-chloromethyl-2′,7′-dichlorodihydrofluorescein diacetate acetyl ester)
CTCF	corrected total cell fluorescence
CVD	cardiovascular diseases
DRP-1	dynamin-related protein 1
ER	endoplasmic reticulum
FAs	fatty acids
FAT/CD36	cluster of differentiation 36/fatty acid translocase
ATR-FTIR	Attenuated total Reflectance-Fourier-transform infrared spectroscopy
GSH	glutathione
GPX	glutathione peroxidase
HF	heart failure
I-PSELT	inert selenoprotein T-derived peptide 43–52
LDH	lactate dehydrogenase
MI/R	myocardial ischemia/reperfusion
OCR	oxygen consumption rate
OPA-1	optic Atrophy 1
OST	oligosaccharyl transferase
PA	palmitate
PGC1-α	peroxisome proliferator-activated receptor-gamma coactivator1-α
PSELT	selenoprotein T-derived peptide 43–52
ROS	reactive oxygen species
Sec, U	selenocysteine
SELENOT	selenoprotein T
siRNA	short interfering RNA

SOD	superoxide dismutase
SSO	sulfo-N-succinimidyl oleate
UPR	unfolded protein response

References

1. Powell-Wiley, T.M.; Poirier, P.; Burke, L.E.; Després, J.P.; Gordon-Larsen, P.; Lavie, C.J.; Lear, S.A.; Ndumele, C.E.; Neeland, I.J.; Sanders, P.; et al. American Heart Association Council on Lifestyle and Cardiometabolic Health; Council on Cardiovascular and Stroke Nursing; Council on Clinical Cardiology; Council on Epidemiology and Prevention; and Stroke Council. Obesity and Cardiovascular Disease: A Scientific Statement from the American Heart Association. *Circulation* **2021**, *143*, e984–e1010. [PubMed]
2. Szczepaniak, L.S.; Dobbins, R.L.; Metzger, G.J.; Sartoni-D'Ambrosia, G.; Arbique, D.; Vongpatanasin, W.; Unger, R.; Victor, R.G. Myocardial triglycerides and systolic function in humans: In vivo evaluation by localized proton spectroscopy and cardiac imaging. *Magn. Reson. Med.* **2003**, *49*, 417–423. [CrossRef] [PubMed]
3. Drosatos, K.; Schulze, P.C. Cardiac lipotoxicity: Molecular pathways and therapeutic implications. *Curr. Heart Fail. Rep.* **2013**, *10*, 109–121. [CrossRef] [PubMed]
4. Ebong, I.A.; Goff, D.C., Jr.; Rodriguez, C.J.; Chen, H.; Bertoni, A.G. Mechanisms of heart failure in obesity. *Obes. Res. Clin. Pract.* **2014**, *8*, e540–e548. [CrossRef]
5. Pasqua, T.; Rocca, C.; Giglio, A.; Angelone, T. Cardiometabolism as an Interlocking Puzzle between the Healthy and Diseased Heart: New Frontiers in Therapeutic Applications. *J. Clin. Med.* **2021**, *10*, 721. [CrossRef]
6. Sletten, A.C.; Peterson, L.R.; Schaffer, J.E. Manifestations and mechanisms of myocardial lipotoxicity in obesity. *J. Intern. Med.* **2018**, *284*, 478–491. [CrossRef]
7. Listenberger, L.L.; Han, X.; Lewis, S.E.; Cases, S.; Farese, R.V., Jr.; Ory, D.S.; Schaffer, J.E. Triglyceride accumulation protects against fatty acid-induced lipotoxicity. *Proc. Natl. Acad. Sci. USA* **2003**, *100*, 3077–3082. [CrossRef]
8. Malhotra, J.D.; Kaufman, R.J. Endoplasmic reticulum stress and oxidative stress: A vicious cycle or a double-edged sword? *Antioxid. Redox Signal.* **2007**, *9*, 2277–2293. [CrossRef]
9. Rocca, C.; Pasqua, T.; Boukhzar, L.; Anouar, Y.; Angelone, T. Progress in the emerging role of selenoproteins in cardiovascular disease: Focus on endoplasmic reticulum-resident selenoproteins. *Cell. Mol. Life Sci.* **2019**, *76*, 3969–3985. [CrossRef]
10. Lu, J.; Holmgren, A. Selenoproteins. *J. Biol. Chem.* **2009**, *284*, 723–727. [CrossRef]
11. Anouar, Y.; Lihrmann, I.; Falluel-Morel, A.; Boukhzar, L. Selenoprotein T is a key player in ER proteostasis, endocrine homeostasis and neuroprotection. *Free Radic. Biol. Med.* **2018**, *127*, 145–152. [CrossRef] [PubMed]
12. Pothion, H.; Jehan, C.; Tostivint, H.; Cartier, D.; Bucharles, C.; Falluel-Morel, A.; Boukhzar, L.; Anouar, Y.; Lihrmann, I. Selenoprotein T: An Essential Oxidoreductase Serving as a Guardian of Endoplasmic Reticulum Homeostasis. *Antioxid. Redox Signal.* **2020**, *33*, 1257–1275. [CrossRef] [PubMed]
13. Hamieh, A.; Cartier, D.; Abid, H.; Calas, A.; Burel, C.; Bucharles, C.; Jehan, C.; Grumolato, L.; Landry, M.; Lerouge, P.; et al. Selenoprotein T is a novel OST subunit that regulates UPR signaling and hormone secretion. *EMBO Rep.* **2017**, *18*, 1935–1946. [CrossRef]
14. Prevost, G.; Arabo, A.; Jian, L.; Quelennec, E.; Cartier, D.; Hassan, S.; Falluel-Morel, A.; Tanguy, Y.; Gargani, S.; Lihrmann, I.; et al. The PACAP-regulated gene selenoprotein T is abundantly expressed in mouse and human β-cells and its targeted inactivation impairs glucose tolerance. *Endocrinology* **2013**, *154*, 3796–3806. [CrossRef]
15. Rocca, C.; Boukhzar, L.; Granieri, M.C.; Alsharif, I.; Mazza, R.; Lefranc, B.; Tota, B.; Leprince, J.; Cerra, M.C.; Anouar, Y.; et al. A selenoprotein T-derived peptide protects the heart against ischaemia/reperfusion injury through inhibition of apoptosis and oxidative stress. *Acta Physiol.* **2018**, *223*, e13067. [CrossRef] [PubMed]
16. Rocca, C.; De Bartolo, A.; Granieri, M.C.; Rago, V.; Amelio, D.; Falbo, F.; Malivindi, R.; Mazza, R.; Cerra, M.C.; Boukhzar, L.; et al. The Antioxidant Selenoprotein T Mimetic, PSELT, Induces Preconditioning-like Myocardial Protection by Relieving Endoplasmic-Reticulum Stress. *Antioxidants* **2022**, *11*, 571. [CrossRef]
17. Alsharif, I.; Boukhzar, L.; Lefranc, B.; Godefroy, D.; Aury-Landas, J.; Rego, J.D.; Rego, J.D.; Naudet, F.; Arabo, A.; Chagraoui, A.; et al. Cell-penetrating, antioxidant SELENOT mimetic protects dopaminergic neurons and ameliorates motor dysfunction in Parkinson's disease animal models. *Redox Biol.* **2021**, *40*, 101839. [CrossRef]
18. Rocca, C.; De Bartolo, A.; Grande, F.; Rizzuti, B.; Pasqua, T.; Giordano, F.; Granieri, M.C.; Occhiuzzi, M.A.; Garofalo, A.; Amodio, N.; et al. Cateslytin abrogates lipopolysaccharide-induced cardiomyocyte injury by reducing inflammation and oxidative stress through toll like receptor 4 interaction. *Int. Immunopharmacol.* **2021**, *94*, 107487. [CrossRef]
19. Grande, F.; De Bartolo, A.; Occhiuzzi, M.A.; Caruso, A.; Rocca, C.; Pasqua, T.; Carocci, A.; Rago, V.; Angelone, T.; Sinicropi, M.S. Carbazole and Simplified Derivatives: Novel Tools toward β-Adrenergic Receptors Targeting. *Appl. Sci.* **2021**, *11*, 5486. [CrossRef]
20. Rocca, C.; Grande, F.; Granieri, M.C.; Colombo, B.; De Bartolo, A.; Giordano, F.; Rago, V.; Amodio, N.; Tota, B.; Cerra, M.C.; et al. The chromogranin A1-373 fragment reveals how a single change in the protein sequence exerts strong cardioregulatory effects by engaging neuropilin-1. *Acta Physiol.* **2021**, *231*, e13570. [CrossRef]
21. Joseph, L.C.; Barca, E.; Subramanyam, P.; Komrowski, M.; Pajvani, U.; Colecraft, H.M.; Hirano, M.; Morrow, J.P. Inhibition of NAPDH Oxidase 2 (NOX2) Prevents Oxidative Stress and Mitochondrial Abnormalities Caused by Saturated Fat in Cardiomyocytes. *PLoS ONE* **2016**, *11*, e0145750. [CrossRef]

22. McQueen, M.J. Optimal Assay of LDH and α-HBD at 37 °C. *Ann. Clin. Biochem.* **1972**, *9*, 21–25. [CrossRef]
23. Rocca, C.; Scavello, F.; Colombo, B.; Gasparri, A.M.; Dallatomasina, A.; Granieri, M.C.; Amelio, D.; Pasqua, T.; Cerra, M.C.; Tota, B.; et al. Physiological levels of chromogranin A prevent doxorubicin-induced cardiotoxicity without impairing its anticancer activity. *FASEB J.* **2019**, *33*, 7734–7747. [CrossRef]
24. Ivan, A.; Herman, H.; Balta, C.; Hadaruga, D.I.; Mihali, C.V.; Ardelean, A.; Hermenean, A. Berberis vulgaris extract/β-cyclodextrin complex increases protection of hepatic cells via suppression of apoptosis and lipogenesis pathways. *Exp. Ther. Med.* **2017**, *13*, 2143–2150. [CrossRef]
25. Chen, C.J.; Kong, B.; Shuai, W.; Liu, Y.; Wang, G.J.; Xu, M.; Zhao, J.J.; Fang, J.; Fu, H.; Jiang, X.B.; et al. Myeloid differentiation protein 1 protected myocardial function against high-fat stimulation induced pathological remodelling. *J. Cell. Mol. Med.* **2019**, *23*, 5303–5316.
26. Nasci, V.L.; Chuppa, S.; Griswold, L.; Goodreau, K.A.; Dash, R.K.; Kriegel, A.J. miR-21-5p regulates mitochondrial respiration and lipid content in H9C2 cells. *Am. J. Physiol. Heart Circ. Physiol.* **2019**, *316*, H710–H721. [CrossRef] [PubMed]
27. Güçlü, O.; Doğanlar, O.; Yüksel, V.; Doğanlar, Z.B. FOLFIRI-Mediated Toxicity in Human Aortic Smooth Muscle Cells and Possible Amelioration with Curcumin and Quercetin. *Cardiovasc. Toxicol.* **2020**, *20*, 139–154. [CrossRef] [PubMed]
28. Scicchitano, M.; Carresi, C.; Nucera, S.; Ruga, S.; Maiuolo, J.; Macrì, R.; Scarano, F.; Bosco, F.; Mollace, R.; Cardamone, A.; et al. Icariin Protects H9c2 Rat Cardiomyoblasts from Doxorubicin-Induced Cardiotoxicity: Role of Caveolin-1 Upregulation and Enhanced Autophagic Response. *Nutrients* **2021**, *13*, 4070. [CrossRef]
29. Benoist, L.; Chadet, S.; Genet, T.; Lefort, C.; Heraud, A.; Danila, M.D.; Muntean, D.M.; Baron, C.; Angoulvant, D.; Babuty, D.; et al. Stimulation of P2Y11 receptor protects human cardiomyocytes against Hypoxia/Reoxygenation injury and involves PKCε signaling pathway. *Sci. Rep.* **2019**, *9*, 11613. [CrossRef] [PubMed]
30. De Francesco, E.M.; Ózsvári, B.; Sotgia, F.; Lisanti, M.P. Dodecyl-TPP Targets Mitochondria and Potently Eradicates Cancer Stem Cells (CSCs): Synergy With FDA-Approved Drugs and Natural Compounds (Vitamin C and Berberine). *Front. Oncol.* **2019**, *9*, 615. [CrossRef]
31. Urso, C.J.; Zhou, H. Role of CD36 in Palmitic Acid Lipotoxicity in Neuro-2a Neuroblastoma Cells. *Biomolecules* **2021**, *11*, 1567. [CrossRef]
32. Rehman, I.U.; Movasaghi, Z.; Rehman, S. *Vibrational Spectroscopy for Tissue Analysis*, 1st ed.; CRC Press: Boca Raton, FL, USA, 2012.
33. Güler, G.; Guven, U.; Oktem, G. Characterization of CD133+/CD44+ human prostate cancer stem cells with ATR-FTIR spectroscopy. *Analyst* **2019**, *144*, 2138–2149. [CrossRef]
34. Tamm, L.K.; Tatulian, S.A. Infrared spectroscopy of proteins and peptides in lipid bilayers. *Q. Rev. Biophys.* **1997**, *30*, 365–429. [CrossRef]
35. Ozek, N.S.; Tuna, S.; Erson-Bensan, A.E.; Severcan, F. Characterization of microRNA-125b expression in MCF7 breast cancer cells by ATR-FTIR spectroscopy. *Analyst* **2010**, *135*, 3094–3102. [CrossRef]
36. Cheon, H.G.; Cho, Y.S. Protection of palmitic acid-mediated lipotoxicity by arachidonic acid via channeling of palmitic acid into triglycerides in C2C12. *J. Biomed. Sci.* **2014**, *21*, 13. [CrossRef]
37. Cho, Y.S.; Kim, C.H.; Kim, K.Y.; Cheon, H.G. Protective effects of arachidonic acid against palmitic acid-mediated lipotoxicity in HIT-T15 cells. *Mol. Cell. Biochem.* **2012**, *364*, 19–28. [CrossRef]
38. Estadella, D.; da Penha Oller do Nascimento, C.M.; Oyama, L.M.; Ribeiro, E.B.; Dâmaso, A.R.; de Piano, A. Lipotoxicity: Effects of dietary saturated and transfatty acids. *Mediat. Inflamm.* **2013**, *2013*, 137579. [CrossRef]
39. Stanley, W.C.; Dabkowski, E.R.; Ribeiro, R.F., Jr.; O'Connell, K.A. Dietary fat and heart failure: Moving from lipotoxicity to lipoprotection. *Circ. Res.* **2012**, *110*, 764–776. [CrossRef]
40. Wen, S.Y.; Velmurugan, B.K.; Day, C.H.; Shen, C.Y.; Chun, L.C.; Tsai, Y.C.; Lin, Y.M.; Chen, R.J.; Kuo, C.H.; Huang, C.Y. High density lipoprotein (HDL) reverses palmitic acid induced energy metabolism imbalance by switching CD36 and GLUT4 signaling pathways in cardiomyocyte. *J. Cell. Physiol.* **2017**, *232*, 3020–3029. [CrossRef]
41. Hescheler, J.; Meyer, R.; Plant, S.; Krautwurst, D.; Rosenthal, W.; Schultz, G. Morphological, biochemical, and electrophysiological characterization of a clonal cell (H9c2) line from rat heart. *Circ. Res.* **1991**, *69*, 1476–1486. [CrossRef]
42. Branco, A.F.; Pereira, S.P.; Gonzalez, S.; Gusev, O.; Rizvanov, A.A.; Oliveira, P.J. Gene Expression Profiling of H9c2 Myoblast Differentiation towards a Cardiac-Like Phenotype. *PLoS ONE* **2015**, *10*, e0129303. [CrossRef]
43. Kong, J.Y.; Rabkin, S.W. Palmitate-induced apoptosis in cardiomyocytes is mediated through alterations in mitochondria: Prevention by cyclosporin A. *Biochim. Biophys. Acta* **2000**, *1485*, 45–55. [CrossRef]
44. Leroy, C.; Tricot, S.; Lacour, B.; Grynberg, A. Protective effect of eicosapentaenoic acid on palmitate-induced apoptosis in neonatal cardiomyocytes. *Biochim. Biophys. Acta* **2008**, *1781*, 685–693. [CrossRef]
45. Heitz, F.; Morris, M.C.; Divita, G. Twenty years of cell-penetrating peptides: From molecular mechanisms to therapeutics. *Br. J. Pharmacol.* **2009**, *157*, 195–206. [CrossRef]
46. Zou, L.; Li, X.; Wu, N.; Jia, P.; Liu, C.; Jia, D. Palmitate induces myocardial lipotoxic injury via the endoplasmic reticulum stress-mediated apoptosis pathway. *Mol. Med. Rep.* **2017**, *16*, 6934–6939. [CrossRef]
47. De Geest, B.; Mishra, M. Role of Oxidative Stress in Diabetic Cardiomyopathy. *Antioxidants* **2022**, *11*, 784. [CrossRef]
48. Remigante, A.; Morabito, R.; Spinelli, S.; Trichilo, V.; Loddo, S.; Sarikas, A.; Dossena, S.; Marino, A. d-Galactose Decreases Anion Exchange Capability through Band 3 Protein in Human Erythrocytes. *Antioxidants* **2020**, *9*, 689. [CrossRef]
49. Crupi, R.; Morabito, R.; Remigante, A.; Gugliandolo, E.; Britti, D.; Cuzzocrea, S.; Marino, A. Susceptibility of erythrocytes from different sources to xenobiotics-induced lysis. *Comp. Biochem. Physiol. C Toxicol. Pharmacol.* **2019**, *221*, 68–72. [CrossRef]

50. Tsushima, K.; Bugger, H.; Wende, A.R.; Soto, J.; Jenson, G.A.; Tor, A.R.; McGlauflin, R.; Kenny, H.C.; Zhang, Y.; Souvenir, R.; et al. Mitochondrial Reactive Oxygen Species in Lipotoxic Hearts Induce Post-Translational Modifications of AKAP121, DRP1, and OPA1 That Promote Mitochondrial Fission. *Circ. Res.* **2018**, *122*, 58–73. [CrossRef]

51. Mangali, S.; Bhat, A.; Udumula, M.P.; Dhar, I.; Sriram, D.; Dhar, A. Inhibition of protein kinase R protects against palmitic acid-induced inflammation, oxidative stress, and apoptosis through the JNK/NF-kB/NLRP3 pathway in cultured H9C2 cardiomyocytes. *J. Cell. Biochem.* **2019**, *120*, 3651–3663. [CrossRef]

52. Kurutas, E.B. The importance of antioxidants which play the role in cellular response against oxidative/nitrosative stress: Current state. *Nutr. J.* **2016**, *15*, 71. [CrossRef]

53. Alnahdi, A.; John, A.; Raza, H. N-acetyl cysteine attenuates oxidative stress and glutathione-dependent redox imbalance caused by high glucose/high palmitic acid treatment in pancreatic Rin-5F cells. *PLoS ONE* **2019**, *14*, e0226696. [CrossRef] [PubMed]

54. Negri, S.; Faris, P.; Moccia, F. Reactive Oxygen Species and Endothelial Ca^{2+} Signaling: Brothers in Arms or Partners in Crime? *Int. J. Mol. Sci.* **2021**, *22*, 9821. [CrossRef]

55. Tiwary, S.; Nandwani, A.; Khan, R.; Datta, M. GRP75 mediates endoplasmic reticulum-mitochondria coupling during palmitate-induced pancreatic β-cell apoptosis. *J. Biol. Chem.* **2021**, *297*, 101368. [CrossRef] [PubMed]

56. Ly, L.D.; Xu, S.; Choi, S.K.; Ha, C.M.; Thoudam, T.; Cha, S.K.; Wiederkehr, A.; Wollheim, C.B.; Lee, I.K.; Park, K.S. Oxidative stress and calcium dysregulation by palmitate in type 2 diabetes. *Exp. Mol. Med.* **2017**, *49*, e291. [CrossRef] [PubMed]

57. Huang, L.; Chen, Z.; Chen, R.; Lin, L.; Ren, L.; Zhang, M.; Liu, L. Increased fatty acid metabolism attenuates cardiac resistance to β-adrenoceptor activation via mitochondrial reactive oxygen species: A potential mechanism of hypoglycemia-induced myocardial injury in diabetes. *Redox Biol.* **2022**, *52*, 102320. [CrossRef]

58. Grumolato, L.; Ghzili, H.; Montero-Hadjadje, M.; Gasman, S.; Lesage, J.; Tanguy, Y.; Galas, L.; Ait-Ali, D.; Leprince, J.; Guérineau, N.C.; et al. Selenoprotein T is a PACAP-regulated gene involved in intracellular Ca^{2+} mobilization and neuroendocrine secretion. *FASEB J.* **2008**, *22*, 1756–1768. [CrossRef] [PubMed]

59. Shu, H.; Peng, Y.; Hang, W.; Nie, J.; Zhou, N.; Wang, D.W. The role of CD36 in cardiovascular disease. *Cardiovasc. Res.* **2022**, *118*, 115–129. [CrossRef]

60. Glatz, J.F.C.; Luiken, J.J.F.P. Dynamic role of the transmembrane glycoprotein CD36 (SR-B2) in cellular fatty acid uptake and utilization. *J. Lipid Res.* **2018**, *59*, 1084–1093. [CrossRef]

61. Glatz, J.F.C.; Luiken, J.J.F.P.; Nabben, M. CD36 (SR-B2) as a Target to Treat Lipid Overload-Induced Cardiac Dysfunction. *J. Lipid Atheroscler.* **2020**, *9*, 66–78. [CrossRef] [PubMed]

62. Boukhzar, L.; Hamieh, A.; Cartier, D.; Tanguy, Y.; Alsharif, I.; Castex, M.; Arabo, A.; El Hajji, S.; Bonnet, J.J.; Errami, M.; et al. Selenoprotein T Exerts an Essential Oxidoreductase Activity That Protects Dopaminergic Neurons in Mouse Models of Parkinson's Disease. *Antioxid. Redox Signal.* **2016**, *24*, 557–574. [CrossRef] [PubMed]

63. Santesmasses, D.; Mariotti, M.; Gladyshev, V.N. Tolerance to Selenoprotein Loss Differs between Human and Mouse. *Mol. Biol. Evol.* **2020**, *37*, 341–354. [CrossRef] [PubMed]

64. Gustafsson, Å.B.; Gottlieb, R.A. Heart mitochondria: Gates of life and death. *Cardiovasc. Res.* **2008**, *77*, 334–343. [CrossRef]

65. Peoples, J.N.; Saraf, A.; Ghazal, N.; Pham, T.T.; Kwong, J.Q. Mitochondrial dysfunction and oxidative stress in heart disease. *Exp. Mol. Med.* **2019**, *51*, 1–13. [CrossRef]

66. Kakimoto, P.A.; Kowaltowski, A.J. Effects of high fat diets on rodent liver bioenergetics and oxidative imbalance. *Redox Biol.* **2016**, *8*, 216–225. [CrossRef] [PubMed]

67. Mishra, P.; Varuzhanyan, G.; Pham, A.H.; Chan, D.C. Mitochondrial Dynamics is a Distinguishing Feature of Skeletal Muscle Fiber Types and Regulates Organellar Compartmentalization. *Cell. Metab.* **2015**, *22*, 1033–1044. [CrossRef] [PubMed]

68. De Sousa, I.F.; Migliaccio, V.; Lepretti, M.; Paolella, G.; Di Gregorio, I.; Caputo, I.; Ribeiro, E.B.; Lionetti, L. Dose- and Time-Dependent Effects of Oleate on Mitochondrial Fusion/Fission Proteins and Cell Viability in HepG2 Cells: Comparison with Palmitate Effects. *Int. J. Mol. Sci.* **2021**, *22*, 9812. [CrossRef]

69. Coll, T.; Jové, M.; Rodríguez-Calvo, R.; Eyre, E.; Palomer, X.; Sánchez, R.M.; Merlos, M.; Laguna, J.C.; Vázquez-Carrera, M. Palmitate-mediated downregulation of peroxisome proliferator-activated receptor-gamma coactivator 1alpha in skeletal muscle cells involves MEK1/2 and nuclear factor-kappaB activation. *Diabetes* **2006**, *55*, 2779–2787. [CrossRef] [PubMed]

70. Lim, J.H.; Gerhart-Hines, Z.; Dominy, J.E.; Lee, Y.; Kim, S.; Tabata, M.; Xiang, Y.K.; Puigserver, P. Oleic acid stimulates complete oxidation of fatty acids through protein kinase A-dependent activation of SIRT1-PGC1α complex. *J. Biol. Chem.* **2013**, *288*, 7117–7126. [CrossRef]

71. Dorn, G.W., 2nd; Vega, R.B.; Kelly, D.P. Mitochondrial biogenesis and dynamics in the developing and diseased heart. *Genes Dev.* **2015**, *29*, 1981–1991. [CrossRef]

72. Youle, R.J.; van der Bliek, A.M. Mitochondrial fission, fusion, and stress. *Science* **2012**, *337*, 1062–1065. [CrossRef] [PubMed]

73. Adaniya, S.M.; O-Uchi, J.; Cypress, M.W.; Kusakari, Y.; Jhun, B.S. Posttranslational modifications of mitochondrial fission and fusion proteins in cardiac physiology and pathophysiology. *Am. J. Physiol. Cell. Physiol.* **2019**, *316*, C583–C604. [CrossRef] [PubMed]

74. Hoppins, S.; Nunnari, J. Cell Biology. Mitochondrial dynamics and apoptosis–the ER connection. *Science* **2012**, *337*, 1052–1054. [CrossRef] [PubMed]

75. Yang, L.; Guan, G.; Lei, L.; Liu, J.; Cao, L.; Wang, X. Oxidative and endoplasmic reticulum stresses are involved in palmitic acid-induced H9c2 cell apoptosis. *Biosci. Rep.* **2019**, *39*, BSR20190225. [CrossRef]

76. He, Y.; Zhou, L.; Fan, Z.; Liu, S.; Fang, W. Palmitic acid, but not high-glucose, induced myocardial apoptosis is alleviated by N-acetylcysteine due to attenuated mitochondrial-derived ROS accumulation-induced endoplasmic reticulum stress. *Cell. Death Dis.* **2018**, *9*, 568. [CrossRef]

77. Xue, R.Q.; Zhao, M.; Wu, Q.; Yang, S.; Cui, Y.L.; Yu, X.J.; Liu, J.; Zang, W.J. Regulation of mitochondrial cristae remodelling by acetylcholine alleviates palmitate-induced cardiomyocyte hypertrophy. *Free Radic. Biol. Med.* **2019**, *145*, 103–117. [CrossRef] [PubMed]

78. Gui, T.; Li, Y.; Zhang, S.; Zhang, N.; Sun, Y.; Liu, F.; Chen, Q.; Gai, Z. Docosahexaenoic acid protects against palmitate-induced mitochondrial dysfunction in diabetic cardiomyopathy. *Biomed. Pharmacother.* **2020**, *128*, 110306. [CrossRef]

79. Wu, K.M.; Hsu, Y.M.; Ying, M.C.; Tsai, F.J.; Tsai, C.H.; Chung, J.G.; Yang, J.S.; Tang, C.H.; Cheng, L.Y.; Su, P.H.; et al. High-density lipoprotein ameliorates palmic acid-induced lipotoxicity and oxidative dysfunction in H9c2 cardiomyoblast cells via ROS suppression. *Nutr. Metab.* **2019**, *16*, 36. [CrossRef]

80. Hu, Q.; Zhang, H.; Gutiérrez Cortés, N.; Wu, D.; Wang, P.; Zhang, J.; Mattison, J.A.; Smith, E.; Bettcher, L.F.; Wang, M.; et al. Increased Drp1 Acetylation by Lipid Overload Induces Cardiomyocyte Death and Heart Dysfunction. *Circ. Res.* **2020**, *126*, 456–470. [CrossRef]

81. Perry, B.D.; Rahnert, J.A.; Xie, Y.; Zheng, B.; Woodworth-Hobbs, M.E.; Price, S.R. Palmitate-induced ER stress and inhibition of protein synthesis in cultured myotubes does not require Toll-like receptor 4. *PLoS ONE* **2018**, *13*, e0191313. [CrossRef]

82. Gudbjarnason, S.; De Schryver, C.; Chiba, C.; Yamanaka, J.; Bing, R. Protein and Nucleic Acid Synthesis During the Reparative Processes Following Myocardial Infarction. *Circ. Res.* **1964**, *15*, 320–326. [CrossRef] [PubMed]

83. Kaneko, M.; Panagia, V.; Paolillo, G.; Majumder, S.; Ou, C.; Dhalla, N.S. Inhibiton of cardiac phosphatidylethanolamine N-methylation by oxygen free radicals. *Biochim. Biophys. Acta* **1990**, *1021*, 33–38. [CrossRef] [PubMed]

84. Yadav, D.K.; Kumar, S.; Choi, E.H.; Chaudhary, S.; Kim, M.H. Molecular dynamic simulations of oxidized skin lipid bilayer and permeability of reactive oxygen species. *Sci. Rep.* **2019**, *14*, 4496. [CrossRef] [PubMed]

85. Juan, C.A.; Pérez de la Lastra, J.M.; Plou, F.J.; Pérez-Lebeña, E. The Chemistry of Reactive Oxygen Species (ROS) Revisited: Outlining Their Role in Biological Macromolecules (DNA, Lipids and Proteins) and Induced Pathologies. *Int. J. Mol. Sci.* **2021**, *22*, 4642. [CrossRef] [PubMed]

cells

Article

Mercury Chloride Affects Band 3 Protein-Mediated Anionic Transport in Red Blood Cells: Role of Oxidative Stress and Protective Effect of Olive Oil Polyphenols

Pasquale Perrone [1],[†], Sara Spinelli [2],[†], Gianluca Mantegna [2], Rosaria Notariale [1], Elisabetta Straface [3], Daniele Caruso [4], Giuseppe Falliti [4], Angela Marino [2], Caterina Manna [1], Alessia Remigante [2],[*] and Rossana Morabito [2]

[1] Department of Precision Medicine, School of Medicine, University of Campania Luigi Vanvitelli, 80138 Naples, Italy
[2] Department of Chemical, Biological, Pharmaceutical and Environmental Sciences, University of Messina, 98122 Messina, Italy
[3] Biomarkers Unit, Center for Gender-Specific Medicine, Istituto Superiore di Sanità, 00161 Rome, Italy
[4] Complex Operational Unit of Clinical Pathology, Papardo Hospital, 98122 Messina, Italy
[*] Correspondence: aremigante@unime.it
[†] These authors contributed equally to this work.

Abstract: Mercury is a toxic heavy metal widely dispersed in the natural environment. Mercury exposure induces an increase in oxidative stress in red blood cells (RBCs) through the production of reactive species and alteration of the endogenous antioxidant defense system. Recently, among various natural antioxidants, the polyphenols from extra-virgin olive oil (EVOO), an important element of the Mediterranean diet, have generated growing interest. Here, we examined the potential protective effects of hydroxytyrosol (HT) and/or homovanillyl alcohol (HVA) on an oxidative stress model represented by human RBCs treated with $HgCl_2$ (10 μM, 4 h of incubation). Morphological changes as well as markers of oxidative stress, including thiobarbituric acid reactive substance (TBARS) levels, the oxidation of protein sulfhydryl (-SH) groups, methemoglobin formation (% MetHb), apoptotic cells, a reduced glutathione/oxidized glutathione ratio, Band 3 protein (B3p) content, and anion exchange capability through B3p were analyzed in RBCs treated with $HgCl_2$ with or without 10 μM HT and/or HVA pre-treatment for 15 min. Our data show that 10 μM HT and/or HVA pre-incubation impaired both acanthocytes formation, due to 10 μM $HgCl_2$, and mercury-induced oxidative stress injury and, moreover, restored the endogenous antioxidant system. Interestingly, $HgCl_2$ treatment was associated with a decrease in the rate constant for SO_4^{2-} uptake through B3p as well as MetHb formation. Both alterations were attenuated by pre-treatment with HT and/or HVA. These findings provide mechanistic insights into benefits deriving from the use of naturally occurring polyphenols against oxidative stress induced by $HgCl_2$ on RBCs. Thus, dietary supplementation with polyphenols might be useful in populations exposed to $HgCl_2$ poisoning.

Keywords: mercury chloride; oxidative stress; band 3 protein; anion exchange; human RBCs

Citation: Perrone, P.; Spinelli, S.; Mantegna, G.; Notariale, R.; Straface, E.; Caruso, D.; Falliti, G.; Marino, A.; Manna, C.; Remigante, A.; et al. Mercury Chloride Affects Band 3 Protein-Mediated Anionic Transport in Red Blood Cells: Role of Oxidative Stress and Protective Effect of Olive Oil Polyphenols. *Cells* **2023**, *12*, 424. https://doi.org/10.3390/cells12030424

Academic Editor: Cord Brakebusch

Received: 25 November 2022
Revised: 19 January 2023
Accepted: 25 January 2023
Published: 27 January 2023

1. Introduction

Human exposure to heavy metals has increased dramatically in the last five decades due to exponential increase in their use in various industries and products [1,2]. Among heavy metal pollutants of the natural environment, mercury is the most common. There are three forms of mercury: elemental (or metallic) mercury, inorganic mercury and organic mercury [3]. Although mercury is present in low concentration in nature, it poses a health risk to human beings and other organisms due to its persistency, ingestion of contaminated fish (such as swordfish, tuna or shark), bioaccumulation, and toxicity [4,5].

Mercury preferentially accumulates in red blood cells (RBCs), thus inducing morphological changes and an increase in the pro-coagulant activity of these cells [6,7]. Red blood cells are unique cells regarding their structural organization and function. Although their primary function is the transportation of the respiratory gases O_2 and CO_2 between lungs and tissues, these circulatory cells are equipped with efficient endogenous anti-oxidative systems that make them mobile free radical scavengers providing antioxidant protection, not only to RBCs themselves but also to other tissues and organs in the body [8]. However, these anucleated cells can be affected, to different extents, by oxidative injuries whenever oxidative stress develops. For example, the incubation of RBCs with varying concentrations of mercury results in the externalization of phosphatidylserine (PS), cell eryptosis, and morphological perturbations in cell shape [9–11]. The underlying mechanism of mercury toxicity is correlated with its high affinity for sulfhydryl groups which severely alter enzyme activities as well as membrane structural proteins, such as Band 3 protein (B3p). Band 3 protein, or anion exchanger 1 (AE1), is encoded by the *SLC4A1* gene and, with more than 1 million copies per cell, is the most abundant membrane protein in RBCs. The crystal structure of B3p has been recently obtained, revealing two domains, an N-terminal cytosolic domain that anchors the cytoskeleton at the plasma membrane and interacts with different proteins and a C-terminal membrane domain mediating the anion exchange [12]. Since an imbalance in the physicochemical properties of RBCs can make them dysfunctional and impede efficient tissue oxygenation, the maintenance of their functionality is of the outmost importance [13]. Mercury exposure has been proven to cause hemolysis, thus representing a risk of anemia [14–16]. In fact, mercury is able to trigger a complex reaction that leads to cell damage and finally cell death. When mercury enters RBCs, a decrease in endogenous antioxidant activity and an increase in pro-oxidant activity are observed, leading to the peroxidative destruction of RBC membranes as well as an increase in osmotic pressure resulting in hemolysis [9,17,18]. Specifically, anemia is associated with a decrease in glutathione peroxidase (GPx) activity [19]. Glutathione peroxidase has an important function in the process of aerobic glycolysis on the pentose phosphate pathway, which is the main energy source for RBCs. Decreasing GPx activity inhibits the glycolysis process, which results in a reduced RBC half-life. Other studies have also demonstrated that mercury exposure increases the production of reactive oxygen species (ROS) on account of the Fenton reaction. Therefore, the hemolytic effects of mercury suggest that RBCs may be an important target of this metal [20,21].

The life functions of RBCs can be potentially improved by functional foods and natural products with antioxidant properties, thus ameliorating the homeostasis of the whole body. The consumption of extra virgin olive oil (EVOO) has been associated with several beneficial healthy effects, partly due to its polyphenol content, known for its important antioxidant activity [22–24]. More than 30 phenolic compounds belonging to several polyphenol classes have been identified in EVOO [25]. At a biological level, these compounds are very important not only for their antioxidant activity but also for their capacity in modelling several cellular signaling pathways which exhibit numerous beneficial effects in addition to those directly due to their free radical scavenging activity [26]. It is well known that the stability of cell membranes can be positively affected by exogenous antioxidants [22]. In this regard, several studies have shown a protective effect in terms of olive oil polyphenols, such as hydroxytyrosol (HT) and its metabolite homovanillyl alcohol (HVA), on RBCs (Figure 1a,b) [27–29].

Since RBCs deliver oxygen to the entire body, the maintenance of a functional and constant amount of RBCs is of pivotal importance for the health of an individual.

In particular, we explored the protective capacity of HT and its metabolite HVA in a model of oxidative stress represented by human RBCs treated with a non-hemolytic concentration of $HgCl_2$ (10 µM). This cell-based model could represent those human pathologic conditions that are hallmarks of chronic mercury toxicity, including hemolytic anemia, and thus affect RBC integrity. Since B3p function has been previously established as a sensitive tool to assess the impact of oxidative stress on the homeostasis of RBCs [30–36], the anion exchange capability through B3p and the effects of oxidative stress on various cellular components were evaluated.

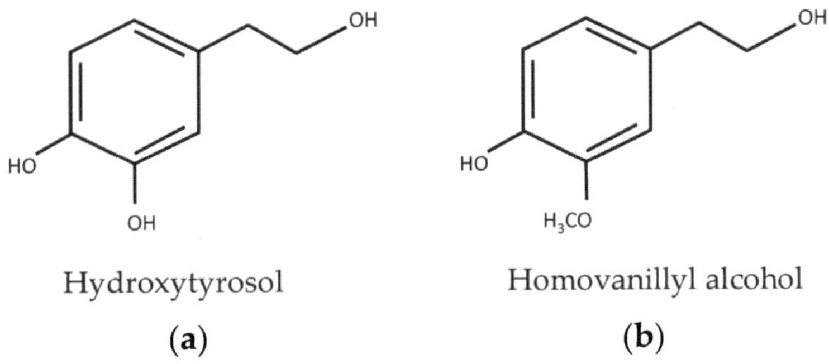

Figure 1. Chemical structures of (**a**) hydroxytyrosol (HT) and (**b**) homovanillyl alcohol (HVA).

2. Materials and Methods

2.1. Solutions and Chemicals

All chemicals were purchased from Sigma (Milan, Italy). 3,4-Dihydroxyphenethyl alcohol (HT, hydroxytyrosol; CAS number: 10597-60-1), 4-hydroxy-3-methoxyphenethyl alcohol (HVA, homovanillyl alcohol; CAS number: 2380-78-1), and 4,4′-diisothiocyanatostilbene-2,2′-disulfonate (DIDS) stock solutions (100 mM for HT and HVA; 10 mM for DIDS) were prepared in dimethyl sulfoxide (DMSO). N-ethylmaleimide (NEM) stock solution (310 mM) was prepared in ethanol. Mercury chloride ($HgCl_2$) was diluted in distilled water from a 100 mM stock solution. Both ethanol and DMSO never exceeded 0.001% *v/v* in the experimental solutions and were previously tested on RBCs to exclude hemolytic damage.

2.2. Preparation of Red Blood Cells

This study was prospectively reviewed and approved by a duly constitute Ethics Committee (prot.52-22, 20-04-2022). Upon informed consent, whole human blood from healthy volunteers was collected in test tubes containing ethylenediaminetetraacetic acid (EDTA). The plasma concentration of glycated hemoglobin (A1c) was less than 5%. Red blood cells were washed in isotonic solution (composition in mM: NaCl 150, 4-(2-hydroxyethyl)-1-piperazineethanesulfonic acid (HEPES) 5, glucose 5, pH 7.4, osmotic pressure 300 mOsm/kgH_2O) and centrifuged thrice (Neya 16R, 1200× *g*, 5 min) to remove plasma and buffy coat. Red blood cells were then suspended at specific hematocrits in isotonic solution and prepared for downstream analysis.

2.3. Analysis of Cell Shape by Scanning Electron Microscopy (SEM)

Samples, which were left untreated or exposed to $HgCl_2$ (4 h at 37 °C) or pre-incubated with 10 µM HT and/or HVA (15 min at 25 °C) and then exposed to 10 µM $HgCl_2$ treatment, were collected, plated on poly-l-lysine-coated slides and fixed with 2.5% glutaraldehyde in 0.1 M cacodylate buffer (pH 7.4) at 25 °C for 20 min. Then, samples were post-fixed with 1% OsO_4 in 0.1 M sodium cacodylate buffer and dehydrated through a graded series of ethanol solutions (from 30% to 100%). Absolute ethanol was gradually substituted by a 1:1 solution of hexamethyldisilazane (HMDS)/absolute ethanol and successively by pure HMDS. Afterwards, HMDS was completely removed, and samples were dried in a desiccator. Dried samples were mounted on stubs, coated with gold (10 nm), and analyzed by a Cambridge 360 scanning electron microscope (Leica Microsystem, Wetzlar, Germany) [37]. The altered shapes of RBCs were evaluated by counting ≥ 500 cells (50 RBCs for each different scanning electron microscopy (SEM) field at a magnification of 3000×) from samples in triplicate.

2.4. Detection of Reactive Oxygen Species (ROS)

To evaluate intracellular reactive oxygen intermediates, RBCs left untreated or exposed to 10 µM $HgCl_2$-containing solutions with or without pre-incubation with 10 µM HT and/or

HVA (15 min at 25 °C) were incubated in Hanks' balanced salt solution, pH 7.4, containing dihydrorhodamine 123 (DHR 123; Molecular Probes, Milan, Italy) and then analyzed with a FACScan flow cytometer (Becton-Dickinson, Mountain View, CA, USA). At least 20,000 events were acquired. The median fluorescence intensity histogram values were used to provide a semi-quantitative analysis of ROS production [38].

2.5. Thiobarbituric-Acid-Reactive Substance (TBARS) Level Measurement

TBARS levels were measured as described by Mendanha and collaborators [39], with minor modifications. TBARSs derive from the reaction between thiobarbituric acid (TBA) and malondialdehyde (MDA), which is the end-product of lipid peroxidation [39,40]. Red blood cells were suspended at 20% hematocrit and incubated with 10 μM $HgCl_2$ (4 h at 37 °C) or pre-incubated with 10 μM HT and/or HVA (15 min at 25 °C, with or without $HgCl_2$ treatment). Then, samples were centrifuged (Neya 16R, 1200× *g*, 5 min) and suspended in isotonic solution. Red blood cells (1.5 mL) were treated with 10% (*w/v*) trichloroacetic acid (TCA) and centrifuged (Neya 16R, 3000× *g*, 10 min). TBA (1% in hot distilled water, 1 mL) was added to the supernatant, and the mixture was incubated at 95 °C for 30 min. Finally, TBARS levels were obtained by subtracting 20% of the absorbance at 453 nm from the absorbance at 532 nm (Onda Spectrophotometer, UV-21). Results are indicated as μM TBARS levels (1.56×10^5 M^{-1} cm^{-1} molar extinction coefficient).

2.6. Total Sulfhydryl Group Content

The measurement of total sulfhydryl group (-SH) groups was carried out according to the method of Aksenov and Markesbery [41], with minor modifications. In short, RBCs (35% hematocrit), left untreated, exposed to 10 μM $HgCl_2$ (4 h at 37 °C), or pre-incubated with 10 μM HT and/or HVA (15 min at 25°C, with or without 10 μM $HgCl_2$ treatment) were centrifuged (Neya 16R, 1200× *g*, 5 min) and a sample of 100 μL was hemolyzed in 1 mL of distilled water. A 50 μL aliquot was added to 1 mL of phosphate-buffered saline (PBS, pH 7.4) containing EDTA (1 mM). 5,5′-Dithiobis (2-nitrobenzoic acid) (DTNB, 10 mM, 30 μL) was added to initiate the reaction, and the samples were incubated for 30 min at 25 °C while protected from light. Control samples, without cell lysate or DTNB, were processed concurrently. After incubation, the sample absorbance was measured at 412 nm (Onda spectrophotometer, UV-21) and 3-thio-2-nitro-benzoic acid (TNB) levels were detected after the subtraction of the blank absorbance (samples containing only DTNB). To achieve the full oxidation of -SH groups, an aliquot of RBCs (positive control) was incubated with 2 mM NEM for 1 h at 25 °C [42,43]. Data were normalized to protein content and results reported as μM TNB/mg protein.

2.7. Determination of Methemoglobin (MetHb) Levels

Methemoglobin levels were determined as reported by Naoum and collaborators [44], with minor modifications. The assay is based on MetHb and (oxy)-hemoglobin (Hb) determination by spectrophotometry at a wavelength of 630 and 540 nm, respectively. After incubation (10 μM $HgCl_2$ for 4 h at 37 °C with or without pre-incubation with 10 μM HT and/or HVA for 15 min at 25 °C), samples were centrifuged (2000× *g*, 5 min, 25 °C; Eppendorf), and 25 μL of RBCs at 40% hematocrit were lysed in 1975 μL hypotonic buffer (composition: 2.5 mM NaH_2PO_4, pH 7.4; 4 °C). Then, samples were centrifuged (13,000× *g*, 15 min, 4 °C; Eppendorf) to eliminate membranes. The absorbance of the supernatant was measured (BioPhotometer Plus; Eppendorf). Incubation with 4 mM $NaNO_2$ (for 1 h at 25 °C), a well-known MetHb-forming agent, was used to obtain complete Hb oxidation [45]. The MetHb percentage (%) was determined as follows: % MetHb = (OD630/OD540) × 100 (OD is optical density).

2.8. Detection of Apoptotic Red Blood Cells

Red blood cells, left untreated or exposed to 10 μM $HgCl_2$-containing solutions with or without pre-incubation with 10 μM HT and/or HVA (15 min at 25 °C), were processed to detect eryptosis by using the FITC-conjugated Annexin V eryptosis detection kit (Biovision, CA, USA.) and Trypan blue staining (0.05% Trypan blue for 15 min at room temperature) [46]. Then, RBCs were analyzed with a FACS scan flow cytometer (Becton-Dickinson, Mountain View, CA, USA) equipped with a 488 nm argon laser.

2.9. Preparation of Red Blood Cell Membranes

Red blood cell membranes were prepared as described by other authors [47], with slight modifications. Briefly, packed RBCs (untreated or treated with 10 µM HgCl$_2$ for 4 h at 37 °C with or without 10 µM or pre-incubated with 10 µM HT and/or HVA (15 min at 37 °C)) were diluted into 1.5 mL of 2.5 mM NaH$_2$PO$_4$ (cold hemolysis solution) containing a cocktail of inhibitors (1 mM PMSF, 1 mM NaF, and 1 mM Na3VO4). Samples were repeatedly centrifuged (Eppendorf, 4 °C, 18,000× g, 15 min) to take out hemoglobin. The obtained membranes were solubilized by sodium dodecyl sulphate (SDS, 1% v/v) and kept for 20 min on ice. After centrifugation (Eppendorf, 4 °C, 13,000× g, 30 min), the supernatant, containing the solubilized membrane proteins, was stored at −80 °C until use.

SDS-PAGE Preparation and Western Blotting Analysis

After thawing, membranes were solubilized in Laemmli buffer (1:1 volume ratio) [48] and heated for 15 min at 95 °C. The protein samples (5 µg/µL), measured by Bradford assay, were separated by 7.5% SDS-polyacrylamide gel electrophoresis and transferred to a polyvinylidene fluoride membrane by applying a constant voltage (75 V) for 2 h at 4°C. Membranes were blocked for 1 h at room temperature in 5% bovine serum albumin (BSA) diluted in Tris-buffered saline (150 mM NaCl, 15 mM Tris-HCl) containing 0.1% Tween-20 (TBST) and incubated overnight at 4 °C with the primary antibody (monoclonal anti-B3p antibody, B9277, Sigma-Aldrich, Milan, Italy, produced in mouse and diluted 1:5000 in TBST). Successively, membranes were incubated for 1 h with peroxidase-conjugated goat anti-mouse IgG secondary antibodies (A9044, Sigma-Aldrich, Milan, Italy) diluted 1:10,000 in TBST solution at room temperature. To assess the presence of equal amounts of protein, a monoclonal anti-β-actin antibody (A1978, Sigma-Aldrich, Milan, Italy), diluted 1:10,000 in TBST solution and produced in mouse, was incubated with the same membrane, as suggested by Yeung and collaborators [49]. A chemiluminescence detection system (Super Signal West Pico Chemiluminescent Substrate, Pierce Thermo Scientific, Rockford, IL, USA) was used to detect signals, whose images were imported to analysis software (Image Quant TL, v2003). The intensities of the corresponding protein bands were determined by densitometry (Bio-Rad ChemiDocTM XRS+).

2.10. SO$_4^{2-}$ Uptake Measurement

2.10.1. Control Condition

An SO$_4^{2-}$ uptake measurement was used to evaluate the anion exchange through B3p, as described elsewhere [50–53]. Briefly, after washing, RBCs were suspended to 3% hematocrit in 35 mL SO$_4^{2-}$ medium (composition in mM: Na$_2$SO$_4$ 118, HEPES 10, glucose 5, pH 7.4, osmotic pressure 300 mOsm/kgH$_2$O) and incubated at 25 °C in this medium. After 5, 10, 15, 30, 45, 60, 90, and 120 min, DIDS (10 µM), which is an inhibitor of B3p activity [54,55], was added to 5 mL sample aliquots, which were kept on ice. Subsequently, samples were washed three times in cold isotonic solution and centrifuged (Neya 16R, 4 °C, 1200× g, 5 min) to eliminate SO$_4^{2-}$ from the external medium. Distilled water (1 mL) was added to induce the osmotic lysis of RBCs, and perchloric acid (4% v/v) was used to precipitate proteins. After centrifugation (Neya 16R, 4 °C, 2500× g, 10 min), the supernatant containing SO$_4^{2-}$ trapped by RBCs was directed to the turbidimetric analysis. Supernatant (500 µL from each sample) was sequentially mixed to 500 µL glycerol diluted (1:1) in distilled water, 1 mL 4 M NaCl, and 500 µL 1.24 M BaCl$_2$•2H$_2$O. Finally, the absorbance of each sample was measured at 425 nm (Onda Spectrophotometer, UV-21). The absorbance was converted to [SO$_4^{2-}$] L cells × 10^{-2} by means of a standard curve previously obtained by precipitating known SO$_4^{2-}$ concentrations. The rate constant of SO$_4^{2-}$ uptake (min^{-1})

was derived from the following equation: $C_t = C_\infty (1 - e^{-rt}) + C_0$, where C_t, C_∞, and C_0 indicate the intracellular SO_4^{2-} concentrations measured at times t, ∞, and 0, respectively, with e representing the Neper number (2.7182818), r indicating the rate constant accounting for the process velocity, and t being the specific time at which the SO_4^{2-} concentration was measured. The rate constant is the inverse of the time needed to reach ~63% of the total SO_4^{2-} intracellular concentration [50], and the $[SO_4^{2-}]$ L cells $\times$ 10^{-2} reported in figures represents the SO_4^{2-} micromolar concentration internalized by 5 mL RBCs suspended at 3% hematocrit.

2.10.2. Experimental Conditions

Samples (3% hematocrit), which were left untreated or exposed to 10 μM HgCl$_2$ (4 h at 37 °C) or pre-incubated with 10 μM HT and/or HVA (15 min at 25 °C, with or without HgCl$_2$ treatment), were centrifuged (Neya 16R, 4 °C, 1200$\times$ g, 5 min) to replace the supernatant with SO_4^{2-} medium. The rate constant of SO_4^{2-} uptake was then determined as described for the control condition.

2.11. Measurement of Reduced Glutathione (GSH) Content

GSH levels were assayed according to Teti and collaborators [56], with slight modifications. Samples (20% hematocrit), which were left untreated or exposed to 10 μM HgCl$_2$ (4 h at 37 °C) or pre-incubated with 10 μM HT and/or HVA (15 min at 25 °C, with or without HgCl$_2$ treatment), were centrifuged (Neya 16R, 4 °C, 1200$\times$ g, 5 min) and resuspended in isotonic solution. After treatments, the content of GSH was measured by Cayman's GSH assay kit using an enzymatic recycling method with glutathione reductase. This assay is based on the oxidation of GSH by Ellman's reagent DTNB, which produces oxidized glutathione (GSSG) and 3-thio-2-nitro-benzoic acid (TNB), absorbing at a wavelength of 412 nm. The amount of GSSG was calculated by the following formula: 1/2 GSSG = GSH total-GSH reduced. Results are expressed as a GSH/ GSSG ratio.

2.12. Experimental Data and Statistics

All data are expressed as arithmetic mean $\pm$ standard error of the mean. For statistical analysis and graphics, GraphPad Prism (version 9.0, GraphPad Software, San Diego, CA, USA) and Excel (Version 2019, Microsoft, Redmond, WA, USA) software were used. Data normality was verified with the D'Agostino and Pearson Omnibus normality test. Significant differences between mean values were determined by one-way analysis of variance (ANOVA), followed by Bonferroni's multiple comparison post-test or ANOVA with Dunnett's post-test, as appropriate. Statistically significant differences were assumed at $p < 0.05$; (n) corresponds to the number of separate measurements.

3. Results

3.1. Evaluation of Red Blood Cell Shape

As depicted in Figure 2, incubation for 4 h at 37°C with 10 μM HgCl$_2$ induced the morphological alteration of RBCs with 44.3% of acanthocytes (RBCs with surface blebs) detected by scanning electron microscopy analysis (SEM). However, in samples pre-treated for 15 min at 25 °C with 10 μM HT and/or HVA and then treated with 10 μM HgCl$_2$, the percentage of morphologically altered cells was reduced to 3.5% and 2.9%, respectively (Table 1).

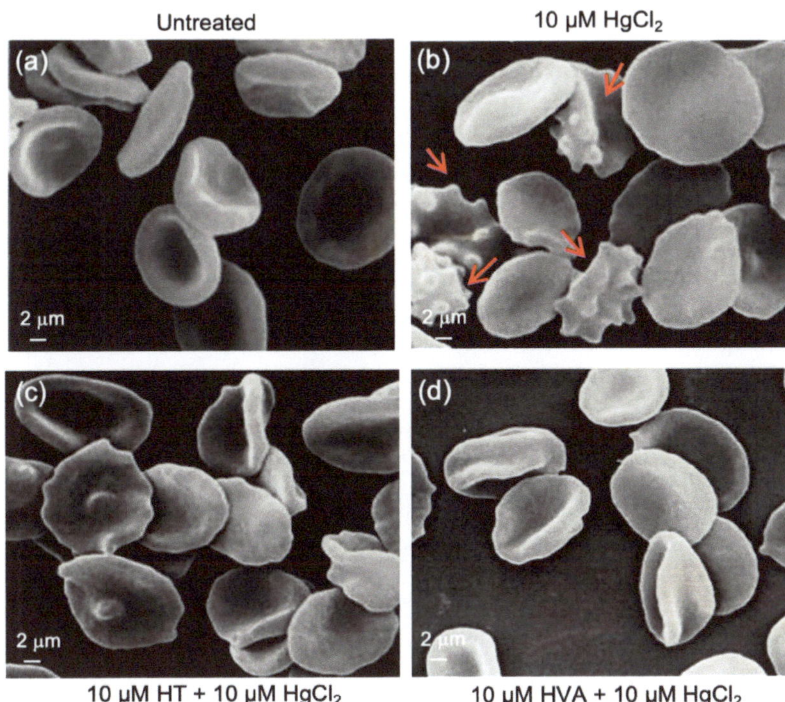

Figure 2. Red blood cell morphology evaluation. Representative scanning electron microscopy images showing RBCs with a typical biconcave form (**a**) left untreated or, alternatively, with surface blebs (**b**) (acanthocytes, red arrows) (10 μM HgCl$_2$). (**c**) and (**d**) Pre-treatment with HT and/or HVA (10 μM) attenuated the morphological changes compared to HgCl$_2$ treatment. Magnification 3500×.

Table 1. Percentage of morphological alterations in RBCs left untreated (control) or treated as indicated. Data are presented as means ± S.E.M. from separate three independent experiments, where ns indicates not statistically significant versus left untreated (biconcave shape, acanthocytes); ***, $p < 0.001$ versus control (biconcave shape); °°°, $p < 0.001$ versus control (acanthocytes), one-way ANOVA followed by Bonferroni's multiple comparison post-hoc test.

Experimental Conditions	Biconcave Shape	Acanthocytes	n
Untreated (control)	94% ± 0.013	3% ± 0.011	5
10 μM HgCl$_2$	55.7% ± 0.011 [***]	44.3% ± 0.012 [°°°]	5
10 μM HT + 10 μM HgCl$_2$	96.5% ± 0.013 [ns]	3.5% ± 0.009 [ns]	5
10 μM HVA + 10 μM HgCl$_2$	97.1% ± 0.007 [ns]	2.9% ± 0.008 [ns]	5

3.2. Oxidative Stress Assessment

3.2.1. Evaluation of Intracellular ROS Levels

The evaluation of ROS species was carried out by flow cytometry in RBCs left untreated or, alternatively, exposed to 10 μM HgCl$_2$ with or without pre-exposure to 10 μM HT and/or HVA for 15 min at 25 °C. Figure 3A shows the intracellular ROS levels at different time points (0, 30, 60, 120, 180, and 240 min after exposure to 10 μM HgCl$_2$).

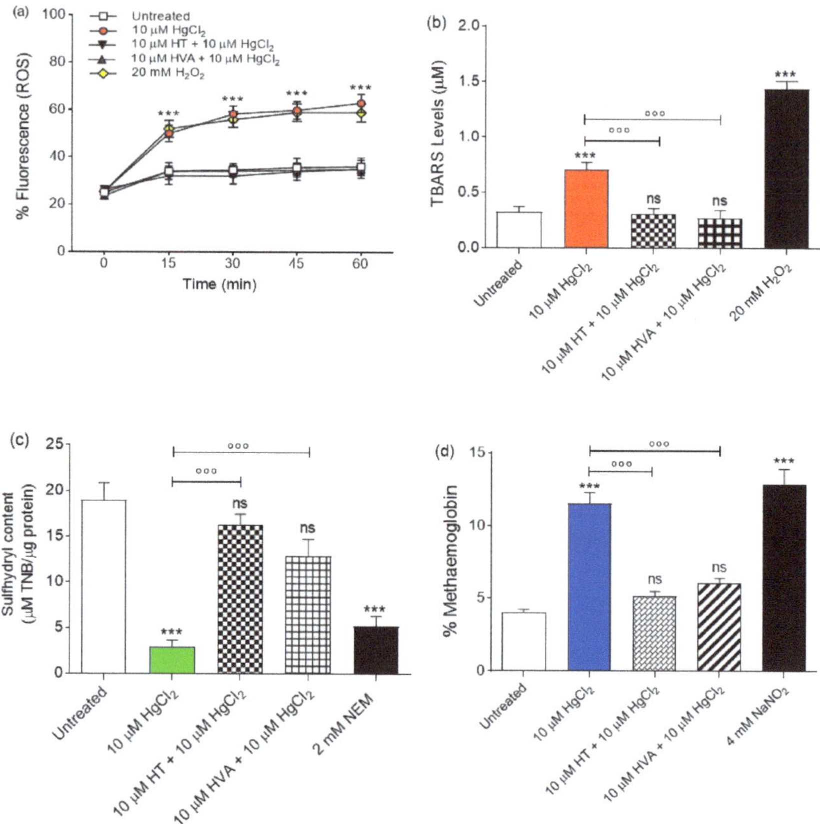

Figure 3. Determination of oxidative stress levels. (**a**) Detection of reactive oxygen species (ROS) levels by flow cytometry. Time course of ROS production in RBCs left untreated (control) or treated for 4 h with 10 μM HgCl$_2$, with or without pre-exposure to 10 μM HT and/or HVA for 15 min. H$_2$O$_2$ was used as a positive control. ns, not statistically significant versus left untreated (control); *** $p < 0.001$ versus control, one-way ANOVA followed by Bonferroni's post-hoc test ($n = 8$). (**b**) Detection of TBARS levels. TBARS levels (μM) in RBCs left untreated (control) or treated for 4 h with 10 μM HgCl$_2$, with or without pre-exposure to 10 μM HT and/or HVA for 15 min. H$_2$O$_2$ was used as a positive control. ns, not statistically significant versus left untreated (control); *** $p < 0.001$ versus left untreated (control); °°° $p < 0.001$ versus 10 μM HgCl$_2$, one-way ANOVA followed by Bonferroni's post-hoc test ($n = 11$). (**c**) Sulfhydryl group content evaluation. Sulfhydryl group content (μM TNB/μg protein) in RBCs left untreated (control) and in RBCs treated for 4 h with HgCl$_2$, with or without pre-exposure to 10 μM HT and/or HVA for 15 min. ns, not statistically significant versus left untreated (control); *** $p < 0.001$ versus left untreated (control); °°° $p < 0.001$ versus 10 μM HgCl$_2$, one-way ANOVA followed by Bonferroni's post-hoc test ($n = 11$). NEM was used as a positive control. ns, not statistically significant versus control; *** $p < 0.001$ versus control, one-way ANOVA followed by Bonferroni's post-hoc test ($n = 10$). (**d**) Methemoglobin (% MetHb) content. Red blood cells were left untreated or incubated with 10 μM HgCl$_2$, with or without pre-exposure to 10 μM HT and/or HVA for 15 min. NaNO$_2$ (4 mM for 1 h) was used as positive control. ns, not statistically significant; *** $p < 0.001$ versus left untreated (control); °°° $p < 0.001$ versus 10 μM HgCl$_2$, one way ANOVA followed by Bonferroni's post-hoc test ($n = 11$).

Samples exposed to 10 μM HgCl$_2$ showed a significant increase in ROS levels compared to the control samples. After 30 min, the levels of ROS increased by 50% in the 10 μM

HgCl$_2$-treated samples and remained unchanged over time. In Figure 4, the effect of olive oil polyphenols is also reported. In samples pre-exposed to 10 μM HT and/or HVA, 10 μM HgCl$_2$ failed to significantly increase ROS levels, with these being unchanged compared to control values (Figure 3a). Of note, olive oil polyphenols alone did not significantly induce an increase in ROS levels.

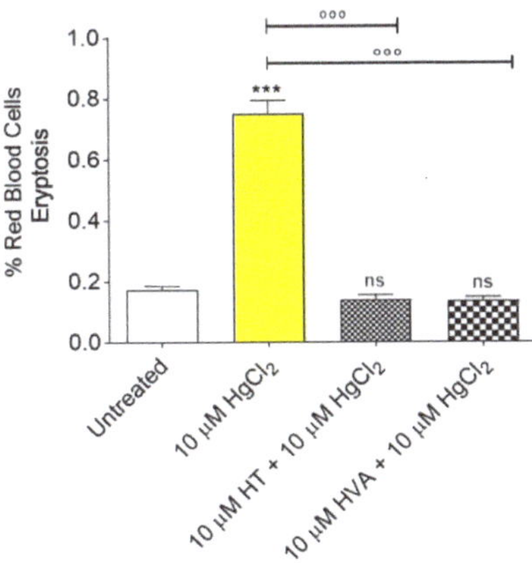

Figure 4. Detection of eryptosis by flow cytometry. Percentage of apoptotic RBCs positive to Annexin 5 and Trypan Blue detected in samples treated with 10 μM HgCl$_2$ (4 h at 37 °C) with or without pre-treatment with 10 μM HT and/or HVA (1 h at 25 °C). ns, not statistically significant versus control; *** $p < 0.001$ versus left untreated (control); °°° $p < 0.001$ versus 10 μM HgCl$_2$, one way ANOVA followed by Bonferroni's post-hoc test ($n = 11$).

3.2.2. Measurement of Thiobarbituric Acid Reactive Substance (TBARS) Levels

The measurement of Thiobarbituric acid reactive substances (TBARSs) in RBCs is reported in Figure 3b. As expected, the TBARS levels of RBCs treated with 20 mM H$_2$O$_2$ (positive control) for 1 h were significantly higher with respect to those of RBCs left untreated (control). Moreover, the TBARS levels of RBCs treated with 10 μM HgCl$_2$ for 4 h were significantly higher than those of RBCs left untreated (control). Importantly, in RBCs pre-treated with 10 μM HT and/or HVA and then exposed to 10 μM HgCl$_2$, TBARS levels were significantly reduced compared to those measured in 10 μM HgCl$_2$-treated RBCs. It is of note that olive oil polyphenols alone did not significantly affect TBARS levels.

3.2.3. Total Sulfhydryl Group Content Measurement

Figure 3c shows the total content of sulfhydryl groups (μM TNB/μg protein) in RBCs left untreated or treated with either the oxidizing compound NEM (2 mM for 1 h, as the positive control), or 10 μM HgCl$_2$ for 4 h with or without HT and/or HVA pre-treatment. As expected, exposure to NEM led to a significant reduction in the sulfhydryl group content. Sulfhydryl groups in 10 μM HgCl$_2$-treated RBCs were also significantly reduced with respect to the control. Importantly, pre-treatment with 10 μM HT and/or HVA significantly restored sulfhydryl group total content in 10 μM HgCl$_2$-treated RBCs (Figure 3b). Olive oil polyphenols alone did not significantly affect the total sulfhydryl group content (Figure 3b).

3.2.4. Evaluation of Methemoglobin (MetHb) Levels

Figure 3d shows MetHb levels (%MetHb) measured in RBCs left untreated or treated with 10 μM HgCl$_2$ (4 h at 37 °C) with or without pre-treatment with 10 μM HT and/or HVA (1 h at 25 °C), or, alternatively, treated with the well-known MetHb-forming agent NaNO$_2$ (4 mM for 1 h at 25 °C). Methemoglobin levels measured after incubation with NaNO$_2$ were significantly higher than those detected in RBCs left untreated (control). In parallel, MetHb levels measured following exposure to 10 μM HgCl$_2$ were significantly higher than those measured in the control (left untreated). Pre-exposure to 10 μM HT and/or HVA significantly reduced the MetHb levels in 10 μM HgCl$_2$-treated RBCs towards values that did not differ from control values. Olive oil polyphenols alone did not significantly affect the %MetHb.

3.3. Determination of Apoptotic Red Blood Cells

Figure 4 shows eryptosis levels measured in RBCs left untreated or treated with 10 μM HgCl$_2$ (4 h at 37 °C) with or without pre-treatment with 10 μM HT and/or HVA (1 h at 25 °C). Regarding the percentage of RBCs in eryptosis, significant differences were detected after treatment with 10 μM HgCl$_2$ for 4 h. However, pre-treatment with 10 μM HT and/or HVA significantly reduced eryptosis levels in 10 μM HgCl$_2$-treated RBCs. Olive oil polyphenols alone did not significantly affect eryptosis.

3.4. Detection of Band 3 Protein Levels

Figure 5 shows B3p levels in RBCs incubated with 10 μM HgCl$_2$ (4 h at 37 °C) with or without pre-treatment with 10 μM HT and/or HVA (1 h at 25 °C). Band 3 protein levels after treatments were not significantly different with respect to those determined in control RBCs.

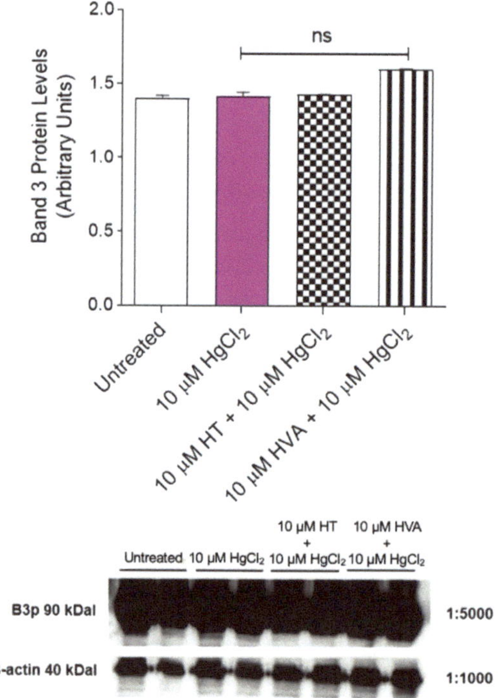

Figure 5. Band 3 protein expression levels measured in RBCs left untreated (control) or treated with 10 μM HgCl$_2$ for 4 h at 37 °C with or without pre-exposure to 10 μM HT and/or HVA for 1 h at 25 °C, detected by Western Blotting analysis. ns, not significant versus left untreated (control), one-way ANOVA followed by Bonferroni's multiple comparison post-hoc test (n = 3).

3.5. Measurement of SO$_4^{2-}$ Uptake through Band 3 Protein

Figure 6 reports the SO$_4^{2-}$ uptake as a function of time in RBCs left untreated (control) and in RBCs treated with 10 μM HgCl$_2$ (4 h at 37 °C) with or without pre-treatment with 10 μM HT and/or HVA (1 h at 25°C). In control conditions, SO$_4^{2-}$ uptake progressively increased and reached equilibrium within 45 min (rate constant of SO_4^{2-} uptake = 0.058 ± 0.001 min^{-1}). Red blood cells treated with 10 µM HT and/or HVA alone showed a rate constant of SO_4^{2-} uptake that was not significantly different with respect to the control. Conversely, the rate constant value (0.043 ± 0.001 min^{-1}) in RBCs treated with 10 µM HgCl$_2$ was significantly lower than in the control (*** $p < 0.001$). In RBCs pre-incubated with 10 µM HT (1 h at 25 °C) and then treated with 10 µM HgCl$_2$ (1 h at 25 °C), the rate constant (0.058 ± 0.001 min^{-1}) was significantly higher than that of RBCs treated with 10 µM HgCl$_2$ (0.043 ± 0.001 min^{-1}) while not being significantly different with respect to the control (Table 2). Similarly, in RBCs pre-incubated with 10 µM HVA (1 h at 25 °C) and then exposed to 10 µM HgCl$_2$ (4 h at 37 °C), the rate constant (0.062 ± 0.001 min^{-1}) was significantly different with respect to that of RBCs treated with 10 µM HgCl$_2$ (0.043 ± 0.001 min^{-1}) (Table 2). SO_4^{2-} uptake was almost completely blocked by 10 µM DIDS applied at the beginning of incubation in SO_4^{2-} medium (0.018 ± 0.001 min^{-1}, *** $p < 0.001$, Table 2). Additionally, the SO_4^{2-} amount internalized by HgCl$_2$-treated RBCs after 45 min of incubation in SO_4^{2-} medium was significantly lower than the control (Table 2), while in RBCs pre-incubated with 10 µM HT and/or HVA and then exposed to 10 µM HgCl$_2$, it was not significant different with respect to control (Table 2). In DIDS-treated cells, the internalized SO_4^{2-} amount (5.49 ± 2.50) was significantly lower than what was determined in both control and treated RBCs (*** $p < 0.001$, Table 2).

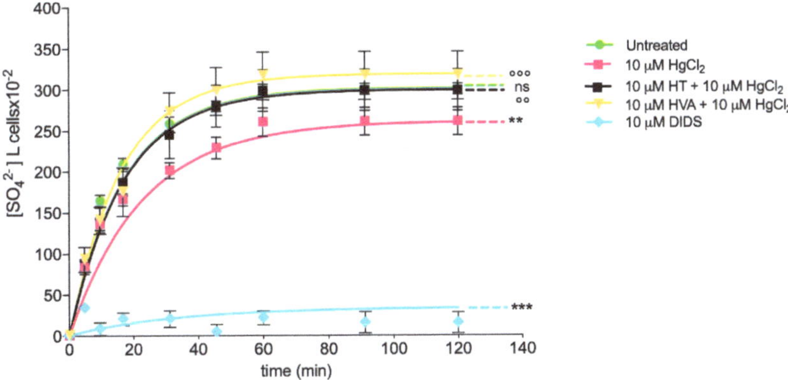

Figure 6. Time course of SO_4^{2-} uptake. Red blood cells were left untreated (control) or treated with 10 µM HgCl$_2$ for 4 h at 37 °C with or without pre-exposure to 10 µM HT and/or HVA for 1 h at 25 °C or exposed to 10 µM DIDS. ns, not statistically significant versus control; ** $p < 0.01$ and *** $p < 0.001$ versus control; °° $p < 0.05$ or °°° $p < 0.001$ versus 10 µM HgCl$_2$, one way ANOVA followed by Bonferroni's post-hoc test.

Table 2. Rate constant of SO_4^{2-} uptake and amount of SO_4^{2-} trapped in RBCs left untreated (control) and RBCs treated as indicated. Results are presented as means $\pm$ S.E.M. from separate (n) experiments, where ns indicates not statistically significant versus left untreated or 10 µM HgCl$_2$; ** $p < 0.01$; *** $p < 0.001$ versus control; °° $p < 0.01$ or °°° $p < 0.001$ versus 10 µM HgCl$_2$, one-way ANOVA followed by Bonferroni's multiple comparison post-hoc test.

Experimental Conditions	Rate Constant (min^{-1})	Time (min)	n	SO_4^{2-} Amount Trapped after 45 min of Incubation in SO_4^{2-} Medium (SO_4^{2-}) L Cells $\times 10^{-2}$
Control	0.058 ± 0.001	16.00	10	279 ± 17.43
10 µM HgCl$_2$	$55.7\% \pm 0.001$ ***	22.23	10	228 ± 15.40 ***
10 µM HT + 10 µM HgCl$_2$	$96.5\% \pm 0.001$ [ns,°°]	16.95	10	280 ± 14.39 [ns]
10 µM HVA + 10 µM HgCl$_2$	$97.1\% \pm 0.001$ [ns,°°°]	15.90	10	298 ± 17.70 [**,ns]
10 µM DIDS	0.018 ± 0.001 ***	63.50	10	5.49 ± 3.60 ***

3.6. GSH/GSSG Ratio Measurement

Figure 7 shows the GSH/GSSG ratio measured in RBCs treated with 10 µM HgCl$_2$ for 4 h at 37 °C with or without pre-treatment with 10 µM HT and/or HVA (15 min at 25 °C). The GSH/GSSG ratio measured after incubation with 10 µM HgCl$_2$ was significantly lower than that detected in control RBCs. This effect can be associated with an increased GSSG abundance and/or decreased GSH concentration, both of which are indicative of an increase in cellular oxidative stress.

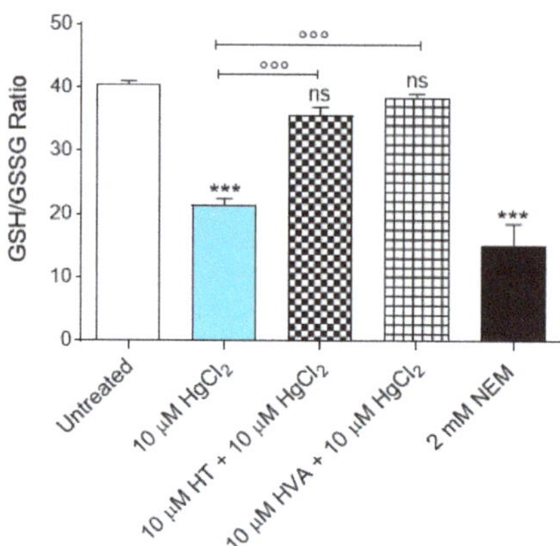

Figure 7. Estimation of the GSH/GSSG ratio measured in RBCs incubated for 4 h at 37 °C with 10 µM HgCl$_2$ with or without pre-treatment with 10 µM HT and/or HVA (15 min at 25 °C). GSH, reduced glutathione; GSSG, oxidized glutathione. *** $p < 0.001$ versus left untreated (control) RBCs, °°° $p < 0.001$ versus 10 µM HgCl$_2$, one-way ANOVA followed by Bonferroni's multiple comparison post-hoc test ($n = 8$).

4. Discussion

In this study, the antioxidant capacity of the most important olive oil polyphenol metabolites (HT and/or HVA) on morphology and function on an HgCl$_2$-induced oxidative stress model in human RBCs were investigated. With regard to the first aspect, the susceptibility of RBCs to 10 µM HgCl$_2$ exposure was verified by scanning electron microscopy (SEM), revealing dramatic changes in the RBC shape. Such a finding is in line with what was already reported by Lim and collaborators [57] when they observed RBC shape changes from acanthocytes to echinocytes to spherocytes accompanied by micro-vesicle generation after prolonged exposure to low-dose HgCl$_2$. In the present case, the typical biconcave shape was lost in a significant number of cells displaying surface blebs (acanthocytes, 10 µM HgCl$_2$). Indeed, the percentage of acanthocytes was higher (44.3%) than that detected in untreated cells (3%). However, pre-treatment with 10 µM HT and/or HVA polyphenol extracts prevented the morphological changes (Figure 2c,d), with a reduction in the percentage of acanthocytes (Table 1). These data also corroborate Paiva-Martins and co-authors [22,58], who reported that treatment with both olive oil polyphenols protected RBC morphology from the oxidative damage induced by the radical initiator 2,2-azo-bis(2-amidinopropane) dihydrochloride or, alternatively, caused by H$_2$O$_2$ exposure.

The study of RBC morphology is of great importance in the field of hemorheology [59,60]. Although RBC membranes are designed to withstand the mechanical stress of blood flow, these cells may respond to any form of damage by changing their morphology following changes in their structure or biochemical composition (the oxidation of sulfhydryl groups of membrane proteins, the oxidation of membrane fatty acid residues, or the oxidation of hemoglobin [61]). In this regard, the exposure to HgCl$_2$ has also been associated with eryptosis [62,63]. By analogy with the apoptosis of nucleated cells, RBCs may undergo programmed cell death, which is characterized by cell shrinkage and cell membrane phospholipid scrambling. Eryptosis machinery includes the activation of redox-sensitive calcium-permeable cation channels, resulting in calcium entry [64]. The elevation of cytosolic calcium further activates RBC scramblase and calpain resulting in PS externalization and membrane blebbing, respectively. In this context, 10 μM $HgCl_2$ exposure induced PS translocation to the external surface of the RBC membranes after 4 h of treatment (Figure 4).

Specifically, $HgCl_2$ treatment inhibited flippase, an enzyme that recovers PS into the inner leaflet of the cell membrane, and activated scramblase, an enzyme that alters lipid asymmetry in the cell plasma membrane [65], resulting in increased pro-coagulant activity and the removal of RBCs from the circulatory stream [57,66]. In this investigation, the antioxidant capacity of HT and/or HVA to prevent $HgCl_2$-induced PS exposure and morphological changes in human RBCs was demonstrated (Figure 4).

Since the oxidation of biological macromolecules, such as lipids and proteins, derives from the deleterious effects of ROS generated during normal cellular metabolism [67,68], intracellular ROS levels after exposure to 10 μM $HgCl_2$ were initially evaluated. Our data showed that pre-treatment with 10 μM HT and/or HVA induced a decrease in ROS levels induced by treatment with 10 μM $HgCl_2$ (Figure 3a).

This finding is supported by many authors and suggests that dietary supplementation with regular polyphenol consumption has an antioxidant activity, with this being generally attributed to their ability to directly neutralize ROS, in particular in populations occupationally exposed to $HgCl_2$ poisoning [22,29,69]. To better explain the mechanisms underlying the morphological changes, certain parameters related to oxidative stress assessment were monitored. Different phenomena, including the oxidation of membrane fatty acid residues, the oxidation of sulfhydryl groups of membrane proteins, or the oxidation of hemoglobin, could alter membrane properties and cell shape, thus causing a decrease in the deformability and a loss in membrane integrity. Oxidative modifications tend to stimulate the accumulation of oxidized lipids and proteins, which, when present in excess amounts, impair RBC architecture and function and result in the shortened life span of these cells. Due to oxidation stress, membrane polyunsaturated fatty acids of RBCs are damaged. This results in a steep increase in malondialdehyde (MDA) and thiobarbituric acid (TBA) levels, biomarkers currently used to reveal the oxidation of lipids under different experimental conditions [70]. The present data show that 15 min of pre-treatment with 10 μM HT and/or HVA prevented lipid peroxidation of membranes induced by $HgCl_2$ treatment for 4 h (Figure 3b). However, ROS attack on lipids initiates a chain reaction, which leads to the generation of more ROS that can harm other cellular components, including proteins [71]. Protein oxidation is characterized by a simultaneous decrease in free amino and sulfhydryl groups. Thus, the measurement of the sulfhydryl group content of total proteins was also evaluated. Exposure to 10 μM HT and/or HVA protected RBC proteins from oxidative injury (Figure 3c). These data are in line with what has been previously demonstrated by other authors. For example, pre-incubation with HT and HVA inhibited H_2O_2-induced lipid-peroxidation in LLC-PK1, a porcine kidney epithelial cell line [72]. In addition, the effects of HT exposure on UVA-induced cell damages were investigated using a human melanoma cell line (M14) as a model system [73]. In UVA-irradiated M14 cells, HT extract prevented the uprise of typical markers of oxidative stress, such as TBARSs, protein oxidation, and an increase in ROS levels.

The biological effects of olive oil are related to its chemical matrix, which includes numerous phytochemicals components, such as flavonoids [74]. These molecules can directly neutralize ROS and/or attenuate lipid peroxidation by scavenging free radicals, thus sustaining the endogenous antioxidant system [26,75]. In a previous investigation [76], the potential protective role of the polyphenolic flavonoid compound quercetin (Q) on an oxidative stress model represented by RBCs treated with 20 mM H_2O_2 was evaluated. In this regard, TBARS levels, augmented following 20 mM H_2O exposure (1 h), were completely restored by 10 μM Q pre-treatment (1 h). Similarly, the oxidation of protein sulfhydryl groups was totally restored after pre-treatment (1 h) with Q in RBCs treated with

20 mM H_2O_2 (1 h). Additionally, the protective effects of freeze-dried Açaì extract—the main chemical components present in the Açaí pulp are polyphenols—in a D-Galactose-induced aging model in human RBCs have been also reported [37].

Another major property of RBC oxidation is the clustering and/or the breakdown of B3p and the binding of oxidized hemoglobin (MetHb) to high affinity sites on B3p. Band 3 protein is an integral membrane protein that accounts for approximately 25% of the RBC membrane. It has several crucial functions and is composed by two rather different domains of a similar size [77]. In RBCs, $HgCl_2$-induced oxidative damage resulted in the oxidation of the ferrous iron of hemoglobin to a ferric form given methemoglobin (MetHb), which is inactive as an oxygen transporter in the blood [7]. To better elucidate the molecular interaction between B3p and oxidized hemoglobin, MetHb levels were also evaluated. Our results indicated that exposure to 10 μM $HgCl_2$ for 4 h increased the levels of MetHb in RBCs (Figure 3c). These modifications can start a cascade of structural and biochemical transformations, including the release of microparticles containing hemicromes and the clustering of B3p regions, as formerly demonstrated by other groups [23,24,69]. When the $HgCl_2$-induced oxidation processes are advanced, these clusters can provide a recognition site for antibodies directed against aging cells, thus triggering the premature removal of senescent RBCs from the circulation at the end of their 120-day life span [78]. Interestingly, pre-treatment with 10 μM HT and/or HVA prevented $HgCl_2$-induced MetHb production (Figure 3d), which was also observed when Q was applied after an oxidative stress increase (20 mM H_2O_2) [76].

One of the most interesting and still poorly investigated implications of $HgCl_2$ toxicity is its impact on membrane transport systems. The Band 3 protein is mainly involved in the chloride–bicarbonate exchange. Firstly, the content of B3p was determined. No change in the protein content was detected under these experimental conditions (Figure 5). Secondly, to define the potential effect of $HgCl_2$ on RBC functional activity, the SO_4^{2-} uptake was measured by means of a validated method to assay anion exchange capability via B3p [79]. In RBCs incubated with 10 μM $HgCl_2$, the rate constant for SO_4^{2-} uptake was reduced compared to the control (Figure 6 Table 2), and, in parallel, the amount of internalized SO_4^{2-} was significantly reduced. However, 15 min pre-treatment with 10 μM of HT and/or HVA completely restored the SO_4^{2-} absorption rate constant as well as the amount of internalized SO_4^{2-}, thus confirming the beneficial effect of the olive oil polyphenols on B3p function.

The evidence that B3p exhibits modifications in the rate constant for SO_4^{2-} uptake following the exposure of RBCs to different oxidative stressors has been previously demonstrated. Specifically, H_2O_2 non-hemolytic (300 μM for 30 min) concentrations induced oxidative stress and provoked a decrease in the rate constant for SO_4^{2-} uptake via B3p [43]. However, pre-treatment with melatonin ameliorated the reduction in the rate constant for SO_4^{2-} uptake as well as the reduction in B3p expression levels observed following treatment with H_2O_2 [34]. Moreover, the reduction in B3p anion exchange efficiency caused by a mild oxidative stress (300 μM for 30 min) was prevented or attenuated by a short-term pre-incubation of RBCs with low H_2O_2 doses (10 μM for 30 min) [35]. This pre-treatment allows RBCs to adapt to a mild and transient oxidative stress and favors an increased tolerance to a successive stronger oxidant condition. Such an adaptive response, termed pre-conditioning and monitored by measuring B3p activity, did not involve B3p-related Tyr-phosphorylation pathways but was mediated by increased catalase activity. In summary, a reduction in the transport rate following H_2O_2 exposure is most likely linked to the formation of oxidizing hemoglobin after oxidative stress. Therefore, the different effect on the SO_4^{2-} transport kinetics observed in a former study is not surprising. Indeed, in other models of oxidative stress obtained by exposing RBCs to high concentrations of D-Galactose and D-Glucose, the acceleration of the anion exchange through B3p, instead of a slowing, was detected [31,32]. It is tempting to speculate that such a two-sided effect on anion exchange velocity depends on the specific structure targeted by the stressors and on the possible underlying pathways. To name just a few examples, the acceleration of the transport rate in two former studies could be most likely linked to mechanisms other than oxidative stress, or, putatively, to the formation of glycated hemoglobin [31,80].

Finally, we directed our focus towards an assay of the endogenous antioxidant system. The underlying mechanism of $HgCl_2$ toxicity is correlated with a high affinity of $HgCl_2$ for sulfhydryl groups that can severely alter enzyme activities, including Glutathione [81]. The $HgCl_2$ ions react with sulfhydryl groups of cysteine residue, thus depleting intracellular thiols [82]. A decrease in total sulfhydryl group content demonstrates redox imbalance and the perturbation of the thiol status of the cell (Figure 3b). Hence the GSH/GSSG ratio was evaluated. The obtained results confirmed that 10 μM $HgCl_2$ treatment reduced the GSH/GSSG ratio. However, the pre-incubation of RBCs with 10 μM HT and/or HVA completely restored the redox balance (Figure 7). To support this evidence, in a rat model of Parkinson's disease, HT extract also resulted in the preservation of the brain GSH/GSSG ratio after MPP^+-induced oxidative injury [83]. Glutathione is an important intracellular non-enzymatic antioxidant that neutralizes many electrophiles and ROS [7]. Its depletion makes cells more prone to oxidative damage. To confirm this finding, a study performed by Lund and co-authors [84] suggests that $HgCl_2$ increases the production of hydrogen peroxide and GSH depletion in renal mitochondria, thus causing oxidative damage in cell membranes, resulting in hemolysis.

5. Conclusions

In summary, exposure to $HgCl_2$ induced increased oxidative stress in human RBCs. Indeed, the damaged RBCs may quickly be removed from circulation, thus leading to the shortened lifespan of RBCs, resulting in anemia. However, pre-treatment with HT and/or HVA avoided the formation of acanthocytes observed after exposure to $HgCl_2$ prevented $HgCl_2$-induced oxidative stress damage, including ROS production, lipid peroxidation, and sulfhydryl group oxidation of the total proteins on the plasma membrane. Moreover, $HgCl_2$ exposure was associated with a reduction in the rate constant of SO_4^{2-} uptake through B3p as well as MetHb production. Both alterations were attenuated by pre-treatment with HT and/or HVA. Importantly, pre-incubation with both polyphenols restored the GSH/GSSG ratio, resulting in the improvement of the endogenous antioxidant system.

In conclusion, the present investigation elucidates benefits deriving from the use of naturally occurring polyphenols against $HgCl_2$-induced oxidative stress damage on a cellular level. Additionally, the results obtained here confirm that the measurement of the B3p anion exchange capability remains a suitable tool for monitoring the impact of oxidative stress on RBC homeostasis. Considering the involvement of oxidative stress damage in $HgCl_2$-induced poisoning, new biomarkers with both diagnostic and monitoring potential are needed. Blood can be obtained from patients with minimally invasive procedures, with it reflecting the physiological status of peripheral tissues, and therefore it might represent a convenient source of biomarkers. In this light, future investigations are needed to clarify the signaling underlying the protective activity of HT and/or HVA on the interaction between B3p and cytoskeletal proteins such as ankyrin and spectrin and their potential post-translation modifications. These results are a step towards a better understanding of the biochemical mode of mercury-induced toxicity and cell damage. This will allow for appropriate methods to overcome the harmful effects of this extremely dangerous metal and widespread environmental pollutant to be devised.

Author Contributions: A.R., C.M. and R.M. conceived and designed the research; P.P., S.S., G.M., R.N., E.S., D.C. and G.F. performed the experiments and analyzed the data; A.M., A.R. and R.M. interpreted the results of the experiments; A.R. prepared the figures; A.R. drafted the manuscript; A.R., A.M. and R.M. edited and revised the manuscript. All authors have read and agreed to the published version of the manuscript.

Funding: This research received no external funding.

Institutional Review Board Statement: The study was conducted in accordance with the Declaration of Helsinki and approved by the Institutional Review Board (or Ethics Committee) of University of Messina (prot.52-22, date of approval 20 April 2022).

Informed Consent Statement: Informed consent was obtained from all subjects involved in the study.

Data Availability Statement: The data that support the findings of this study are available from the corresponding author upon reasonable request.

Conflicts of Interest: The authors declare no conflict of interest.

References

1. Hassanin, M.; Kerek, E.; Chiu, M.; Anikovskiy, M.; Prenner, E.J. Binding Affinity of Inorganic Mercury and Cadmium to Biomimetic Erythrocyte Membranes. *J. Phys. Chem. B* **2016**, *120*, 12872–12882. [CrossRef] [PubMed]
2. Perrone, P.; Lettieri, G.; Marinaro, C.; Longo, V.; Capone, S.; Forleo, A.; Pappalardo, S.; Montano, L.; Piscopo, M. Molecular Alterations and Severe Abnormalities in Spermatozoa of Young Men Living in the "Valley of Sacco River" (Latium, Italy): A Preliminary Study. *Int. J. Environ. Res. Public Health* **2022**, *19*, 11023. [CrossRef] [PubMed]
3. Guzzi, G.; La Porta, C.A. Molecular mechanisms triggered by mercury. *Toxicology* **2008**, *244*, 1–12. [CrossRef] [PubMed]
4. Ackerman, J.T.; Fleck, J.A.; Eagles-Smith, C.A.; Marvin-DiPasquale, M.; Windham-Myers, L.; Herzog, M.P.; McQuillen, H.L. Wetland Management Strategy to Reduce Mercury in Water and Bioaccumulation in Fish. *Environ. Toxicol. Chem.* **2019**, *38*, 2178–2196. [CrossRef]
5. Morabito, R.; Dossena, S.; La Spada, G.; Marino, A. Heavy metals affect nematocysts discharge response and biological activity of crude venom in the jellyfish *Pelagia noctiluca* (Cnidaria, Scyphozoa). *Cell. Physiol. Biochem.* **2014**, *34*, 244–254. [CrossRef]
6. Piscopo, M.; Notariale, R.; Tortora, F.; Lettieri, G.; Palumbo, G.; Manna, C. Novel Insights into Mercury Effects on Hemoglobin and Membrane Proteins in Human Erythrocytes. *Molecules* **2020**, *25*, 3278. [CrossRef]
7. Ahmad, S.; Mahmood, R. Mercury chloride toxicity in human erythrocytes: Enhanced generation of ROS and RNS, hemoglobin oxidation, impaired antioxidant power, and inhibition of plasma membrane redox system. *Environ. Sci. Pollut. Res. Int.* **2019**, *26*, 5645–5657. [CrossRef]
8. Remigante, A.; Morabito, R.; Marino, A. Band 3 protein function and oxidative stress in erythrocytes. *J. Cell. Physiol.* **2021**, *236*, 6225–6234. [CrossRef]
9. He, J.; Zhao, Y.; Zhu, T.; Xue, P.; Zheng, W.; Yao, Y.; Qu, W.; Jia, X.; Lu, R.; He, M.; et al. Mercury Chloride Impacts on the Development of Erythrocytes and Megakaryocytes in Mice. *Toxics* **2021**, *9*, 252. [CrossRef]
10. Notariale, R.; Langst, E.; Perrone, P.; Crettaz, D.; Prudent, M.; Manna, C. Effect of Mercury on Membrane Proteins, Anionic Transport and Cell Morphology in Human Erythrocytes. *Cell. Physiol. Biochem.* **2022**, *56*, 500–513. [CrossRef]
11. Crupi, R.; Morabito, R.; Remigante, A.; Gugliandolo, E.; Britti, D.; Cuzzocrea, S.; Marino, A. Susceptibility of erythrocytes from different sources to xenobiotics-induced lysis. *Comp. Biochem. Physiol. C Toxicol. Pharmacol.* **2019**, *221*, 68–72. [CrossRef]
12. Hatae, H.; Inaka, K.; Okamura, R.; Furubayashi, N.; Kamo, M.; Kobayashi, T.; Abe, Y.; Iwata, S.; Hamasaki, N. Crystallization of Human Erythrocyte Band 3, the anion exchanger, at the International Space Station "KIBO". *Anal. Biochem.* **2018**, *559*, 91–93. [CrossRef]
13. Jensen, F.B. The dual roles of red blood cells in tissue oxygen delivery: Oxygen carriers and regulators of local blood flow. *J. Exp. Biol.* **2009**, *212*, 3387–3393. [CrossRef]
14. Vianna, A.D.S.; Camara, V.M.; Barbosa, M.C.M.; Santos, A.S.E.; Asmus, C.; Luiz, R.R.; Jesus, I.M. Mercury exposure and anemia in children and adolescents from six riverside communities of Brazilian Amazon. *Cien. Saude Colet.* **2022**, *27*, 1859–1871. [CrossRef]
15. Marino, A.; Morabito, R.; La Spada, G. Factors altering the haemolytic power of crude venom from Aiptasia mutabilis (Anthozoa) nematocysts. *Comp. Biochem. Physiol. A Mol. Integr. Physiol.* **2009**, *152*, 418–422. [CrossRef]
16. Tsamesidis, I.; Pouroutzidou, G.K.; Lymperaki, E.; Kazeli, K.; Lioutas, C.B.; Christodoulou, E.; Perio, P.; Reybier, K.; Pantaleo, A.; Kontonasaki, E. Effect of ion doping in silica-based nanoparticles on the hemolytic and oxidative activity in contact with human erythrocytes. *Chem. Biol. Interact.* **2020**, *318*, 108974. [CrossRef]
17. Zabinski, Z.; Dabrowski, Z.; Moszczynski, P.; Rutowski, J. The activity of erythrocyte enzymes and basic indices of peripheral blood erythrocytes from workers chronically exposed to mercury vapours. *Toxicol. Ind. Health* **2000**, *16*, 58–64. [CrossRef]
18. Lupescu, A.; Bissinger, R.; Goebel, T.; Salker, M.S.; Alzoubi, K.; Liu, G.; Chirigiu, L.; Mack, A.F.; Qadri, S.M.; Lang, F. Enhanced suicidal erythrocyte death contributing to anemia in the elderly. *Cell. Physiol. Biochem.* **2015**, *36*, 773–783. [CrossRef]
19. Balali-Mood, M.; Naseri, K.; Tahergorabi, Z.; Khazdair, M.R.; Sadeghi, M. Toxic Mechanisms of Five Heavy Metals: Mercury, Lead, Chromium, Cadmium, and Arsenic. *Front. Pharmacol.* **2021**, *12*, 643972. [CrossRef]
20. Vianna, A.D.S.; Matos, E.P.; Jesus, I.M.; Asmus, C.; Camara, V.M. Human exposure to mercury and its hematological effects: A systematic review. *Cad. Saude Publica* **2019**, *35*, e00091618. [CrossRef]
21. Notariale, R.; Infantino, R.; Palazzo, E.; Manna, C. Erythrocytes as a Model for Heavy Metal-Related Vascular Dysfunction: The Protective Effect of Dietary Components. *Int. J. Mol. Sci.* **2021**, *22*, 6604. [CrossRef] [PubMed]
22. Fernandes, S.; Ribeiro, C.; Paiva-Martins, F.; Catarino, C.; Santos-Silva, A. Protective effect of olive oil polyphenol phase II sulfate conjugates on erythrocyte oxidative-induced hemolysis. *Food Funct.* **2020**, *11*, 8670–8679. [CrossRef] [PubMed]

23. Officioso, A.; Alzoubi, K.; Lang, F.; Manna, C. Hydroxytyrosol inhibits phosphatidylserine exposure and suicidal death induced by mercury in human erythrocytes: Possible involvement of the glutathione pathway. *Food Chem. Toxicol.* **2016**, *89*, 47–53. [CrossRef] [PubMed]

24. Manna, C.; Migliardi, V.; Sannino, F.; De Martino, A.; Capasso, R. Protective effects of synthetic hydroxytyrosol acetyl derivatives against oxidative stress in human cells. *J. Agric. Food Chem.* **2005**, *53*, 9602–9607. [CrossRef] [PubMed]

25. Romani, A.; Ieri, F.; Urciuoli, S.; Noce, A.; Marrone, G.; Nediani, C.; Bernini, R. Health Effects of Phenolic Compounds Found in Extra-Virgin Olive Oil, By-Products, and Leaf of *Olea europaea* L. *Nutrients* **2019**, *11*, 1176. [CrossRef] [PubMed]

26. Bucciantini, M.; Leri, M.; Nardiello, P.; Casamenti, F.; Stefani, M. Olive Polyphenols: Antioxidant and Anti-Inflammatory Properties. *Antioxidants* **2021**, *10*, 1044. [CrossRef]

27. Notariale, R.; Perrone, P.; Mele, L.; Lettieri, G.; Piscopo, M.; Manna, C. Olive Oil Phenols Prevent Mercury-Induced Phosphatidylserine Exposure and Morphological Changes in Human Erythrocytes Regardless of Their Different Scavenging Activity. *Int. J. Mol. Sci.* **2022**, *23*, 5693. [CrossRef]

28. Tagliafierro, L.; Officioso, A.; Sorbo, S.; Basile, A.; Manna, C. The protective role of olive oil hydroxytyrosol against oxidative alterations induced by mercury in human erythrocytes. *Food Chem. Toxicol.* **2015**, *82*, 59–63. [CrossRef]

29. Officioso, A.; Panzella, L.; Tortora, F.; Alfieri, M.L.; Napolitano, A.; Manna, C. Comparative Analysis of the Effects of Olive Oil Hydroxytyrosol and Its 5-S-Lipoyl Conjugate in Protecting Human Erythrocytes from Mercury Toxicity. *Oxid. Med. Cell. Longev.* **2018**, *2018*, 9042192. [CrossRef]

30. Morabito, R.; Remigante, A.; Cavallaro, M.; Taormina, A.; La Spada, G.; Marino, A. Anion exchange through band 3 protein in canine leishmaniasis at different stages of disease. *Pflugers Arch.* **2017**, *469*, 713–724. [CrossRef]

31. Remigante, A.; Morabito, R.; Spinelli, S.; Trichilo, V.; Loddo, S.; Sarikas, A.; Dossena, S.; Marino, A. d-Galactose Decreases Anion Exchange Capability through Band 3 Protein in Human Erythrocytes. *Antioxidants* **2020**, *9*, 689. [CrossRef]

32. Morabito, R.; Remigante, A.; Spinelli, S.; Vitale, G.; Trichilo, V.; Loddo, S.; Marino, A. High Glucose Concentrations Affect Band 3 Protein in Human Erythrocytes. *Antioxidants* **2020**, *9*, 365. [CrossRef]

33. Morabito, R.; Remigante, A.; Cordaro, M.; Trichilo, V.; Loddo, S.; Dossena, S.; Marino, A. Impact of acute inflammation on Band 3 protein anion exchange capability in human erythrocytes. *Arch. Physiol. Biochem.* **2020**, *128*, 1242–1248. [CrossRef]

34. Morabito, R.; Remigante, A.; Marino, A. Melatonin Protects Band 3 Protein in Human Erythrocytes against H2O2-Induced Oxidative Stress. *Molecules* **2019**, *24*, 2741. [CrossRef]

35. Morabito, R.; Remigante, A.; Di Pietro, M.L.; Giannetto, A.; La Spada, G.; Marino, A. SO4(=) uptake and catalase role in preconditioning after H2O2-induced oxidative stress in human erythrocytes. *Pflugers Arch.* **2017**, *469*, 235–250. [CrossRef]

36. Remigante, A.; Morabito, R.; Marino, A. Natural Antioxidants Beneficial Effects on Anion Exchange through Band 3 Protein in Human Erythrocytes. *Antioxidants* **2019**, *9*, 25. [CrossRef]

37. Remigante, A.; Spinelli, S.; Straface, E.; Gambardella, L.; Caruso, D.; Falliti, G.; Dossena, S.; Marino, A.; Morabito, R. Acai (*Euterpe oleracea*) Extract Protects Human Erythrocytes from Age-Related Oxidative Stress. *Cells* **2022**, *11*, 2391. [CrossRef]

38. Lucantoni, G.; Pietraforte, D.; Matarrese, P.; Gambardella, L.; Metere, A.; Paone, G.; Bianchi, E.L.; Straface, E. The red blood cell as a biosensor for monitoring oxidative imbalance in chronic obstructive pulmonary disease: An ex vivo and in vitro study. *Antioxid. Redox Signal.* **2006**, *8*, 1171–1182. [CrossRef]

39. Mendanha, S.A.; Anjos, J.L.; Silva, A.H.; Alonso, A. Electron paramagnetic resonance study of lipid and protein membrane components of erythrocytes oxidized with hydrogen peroxide. *Braz. J. Med. Biol. Res.* **2012**, *45*, 473–481. [CrossRef]

40. Akki, R.; Siracusa, R.; Morabito, R.; Remigante, A.; Campolo, M.; Errami, M.; La Spada, G.; Cuzzocrea, S.; Marino, A. Neuronal-like differentiated SH-SY5Y cells adaptation to a mild and transient H2 O2 -induced oxidative stress. *Cell. Biochem. Funct.* **2018**, *36*, 56–64. [CrossRef]

41. Aksenov, M.Y.; Markesbery, W.R. Changes in thiol content and expression of glutathione redox system genes in the hippocampus and cerebellum in Alzheimer's disease. *Neurosci. Lett.* **2001**, *302*, 141–145. [CrossRef] [PubMed]

42. Morabito, R.; Falliti, G.; Geraci, A.; Spada, G.L.; Marino, A. Curcumin Protects -SH Groups and Sulphate Transport after Oxidative Damage in Human Erythrocytes. *Cell. Physiol. Biochem.* **2015**, *36*, 345–357. [CrossRef] [PubMed]

43. Morabito, R.; Romano, O.; La Spada, G.; Marino, A. H2O2-Induced Oxidative Stress Affects SO4= Transport in Human Erythrocytes. *PLoS ONE* **2016**, *11*, e0146485. [CrossRef] [PubMed]

44. Naoum, P.C.R., Jr.; Magaly da Silva, M. Spectrometric measurement of methemoglobin without interference of chemical or enzymatic reagents. *Rev. Bras. Hematol. Hemoter.* **2004**, *26*, 19–22.

45. Zavodnik, I.B.; Lapshina, E.A.; Rekawiecka, K.; Zavodnik, L.B.; Bartosz, G.; Bryszewska, M. Membrane effects of nitrite-induced oxidation of human red blood cells. *Biochim. Biophys. Acta* **1999**, *1421*, 306–316. [CrossRef]

46. Kupcho, K.; Shultz, J.; Hurst, R.; Hartnett, J.; Zhou, W.; Machleidt, T.; Grailer, J.; Worzella, T.; Riss, T.; Lazar, D.; et al. A real-time, bioluminescent annexin V assay for the assessment of apoptosis. *Apoptosis* **2019**, *24*, 184–197. [CrossRef]

47. Pantaleo, A.; Ferru, E.; Pau, M.C.; Khadjavi, A.; Mandili, G.; Matte, A.; Spano, A.; De Franceschi, L.; Pippia, P.; Turrini, F. Band 3 Erythrocyte Membrane Protein Acts as Redox Stress Sensor Leading to Its Phosphorylation by p (72) Syk. *Oxid. Med. Cell. Longev.* **2016**, *2016*, 6051093. [CrossRef]

48. Laemmli, U.K. Cleavage of structural proteins during the assembly of the head of bacteriophage T4. *Nature* **1970**, *227*, 680–685. [CrossRef]

49. Yeung, Y.G.; Stanley, E.R. A solution for stripping antibodies from polyvinylidene fluoride immunoblots for multiple reprobing. *Anal. Biochem.* **2009**, *389*, 89–91. [CrossRef]

50. Romano, L.; Peritore, D.; Simone, E.; Sidoti, A.; Trischitta, F.; Romano, P. Chloride-sulphate exchange chemically measured in human erythrocyte ghosts. *Cell. Mol. Biol.* **1998**, *44*, 351–355.

51. Romano, L.; Passow, H. Characterization of anion transport system in trout red blood cell. *Am. J. Physiol.* **1984**, *246*, C330–C338. [CrossRef]

52. Morabito, R.; Remigante, A.; Marino, A. Protective Role of Magnesium against Oxidative Stress on SO4(=) Uptake through Band 3 Protein in Human Erythrocytes. *Cell. Physiol. Biochem.* **2019**, *52*, 1292–1308. [CrossRef]

53. Morabito, R.R.A.; Arcuri, B.; Marino, A.; Giammanco, M.; La Spada, G. Effect of cadmium on anion exchange capability through Band 3 protein in human erythrocytes. *J. Biol. Res.* **2018**, *91*. [CrossRef]

54. Jessen, F.; Sjoholm, C.; Hoffmann, E.K. Identification of the anion exchange protein of Ehrlich cells: A kinetic analysis of the inhibitory effects of 4,4′-diisothiocyano-2,2′-stilbene-disulfonic acid (DIDS) and labeling of membrane proteins with 3H-DIDS. *J. Membr. Biol.* **1986**, *92*, 195–205. [CrossRef]

55. Knauf, P.A.; Law, F.Y.; Hahn, K. An oxonol dye is the most potent known inhibitor of band 3-mediated anion exchange. *Am. J. Physiol.* **1995**, *269*, C1073–C1077. [CrossRef]

56. Teti, D.; Crupi, M.; Busa, M.; Valenti, A.; Loddo, S.; Mondello, M.; Romano, L. Chemical and pathological oxidative influences on band 3 protein anion-exchanger. *Cell. Physiol. Biochem.* **2005**, *16*, 77–86. [CrossRef]

57. Lim, K.M.; Kim, S.; Noh, J.Y.; Kim, K.; Jang, W.H.; Bae, O.N.; Chung, S.M.; Chung, J.H. Low-level mercury can enhance procoagulant activity of erythrocytes: A new contributing factor for mercury-related thrombotic disease. *Environ. Health Perspect.* **2010**, *118*, 928–935. [CrossRef]

58. Paiva-Martins, F.; Silva, A.; Almeida, V.; Carvalheira, M.; Serra, C.; Rodrigues-Borges, J.E.; Fernandes, J.; Belo, L.; Santos-Silva, A. Protective activity of hydroxytyrosol metabolites on erythrocyte oxidative-induced hemolysis. *J. Agric. Food Chem.* **2013**, *61*, 6636–6642. [CrossRef]

59. Fujii, J.; Homma, T.; Kobayashi, S.; Warang, P.; Madkaikar, M.; Mukherjee, M.B. Erythrocytes as a preferential target of oxidative stress in blood. *Free Radic. Res.* **2021**, *55*, 562–580. [CrossRef]

60. De Franceschi, L.; Scardoni, G.; Tomelleri, C.; Danek, A.; Walker, R.H.; Jung, H.H.; Bader, B.; Mazzucco, S.; Dotti, M.T.; Siciliano, A.; et al. Computational identification of phospho-tyrosine sub-networks related to acanthocyte generation in neuroacanthocytosis. *PLoS ONE* **2012**, *7*, e31015. [CrossRef]

61. Adjobo-Hermans, M.J.; Cluitmans, J.C.; Bosman, G.J. Neuroacanthocytosis: Observations, Theories and Perspectives on the Origin and Significance of Acanthocytes. *Tremor Other Hyperkinet. Mov.* **2015**, *5*, 328. [CrossRef]

62. Gyawali, P.; Richards, R.S.; Bwititi, P.T.; Nwose, E.U. Association of abnormal erythrocyte morphology with oxidative stress and inflammation in metabolic syndrome. *Blood Cells Mol. Dis.* **2015**, *54*, 360–363. [CrossRef] [PubMed]

63. Gyawali, P.; Richards, R.S.; Uba Nwose, E. Erythrocyte morphology in metabolic syndrome. *Expert Rev. Hematol.* **2012**, *5*, 523–531. [CrossRef] [PubMed]

64. Lang, K.S.; Duranton, C.; Poehlmann, H.; Myssina, S.; Bauer, C.; Lang, F.; Wieder, T.; Huber, S.M. Cation channels trigger apoptotic death of erythrocytes. *Cell Death Differ.* **2003**, *10*, 249–256. [CrossRef] [PubMed]

65. Eisele, K.; Lang, P.A.; Kempe, D.S.; Klarl, B.A.; Niemoller, O.; Wieder, T.; Huber, S.M.; Duranton, C.; Lang, F. Stimulation of erythrocyte phosphatidylserine exposure by mercury ions. *Toxicol. Appl. Pharmacol.* **2006**, *210*, 116–122. [CrossRef]

66. Segawa, K.; Nagata, S. An Apoptotic 'Eat Me' Signal: Phosphatidylserine Exposure. *Trends Cell. Biol.* **2015**, *25*, 639–650. [CrossRef]

67. Tsamesidis, I.; Perio, P.; Pantaleo, A.; Reybier, K. Oxidation of Erythrocytes Enhance the Production of Reactive Species in the Presence of Artemisinins. *Int. J. Mol. Sci.* **2020**, *21*, 4799. [CrossRef]

68. Cyboran-Mikolajczyk, S.; Meczarska, K.; Solarska-Sciuk, K.; Ratajczak-Wielgomas, K.; Oszmianski, J.; Jencova, V.; Bonarska-Kujawa, D. Protection of Erythrocytes and Microvascular Endothelial Cells against Oxidative Damage by *Fragaria vesca* L. and *Rubus idaeus* L. Leaves Extracts-The Mechanism of Action. *Molecules* **2022**, *27*, 5865. [CrossRef]

69. Kouka, P.; Priftis, A.; Stagos, D.; Angelis, A.; Stathopoulos, P.; Xinos, N.; Skaltsounis, A.L.; Mamoulakis, C.; Tsatsakis, A.M.; Spandidos, D.A.; et al. Assessment of the antioxidant activity of an olive oil total polyphenolic fraction and hydroxytyrosol from a Greek Olea europea variety in endothelial cells and myoblasts. *Int. J. Mol. Med.* **2017**, *40*, 703–712. [CrossRef]

70. Akki, R.; Siracusa, R.; Cordaro, M.; Remigante, A.; Morabito, R.; Errami, M.; Marino, A. Adaptation to oxidative stress at cellular and tissue level. *Arch. Physiol. Biochem.* **2020**, *128*, 521–531. [CrossRef]

71. Remigante, A.; Morabito, R. Cellular and Molecular Mechanisms in Oxidative Stress-Related Diseases. *Int. J. Mol. Sci.* **2022**, *23*, 8017. [CrossRef]

72. Deiana, M.; Incani, A.; Rosa, A.; Corona, G.; Atzeri, A.; Loru, D.; Paola Melis, M.; Assunta Dessi, M. Protective effect of hydroxytyrosol and its metabolite homovanillic alcohol on H(2)O(2) induced lipid peroxidation in renal tubular epithelial cells. *Food Chem. Toxicol.* **2008**, *46*, 2984–2990. [CrossRef]

73. D'Angelo, S.; Ingrosso, D.; Migliardi, V.; Sorrentino, A.; Donnarumma, G.; Baroni, A.; Masella, L.; Tufano, M.A.; Zappia, M.; Galletti, P. Hydroxytyrosol, a natural antioxidant from olive oil, prevents protein damage induced by long-wave ultraviolet radiation in melanoma cells. *Free Radic. Biol. Med.* **2005**, *38*, 908–919. [CrossRef]

74. Carrasco-Pancorbo, A.; Gomez-Caravaca, A.M.; Cerretani, L.; Bendini, A.; Segura-Carretero, A.; Fernandez-Gutierrez, A. A simple and rapid electrophoretic method to characterize simple phenols, lignans, complex phenols, phenolic acids, and flavonoids in extra-virgin olive oil. *J. Sep. Sci.* **2006**, *29*, 2221–2233. [CrossRef]

75. Remigante, A.S.S.; Basile, N.; Caruso, D.; Falliti, G.; Dossena, S.; Marino, A.; Morabito, R. Oxidation Stress as a Mechanism of Aging in Human Erythrocytes: Protective Effect of Quercetin. *Int. J. Mol. Sci.* **2022**, *23*, 7781. [CrossRef] [PubMed]

76. Remigante, A.; Spinelli, S.; Straface, E.; Gambardella, L.; Caruso, D.; Falliti, G.; Dossena, S.; Marino, A.; Morabito, R. Antioxidant Activity of Quercetin in a H_2O_2-Induced Oxidative Stress Model in Red Blood Cells: Functional Role of Band 3 Protein. *Int. J. Mol. Sci.* **2022**, *23*, 10991. [CrossRef]

77. Arakawa, T.; Kobayashi-Yurugi, T.; Alguel, Y.; Iwanari, H.; Hatae, H.; Iwata, M.; Abe, Y.; Hino, T.; Ikeda-Suno, C.; Kuma, H.; et al. Crystal structure of the anion exchanger domain of human erythrocyte band 3. *Science* **2015**, *350*, 680–684. [CrossRef]

78. Hogeback, J.; Schwarzer, M.; Wehe, C.A.; Sperling, M.; Karst, U. Investigating the adduct formation of organic mercury species with carbonic anhydrase and hemoglobin from human red blood cell hemolysate by means of LC/ESI-TOF-MS and LC/ICP-MS. *Metallomics* **2016**, *8*, 101–107. [CrossRef]

79. Remigante, A.; Spinelli, S.; Pusch, M.; Sarikas, A.; Morabito, R.; Marino, A.; Dossena, S. Role of SLC4 and SLC26 solute carriers during oxidative stress. *Acta Physiol.* **2022**, *235*, e13796. [CrossRef]

80. Remigante, A.; Spinelli, S.; Trichilo, V.; Loddo, S.; Sarikas, A.; Pusch, M.; Dossena, S.; Marino, A.; Morabito, R. d-Galactose induced early aging in human erythrocytes: Role of band 3 protein. *J. Cell Physiol.* **2022**, *237*, 1586–1596. [CrossRef]

81. Ajsuvakova, O.P.; Tinkov, A.A.; Aschner, M.; Rocha, J.B.T.; Michalke, B.; Skalnaya, M.G.; Skalny, A.V.; Butnariu, M.; Dadar, M.; Sarac, I.; et al. Sulfhydryl groups as targets of mercury toxicity. *Coord. Chem. Rev.* **2020**, *417*, R713–R715. [CrossRef] [PubMed]

82. Tokumoto, M.; Lee, J.Y.; Shimada, A.; Tohyama, C.; Satoh, M. Glutathione has a more important role than metallothionein-I/II against inorganic mercury-induced acute renal toxicity. *J. Toxicol. Sci.* **2018**, *43*, 275–280. [CrossRef] [PubMed]

83. Perez-Barron, G.; Montes, S.; Aguirre-Vidal, Y.; Santiago, M.; Gallardo, E.; Espartero, J.L.; Rios, C.; Monroy-Noyola, A. Antioxidant Effect of Hydroxytyrosol, Hydroxytyrosol Acetate and Nitrohydroxytyrosol in a Rat MPP(+) Model of Parkinson's Disease. *Neurochem. Res.* **2021**, *46*, 2923–2935. [CrossRef]

84. Lund, B.O.; Miller, D.M.; Woods, J.S. Studies on Hg(II)-induced H_2O_2 formation and oxidative stress in vivo and in vitro in rat kidney mitochondria. *Biochem. Pharmacol.* **1993**, *45*, 2017–2024. [CrossRef]

Article

Fresh Medium or L-Cystine as an Effective Nrf2 Inducer for Cytoprotection in Cell Culture

Wujing Dai and **Qin M. Chen** *

Department of Pharmacy Practice and Science, College of Pharmacy, University of Arizona, 1295 N Martin Ave, Tucson, AZ 85721, USA

* Correspondence: qchen@arizona.edu; Tel.: +1-520-626-9126

Abstract: The Nrf2 gene encodes a transcription factor best known for regulating the expression of antioxidant and detoxification genes. A long list of small molecules has been reported to induce Nrf2 protein via Keap1 oxidation or alkylation. Many of these Nrf2 inducers exhibit off-target or toxic effects due to their nature as electrophiles. In searching for non-toxic Nrf2 inducers, we found that a culture medium change to fresh DMEM is capable of inducing Nrf2 protein in HeLa, HEK293, AC16 and MCF7 cells. Testing the components of DMEM led to the discovery of L-Cystine as an effective Nrf2 inducer. L-Cystine induces a dose-dependent increase of Nrf2 protein, from 0.1 to 1.6 mM. RNA-seq analyses and RT-PCR revealed an induction of multiple Nrf2 downstream genes, including NQO1, HMOX1, GCLC, GCLM, SRXN1, TXNRD1, AKR1C and OSGIN1 by 0.8 mM L-Cystine. The induction of Nrf2 protein was dependent on L-Cystine entering cells via the cystine/glutamate antiporter and the presence of Keap1. The half-life of Nrf2 protein increased from 19.4 min to 30.9 min with 0.8 mM L-Cystine treatment. L-Cystine was capable of eliciting cytoprotection by reducing ROS generation and protecting against oxidant- or doxorubicin-induced apoptosis. As an amino acid derivative, L-Cystine is considered a non-toxic Nrf2 inducer that exhibits the potential for protection against oxidative stress and tissue injury.

Keywords: amino acid; cytoprotection; non-toxic; transcription; antioxidant and detoxification genes; oxidative stress

Citation: Dai, W.; Chen, Q.M. Fresh Medium or L-Cystine as an Effective Nrf2 Inducer for Cytoprotection in Cell Culture. *Cells* **2023**, *12*, 291. https://doi.org/10.3390/cells12020291

Academic Editors: Alessia Remigante and Rossana Morabito

Received: 9 December 2022
Revised: 6 January 2023
Accepted: 7 January 2023
Published: 12 January 2023

1. Introduction

The Nrf2 gene has attracted intense interest since its discovery in 1994 due to its ability to elicit a myriad of cytoprotective responses [1–6]. This gene encodes a transcription factor best known for regulating the expression of genes containing the Antioxidant Response Element (ARE) in the promoters [7,8]. A long list of antioxidant and detoxification genes are downstream targets of Nrf2 [5,6]. Typical examples include those encoding key components of redox systems, such as NAD(P)H quinone oxidoreductase 1 (NQO1), heme oxygenase 1 (HMOX1), glutamate-cysteine ligase catalytic subunit or regulatory subunit (GCLC, GCLM), glutathione peroxidases, glutathione reductase, peroxiredoxin, sulfiredoxin 1 (SRXN1), thioredoxin and thioredoxin reductase 1 (TXNRD1). Genomic or transcriptomic profiling has led to the discovery of Nrf2 impact on multiple cellular and molecular pathways, from autophagy to mitochondrial turnover, cell proliferation or extracellular remodeling [5,6]. These findings are consistent with the observations that knocking out Nrf2 leads to an increased susceptibility of mice to a variety of physical or chemical stresses and disease inducers.

Oxidative stress causes the activation of Nrf2 due to increased protein stability, nuclear translocation and de novo protein translation [5,6]. Nrf2 protein normally binds to Keap1 in the cytoplasm, where Keap1 recruits ubiquitin ligases for Nrf2 ubiquitination, contributing to a short half-life of Nrf2 protein, 10–20 min. When Keap1 is modified by oxidation or alkylation, or removed due to autophagy, Nrf2 is relieved from ubiquitination, causing an increase in Nrf2 protein half-life to 40–60 min [3,4,9–11]. In addition to Keap1, two additional mechanisms, HMG-CoA Reductase Degradation protein 1 (Hrd1), also known as Synoviolin, and β-transducin repeat-containing protein (β-TrCP), can carry out Nrf2 protein ubiquitination. Hrd1/Synoviolin responds to ER stress, whereas β-TrCP is activated downstream of GSK3β activity [2,12]. Recently, we and others have found that Nrf2 is among a few genes that undergo de novo protein translation when mammalian cells encounter acute oxidative stress [13–18]. De novo Nrf2 protein translation may cooperate with Nrf2 protein stabilization for the ultimate activation of Nrf2 as a transcription factor.

A number of small molecules have been developed for Nrf2 induction. Many of these molecules alkylate or oxidize Keap1, therefore removing the negative force of Keap1 on Nrf2 [4,19,20]. The most commonly studied Nrf2 inducers include sulforaphane (SFN) and bardoxolone methyl (CDDO-Me). In addition, multiple small molecules derived from herb medicine, spices and dietary plants are capable of inducing Nrf2 [5,9,21,22]. The chemical nature of these compounds as electrophiles suggests off-target effects and toxicity. Given these caveats, it is imperative to identify Nrf2 inducers that can bypass the issue of toxicity.

In an effort of searching for non-toxic Nrf2 inducers, we tested the components in a cell culture medium. Several growth factors have been reported to induce oxidative stress. Examples include Epidermal Growth Factor, Insulin-like Growth Factor-1, Transforming Growth Factor beta 1 and Vascular Endothelial Growth Factors [23–25]. Given the non-toxic nature of growth factors, we asked whether growth factors could induce Nrf2. With cell culture systems, we tested whether the replenishment of a fresh culture medium with the growth factor-containing serum induced Nrf2. Although fetal bovine serum did not induce Nrf2, we found that an amino acid derivative in a cell culture medium, i.e., L-Cystine, is an effective Nrf2 inducer.

2. Materials and Methods

Cell culture: HeLa (CCL-2) cells were obtained from ATCC. AC16 human cardiomyocytes (SCC109) were purchased from Millipore Sigma. HEK293 and MCF7 cells were obtained from Dr. Donna Zhang or Andrew Paek's laboratory at the University of Arizona. Cells were cultured in high-glucose Dulbecco's Modified Eagle's medium (DMEM) with 10% fetal bovine serum (FBS), 100 U/mL penicillin and 100 µg/mL streptomycin at 37 °C with 5% CO_2. Cells at passage 4 to 20 were seeded in tissue culture plates and treated with L-Cystine when the confluence reached 80%. L-Cystine (Research Products International, Mount Prospect, IL, USA) was dissolved in a 500 mM HCl or 200 mM NaOH solution to make a 0.1 M stock solution for its addition to cells at the final concentration of 0.1 to 1.6 mM.

Western blot to detect protein level changes: Cells in culture dishes were harvested in a 1× Laemmli sample buffer containing 5% 2-mercaptoethanol. After boiling and sonicating, samples with an equal cell number or protein concentration were loaded onto a gel for SDS-PAGE. Proteins on the gel were transferred to a PVDF membrane using the Bio-Rad (Hercules, CA, USA) Trans-Blot Turbo Transfer System and incubated with the following antibodies: anti-Nrf2 (sc-13032), anti-Keap1 (sc-365626), anti-HMOX1 (sc-136960), anti-SRXN1 (sc-514940), anti-TXNRD1 (sc-28321), anti-AKR1C (sc-166297), anti-GAPDH (sc-32233) or anti-Ubiquitin (sc-8017), which were purchased from Santa Cruz Biotechnology. Anti-caspase-3 (#9662) was purchased from Cell Signaling Technology (Danvers, MA, USA) for Western blot. The bound antibodies were recognized by anti-mouse IgG-HRP (A9044-2ML) or anti-rabbit IgG-HRP (A9169-2ML), which were purchased from Sigma Aldrich (St. Louis, MO, USA).

RNA-seq for transcriptomics: Total RNAs were extracted from 3 independent sample preparations with control or 0.8 mM L-Cystine-treated for 16 h by RNeasy Kits (Qiagen, Germantown, MD, USA). Sample quality control, transcriptome library preparation, library quality control, sequencing, raw data output, data quality control and bioinformatics analysis were performed by BGI Genomics Americas Corporation (San Jose, CA, USA). Briefly, raw reads were filtered to remove the reads containing the adaptor sequence, unknown nucleotides greater than 5%, or a quality score less than 15 by SOAPnuke software [26].

The clean reads were aligned to a reference genome GRCh38.p12 using HISAT [27] or to a reference transcriptome using Bowtie 2 [28]. RSEM was used to calculate the gene expression level of each sample [29]. For detecting differential genes, the DESeq2 method was employed with Qvalue (Adjusted Pvalue) $\leq$ 0.05 [29].

RT-qPCR to measure the levels of transcripts: Total RNA, 1 µg of each sample, was converted to cDNA with qScript™ cDNA SuperMix (QuantaBio, Beverly, MA, USA) for subsequent qPCR analysis of human genes. The qPCR primer pairs were: forward 5′-TTCCCGGTCACATCGAGAG-3′ and reverse 5′-TCCTGTTGCATACCGTCTAAATC-3′ for Nrf2, forward 5′-ACCACAACAGTGTGGAGAGGT-3′ and reverse 5′-CGATCCTTCGTGT CAGCAT-3′ for Keap1, forward 5′-ATGTATGACAAAGGACCCTTCC-3′ and reverse 5′-TCCCTTGCAGAGAGTACATGG-3′ for NQO1, forward 5′-AACTTTCAGAAGGGCCAGGT-3′ and reverse 5′-CTGGGCTCTCCTTGTTGC-3′ for HMOX1, forward 5′-GACAAAACACA G TTGGAACAGC-3′ and reverse 5′-CAGTCAAATCTGGTGGCATC-3′ for GCLM, forward 5′-ATATGGCAAGAAGGTGATGGTCC-3′ and reverse 5′-GGGCTTGTCCTAACAAAGCTG-3′ for TXNRD1, forward 5′-CAGGGAGGTGACTACTTCTACTC-3′ and reverse 5′-CAGGTA CACCCTTAGGTCTGA-3′ for SRXN1, forward 5′-TCCAGTGTCTGTAAAGCCAGG-3′ and reverse 5′-CCAGCAGTTTTCTCTGGTTGAA-3′ for AKR1C1, forward 5′-CCCGGTCATCA TTGTGGGTAA-3′ and reverse 5′-GCTTCGTGTAGGGTGTGTAGC-3′ for OSGIN1, forward 5′-GGAGCGAGATCCCTCCAAAAT-3′ and reverse 5′-GGCTGTTGTCATACTTCTCATGG-3′ for GAPDH, or forward 5′-TCAACTTTCGATGGTAGTCGCCGT-3′ and reverse 5′-TCCTTGGATGTGGTAGCCGTTTCG-3′ for 18S rRNA. The qPCR was performed with PowerUp™ SYBR™ Green Master Mix (Applied Biosystems™, Waltham, MA, USA) on a BioRad CFX96 thermal cycler and the mRNA abundance was calculated with the ΔCT method by comparing it to that of GAPDH or 18S rRNA.

ARE reporter assay for the activity of Nrf2 transcription factor: pGL4.37(*luc2P*/ARE/Hygro) vector was purchased from Promega (Madison, WI, USA). This construct contains 4 copies of ARE sequence derived from the mouse glutathione S-transferase Ya gene (mGST-ARE). To generate a second reporter under the control of the human NQO1 gene, pGL4.37 was modified by replacing the mGST-ARE sequence with 41 bp human NQO1 ARE sequence (AATCCGCAGTCACAGTGACTCAGCAGAATCTGAGCCTAGGG). For measurement of ARE activation, either pGL4.37-mGST-ARE *luc* or pGL4.37-hNQO1-ARE *luc* vector together with pCMV *Renilla* luciferase construct were transfected into HeLa cells using the Lipofectamine™ 3000 Transfection Reagent (Invitrogen, Waltham, MA, USA). At 24 h after transfection, HeLa cells were treated with L-Cystine or SFN (Santa Cruz Biotechnology, Dallas, TX, USA) for 16 h and were harvested in 1× Passive Lysis Buffer (Promega). The cell lysate was used for detecting the *Firefly* and *Renilla* luciferase activities using Dual-Glo® Luciferase Assay System (Promega).

Ubiquitylation assay: HeLa cells were cultured in 100 mm dishes for treatment with either 0.8 mM L-Cystine or 5 µM SFN along with 10 µM MG132 for 4 h. Cells were lysed in 200 µL TBS buffer [150 mM NaCl, 10 mM Tris-HCl (pH 8.0)] containing 1% SDS and 1 mM DTT. Following 10 min boiling, soluble cell lysates were collected by centrifugation at 15,000× *g*, 4 °C for 10 min and were diluted with 800 µL TBS containing 1% Triton X-100 and 1 mM DTT. Protein A/G agarose beads (20 µL/sample, Thermo Scientific™ 78609) were added to the cell lysate by rotating at 4 °C for 1 h for removal of non-specific binding. Nrf2 antibody (1 µg, 16396-1-AP, Proteintech, Rosemont, IL, USA) was added to the cell lysate for overnight incubation at 4 °C, followed by incubation with 30 µL protein A/G agarose beads for 4 h at 4 °C. The beads were washed 4 times with TBS containing 1% Triton X-100 and 1 mM DTT. Immunoprecipitated proteins were eluted by boiling in 50 µL 1 × Laemmli sample buffer containing 5% 2-mercaptoethanol for Western blot to detect Ubiquitin.

CRISPR-Cas9 knockout of Keap1 gene: Keap1 CRISPR/Cas9 KO kit (sc-400190-KO-2) was purchased from Santa Cruz Biotechnology. This kit contains 3 plasmids, each encoding the Cas9 nuclease and a target-specific 20 nt guide RNA designed for maximal knockout efficiency. After transfecting into HeLa cells, GFP-positive cells were sorted by flow cytometry (BD FACSAria™ III, San Diego, CA, USA) and were seeded at single-cell density in 96-well plates for colony formation. Keap1 knockout was validated by measuring the protein using Western blot.

ROS detection: HeLa cells were seeded in a clear bottom 96-well dark plate and treated with 0.8 mM L-Cystine for 16 h. Following PBS wash, the cells were incubated in the dark with 10 μM 2′,7′–dichlorofluorescein diacetate (DCFDA) for 30 min at 37 °C in a 5% CO_2 incubator. After the removal of DCFDA, 100 μL PBS was added to the cells along with H_2O_2 for measurement of reactive oxygen species (ROS) according to the manufacturer's instruction (Abcam ROS detection kit, Cambridge, UK). DCF fluorescence was read at Ex485/Em535 nm by a plate reader (CLARIOstar Plus, BMG Labtech, Ortenberg, Germany).

Cell viability and toxicity assay: HeLa cells seeded in 96-well plates were treated with various doses of L-Cystine for 8 h before treatment with 2 μM doxorubicin for another 24 h. 3-(4,5-Dimethylthiazol-2-yl)-2,5-diphenyltetrazolium bromide (MTT, 20 μL of stock 2 mg/mL in PBS) was added to cells in 96-well plate for 2-h incubation under cell culture conditions at 37 °C in 5% CO_2 incubator. Upon removal of the culture medium, 100 μL acidified isopropanol (53 μL of concentrated HCl in 50 mL isopropanol) was added to the cells to dissolve formazan. After 10 min of solubilization, absorbance at 570 nm was measured by a microplate reader (CLARIOstar Plus).

Caspase activity assay for quantification of apoptosis: Cells seeded in 24-well plates were pre-treated with L-Cystine for 8 h before treatment with 2 μM doxorubicin (Calbiochem) for 24 h. Detached cells were collected by centrifugation at 3000 rpm for 3 min and washed once with PBS before combining with corresponding adherent cells from the same well. Cells were lysed in 120 μL of lysis buffer (0.5% Nonidet P-40, 0.5 mM EDTA, 150 mM NaCl, and 50 mM Tris, pH 7.5). Caspase 3/7 activity was measured using a fluorogenic substrate Acetyl Asp-Glu-Val-Asp-7-amino-4-methylcoumarin (Ac-DEVD-AMC) [30]. Cell lysates (50 μL/well) were incubated at 37 °C for 1 h with 80 μM of synthetic peptide substrate Ac-DEVD-AMC in a total volume of 200 μL in a 96-well plate. Caspase-3/7 activity was expressed as fluorescent intensity detected at Ex365/Em450 nm by a fluorescence plate reader (CLARIOstar Plus).

Transfection of siRNA for gene silencing: HeLa cells grown to 80% confluence in 12-well plates were transfected with either control siRNA (sc-37007) or Nrf2 siRNA (sc-37030, Santa Cruz Biotechnology). In each reaction, 50 μL opti-MEM containing 40 pmol siRNA was mixed with 4 μL Lipofectamine 3000 and incubated for 30 min. The mixture was added to the HeLa cells, which were placed in 400 μL opti-MEM for 6-h incubation at 37 °C in a CO_2 incubator. DMEM (500 μL) containing 20% FBS was added to cells for an 18-h incubation, followed by replacement with fresh growth medium (10% FBS in DMEM) for 48 h of culture before L-Cystine treatment.

Statistics: Statistical analyses were performed by GraphPad Prism 9 software. Data with one grouping variable were analyzed with one-way ANOVA and corrected by Dunnett's multiple comparisons test. Data with two grouping variables were analyzed with two-way ANOVA and corrected by Turkey's multiple comparisons test. Means of two groups were compared by two-tailed *t*-test. *p* value or adjusted *p* value < 0.05 was set as the threshold for significant difference.

3. Results

3.1. L-Cystine Induces Nrf2 Protein Elevation and Activates Nrf2 Transcription Factor

In addressing whether the culture medium with growth factors induces Nrf2, we discovered that changing the culture medium to fresh DMEM containing 10% FBS indeed induced Nrf2 protein in cultured HeLa cells (Figure 1A). While tracing the components, we found that fresh DMEM but not FBS induced Nrf2 elevation (Figure 1A). DMEM contains high concentrations of D-glucose (4.5 g/L) and L-glutamine (0.584 g/L or 4 mM). When adding D-glucose or L-glutamine to the cells without a medium change, there was no Nrf2 induction (Figure 1B). Induction of Nrf2 by a culture medium change was also observed in

HEK293 human embryonic kidney cells, AC16 human cardiomyocytes and MCF7 human breast cancer cells (Figure 1C), implying that the phenomenon is not unique to HeLa cells.

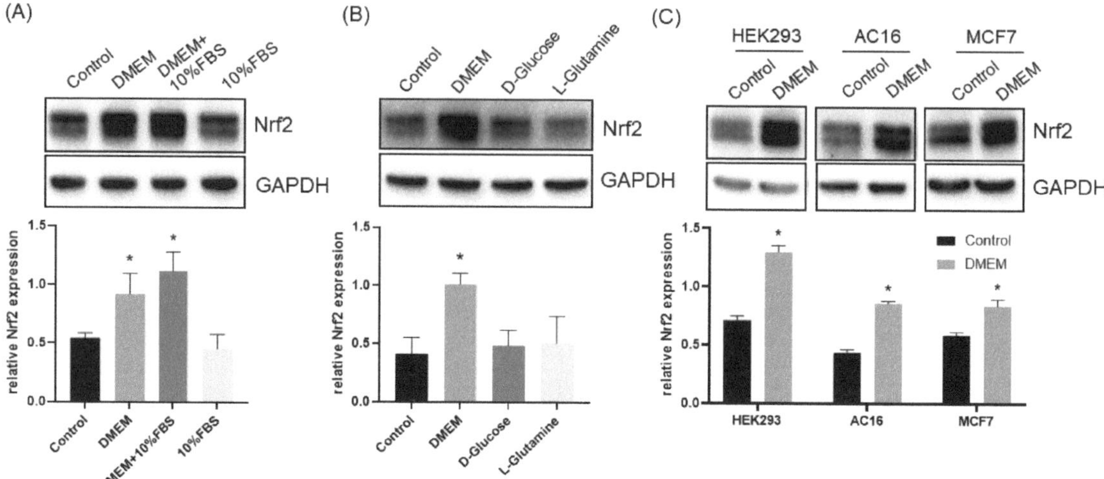

Figure 1. Fresh DMEM Induces Nrf2 Protein. HeLa cells at day 3 after subculture in DMEM with 10% FBS were either left untreated (control) or had a culture medium change to fresh DMEM or fresh DMEM with 10% FBS (**A**). Without changing the medium, cells were treated with 10% FBS (**A**), 25 mM glucose or 4 mM glutamine (**B**). Alternatively, HEK293, AC16 or MCF7 cells were left untreated (control) or had a culture medium change to fresh DMEM (**C**). The cells were collected at 4 h after harvesting for measuring Nrf2 protein levels by Western Blot, using GAPDH as a loading control. The intensities of the Nrf2 or GAPDH band were quantified by NIH ImageJ and presented as means $\pm$ SD of Nrf2/GAPDH ratio from three independent experiments. * indicates adjusted p value < 0.05 compared to the control by one-way ANOVA, corrected by Dunnett's multiple comparisons test (**A,B**) or $p < 0.05$ compared to the control by student's t test (**C**).

We further investigated the component responsible for Nrf2 induction. DMEM contains 14 amino acids or derivatives in addition to L-glutamine. Each of these amino acids was added to cells at its concentration corresponding to that in DMEM to test for Nrf2 induction (Figure 2). We found that 0.2 mM L-Cystine alone is sufficient for producing Nrf2 protein elevation (Figure 2).

To characterize the effect of L-Cystine for Nrf2 induction, we performed the dose-response curves. Since L-Cystine has low solubility in neutral pH, we dissolved it in 500 mM HCl or 200 mM NaOH to add it to the culture medium. With either method of solubilization, the elevation of Nrf2 protein started at 0.1 mM, and the degree of Nrf2 elevation increased with higher doses of L-Cystine (Figure 3A,B). The dose-response curves showed that L-Cystine at 0.8 mM was most effective for Nrf2 induction (Figure 3C). With 0.8 mM L-Cystine, time course studies revealed an induction of Nrf2 by 30 min (Figure 4A,B), which reached the peak at 4 h, before declining after 10 h (Figure 4C,D).

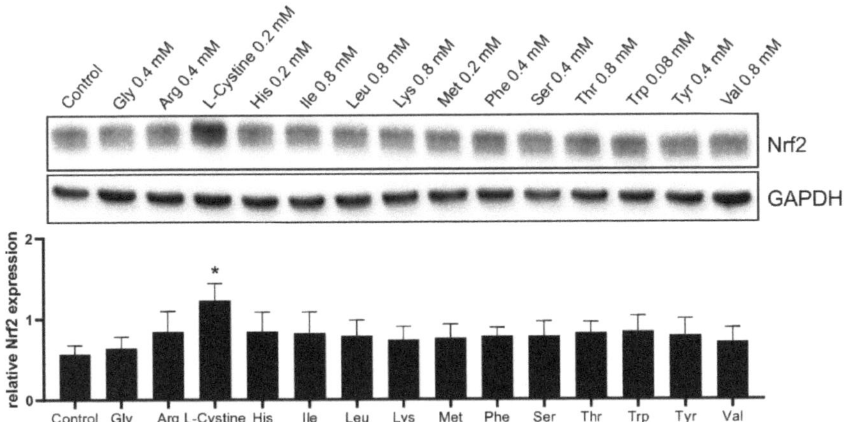

Figure 2. L-Cystine Induces Nrf2 Protein Elevation. The amino acid at its corresponding concentration in DMEM was added to HeLa cells at 3 days after subculture for 4 h of incubation, before harvesting for measuring Nrf2 protein levels by the Western Blot, using GAPDH as a loading control. The band intensities were quantified by NIH ImageJ to be presented as means ± SD of Nrf2/GAPDH ratio from three independent experiments. * indicates statistically significant as the adjusted p value < 0.05 compared to the control by one-way ANOVA, corrected by Dunnett's multiple comparisons test.

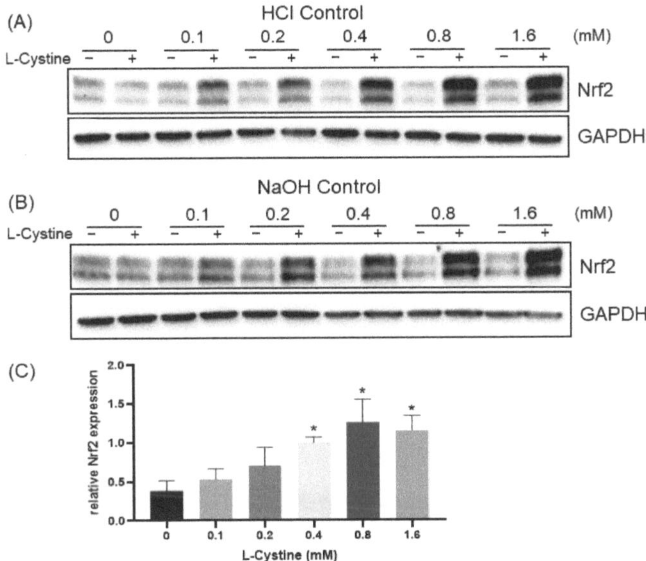

Figure 3. L-Cystine Dose Dependent Induction of Nrf2 Protein. L-Cystine was dissolved in either 500 mM HCl (**A,C**) or 200 mM NaOH (**B**) at a stock concentration of 0.1 M, before adding to HeLa cells on day 3 after subculture at the concentration indicated for 4 h of incubation. An equal volume of solvents was used for controls (**A,B**). Cells were harvested for the Western Blot to detect the levels of Nrf2 protein, using GAPDH as a loading control. The band intensities were quantified by NIH ImageJ to be presented as means ± SD of Nrf2/GAPDH ratio from three independent experiments (**C**). * indicates statistically significant compared to 0 mM L-Cystine as the adjusted p value < 0.05 by one-way ANOVA, corrected by Dunnett's multiple comparisons test (**C**).

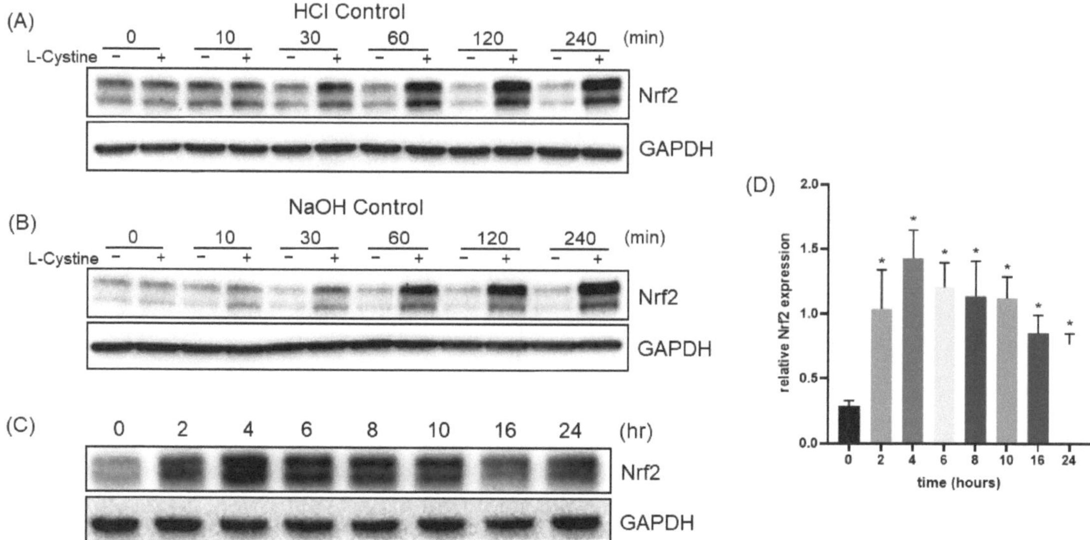

Figure 4. Time-Dependent Nrf2 Protein Induction by L-Cystine. HeLa cells at 3 days after subculture was treated with 0.8 mM L-Cystine from its stock dissolved in 500 mM HCL (**A,C,D**) or 200 mM NaOH (**B**) for indicated time points. An equal volume of solvents was added to the cells in the control groups. The cells were harvested at the end of the time point for Western Blot to detect the level of Nrf2 protein, using GAPDH as a loading control. The band intensities were quantified by NIH ImageJ to be presented as means ± SD of Nrf2/GAPDH ratio from three independent experiments (**D**). * indicates statistically significant compared to the 0-h time point as the adjusted p value < 0.05 by one-way ANOVA, corrected by Dunnett's multiple comparisons test (**D**).

RNA-sequencing technology was employed to determine the transcriptomic landscape change due to L-Cystine treatment. The next generation RNA-seq of DNBseq™ technology, provided by a commercial source (BGI Genomics, San Jose, CA), revealed that 314 genes were up-regulated and 181 genes were down regulated, due to the L-Cystine treatment. The volcano mapping of differentially expressed genes revealed prominent increases of several classical Nrf2 downstream genes, including HMOX1, GCLC, GCLM, SRXN1 and Aldo-keto Reductase 1 family member C1 (AKR1C1, Figure 5A). RNA-seq data revealed that Oxidative Stress-Induced Growth Inhibitor 1 (OSGIN1) had the highest level of induction. OSGIN1 has been reported as an Nrf2 downstream gene [31]. Gene ontology biological process enrichment analysis showed that the Oxidation-Reduction Process has the highest ratio of genes altered by L-Cystine treatment (Figure 5B). The heatmap showed that most of the genes under this category were Nrf2 targets (Figure 5C).

Induction of NQO1, HMOX1, GCLM, SRXN1, TXNRD1, AKR1C1 and OSGIN1 transcripts was validated by real-time RT-PCR in a time course of L-Cystine treatment. All of these genes showed elevated transcripts at 8–24 h (Figure 6A–F). Consistent with RNA-seq data, OSGIN1 had the highest level of elevation by L-Cystine treatment (Figure 6G). L-Cystine did not cause changes in Nrf2 or Keap1 mRNA as measured by quantitative RT-PCR (Figure 6H). These data support the concept that activation of Nrf2 serves as the main molecular event with L-Cystine treatment.

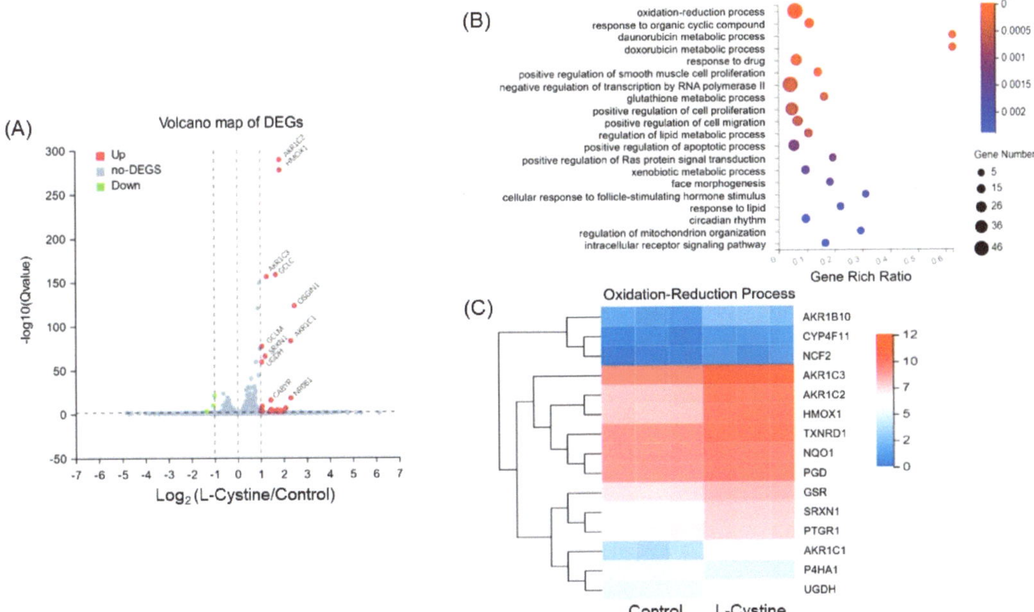

Figure 5. L-Cystine Induces Nrf2 Downstream Genes. HeLa cells at day 3 after subculture were treated with 0.8 mM L-Cystine for 16 h before harvesting total RNA for RNA-seq. (**A**) The volcano plot shows differentially expressed genes (DEGs) by L-Cystine treatment. Red dots indicate significantly up-regulated DEGs with $-\log_{10}$ (Q value) > 1.30 and $\log_2$ (fold change) > 1, which correspond to adjusted *p* value < 0.05 and fold change > 2. (**B**) Gene ontology (GO) for Biological Process Enrichment chart, showing the GO terms significantly changed by L-Cystine treatment. The X-axis is the enrichment ratio, calculated as the number of DEGs over the total number of the genes annotated to the GO term. The size of the dots reflects the ratio of DEGs annotated to the GO Term. The color represents the level of significance in Qvalue, with a threshold of ≤0.05. (**C**) Heatmap representing the variance and expression levels of up-regulated genes (log2 ≥ 0.8, adjusted *p* value ≤ 0.05) annotated in the oxidation-reduction process (GO:0055114) from the RNA-seq analyses. The data were obtained from RNA-seq of three independent sample sets.

Activation of the Nrf2 transcription factor was further confirmed using two different ARE reporters. We have obtained a luciferase reporter construct with the reporter gene under the control of two repeats of 41 nucleotide long promoter sequences from the mouse glutathione S-transferase Ya gene, each containing two copies of ARE sequence. ARE sequence is polymorphic. Therefore, we generated ARE reporter constructs with one copy of a 41-nucleotide-long sequence from the human NQO1 gene. Following transient transfection of the ARE reporter constructs, cells were treated with L-Cystine at various doses. With either form of the ARE reporter constructs, the detected activity of ARE is consistent with the dose-response of Nrf2 induction (Figure 7A). The level of ARE activation by L-Cystine exceeded or was comparable to that of SFN, a well-established widely used Nrf2 inducer (Figure 7A).

Induction of Nrf2 and its downstream targets by L-Cystine treatment was validated at the protein level using the Western blot. Figures 7B,C and 8 show L-Cystine dose- and treatment-time-dependent elevation of the Nrf2 downstream genes, HMOX1, SRXN1, TXNRD1 and AKR1C at the protein level. As expected, Nrf2 protein elevation preceded the induction of its downstream genes (Figure 8A).

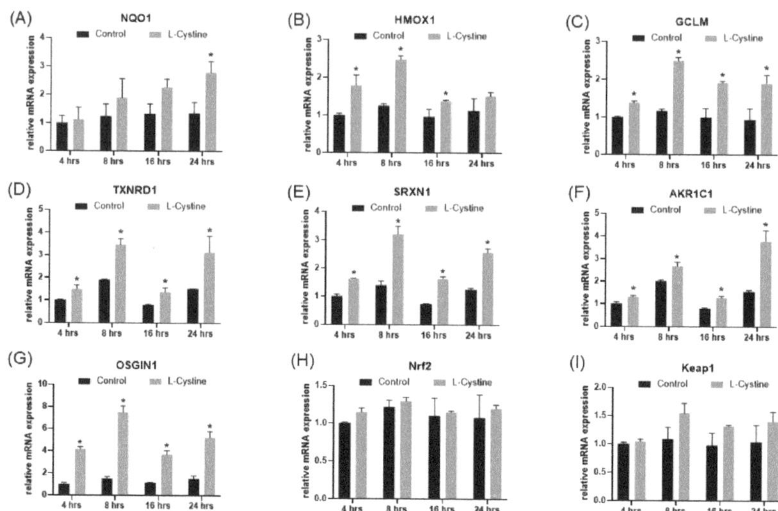

Figure 6. RT-PCR Measurements of Nrf2 Downstream Genes by L-Cystine Treatment. HeLa cells without or with 0.8 mM L-Cystine treatment were harvested at the indicated time for extraction of total RNA. The RNA (1 μg) was reversely transcribed into cDNA for qPCR to measure the levels of genes. The data represent means ± SD from triplicates for the abundance of the gene transcript over that of 18S rRNA (**A–I**). * indicates statistically significant compared to the control at the same time point as the adjusted *p* value < 0.05 by two-way ANOVA, corrected by Turkey's multiple comparisons test (**A–I**).

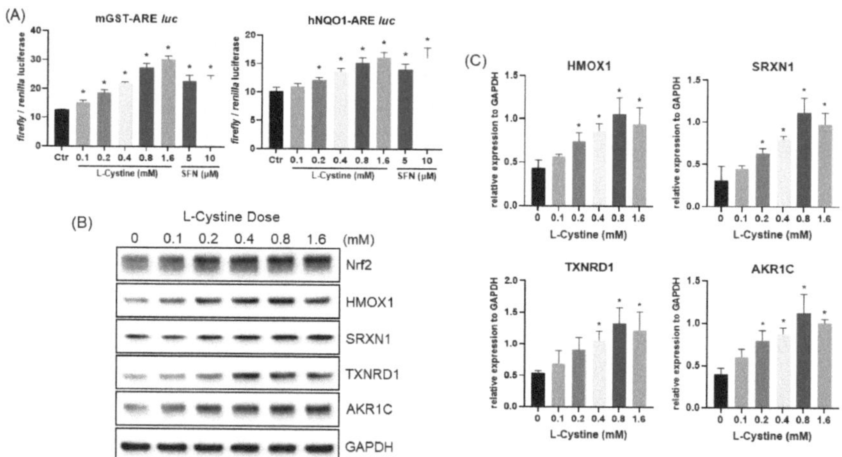

Figure 7. L-Cystine Dose Dependent Activation of ARE and Up-regulation of Nrf2 Downstream Genes at the Protein Level. HeLa cells were transfected with pGL4.37-mGST-ARE *luc* or pGL4.37-hNQO1-ARE *luc* vector together with pCMV of *Renilla* luciferase construct (**A**). At 48 h after transfection, cells were treated with various doses of L-Cystine or SFN as indicated for 16 h for measurements of luciferase activities. The data show means ± SD of the ratio of *Firefly/Renilla* luciferase from triplicate samples (**A**). Without transfection, HeLa cells were treated with L-Cystine at indicated doses for 16 h before harvesting to measure the levels of the proteins by Western Blot, using GAPDH as a loading control (**B**). Quantification of band intensities by NIH ImageJ was shown as means ± SD from three independent experiments (**C**). * indicates statistically significant compared to the control as the adjusted *p* value < 0.05 by one-way ANOVA, corrected by Dunnett's multiple comparisons test (**A,C**).

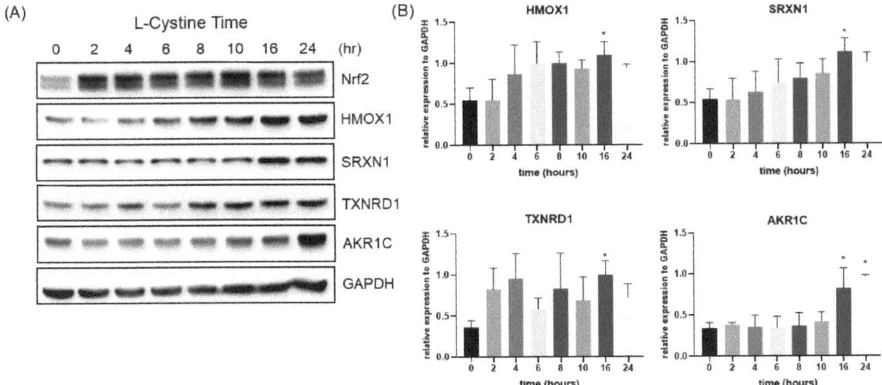

Figure 8. Time-Dependent Elevation of Nrf2 Downstream Genes at the Protein Level. HeLa cells were treated with 0.8 mM L-Cystine and harvested at indicated time points to measure the levels of the proteins by Western Blot, using GAPDH as a loading control (**A**). Quantification of band intensities by NIH ImageJ was shown as means ± SD from three independent experiments (**B**). * indicates statistically significant compared to the 0 h as the adjusted *p* value < 0.05 by one-way ANOVA, corrected by Dunnett's multiple comparisons test (**B**).

3.2. Mechanism of L-Cystine Induced Nrf2 Protein Elevation

L-Cystine is a disulfide conjugate of two L-Cysteine molecules and enters into cells through a cystine/glutamate antiporter encoded by the SLC7A11 gene [32]. Two inhibitors of the antiporter, erastin and sulfasalazine [33], were used to verify the dependency of the cellular entry for Nrf2 induction. Prevention of Nrf2 induction by either inhibitor confirmed that L-Cystine indeed entered into cells to induce Nrf2 (Figure 9A). Upon entering cells, L-Cystine is reduced to L-Cysteine. We treated cells in parallel, either with L-Cysteine or L-Cystine. The results showed that L-Cysteine was incapable of inducing Nrf2 (Figure 9B).

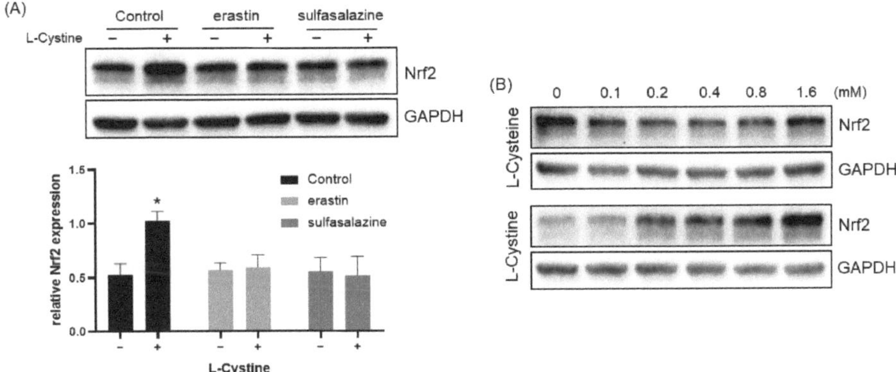

Figure 9. Inhibitors of Cystine/Glutamate Antiporter Block L-Cystine from Inducing Nrf2 Protein. HeLa cells at 2 days after subculture were treated with 5 μM erastin or 500 μM sulfasalazine for 24 h before treatment with 0.8 mM L-Cystine for 4 h (**A**). At day 3 after subculture, HeLa cells were treated with L-Cysteine or L-Cystine in parallel at the indicated doses for 4 h (**B**). Cells were harvested to measure Nrf2 protein levels by Western Blot, using GAPDH as a loading control. Quantification of Nrf2 over GAPDH protein band intensities by NIH ImageJ is shown as means ± SD from three independent experiments. * indicates statistically significant from the control group without L-Cystine treatment as the adjusted *p* value < 0.05 compared to the control by two-way ANOVA, corrected by Turkey's multiple comparisons test (**A**).

Keap1 is known as the main modulator of Nrf2 protein stability due to its binding to Nrf2 and recruitment of ubiquitin E3 ligase for Nrf2 ubiquitination. The Nrf2 protein stability measurements showed that L-Cystine treatment caused an increase in Nrf2 protein half-life from 19.4 min to 30.9 min (Figure 10A,B). The ubiquitylation assay revealed that similar to SFN, L-Cystine treatment decreased Nrf2 protein ubiquitylation (Figure 10C). Using CRISPR-Cas9 to knock out the Keap1 gene, we tested whether this abolished Nrf2 induction by L-Cystine. Among five clones of Keap1 complete knockout cells, KO1, KO2, KO3, KO4 and KO5, L-Cystine was not capable of inducing Nrf2 in four clones, KO1, KO2, KO3 and KO5 (Figure 11A,B). A summary of all five clones supports the lack of Nrf2 induction by Keap1 knockout (Figure 11C). This suggests that Nrf2 induction by L-Cystine is Keap1-dependent.

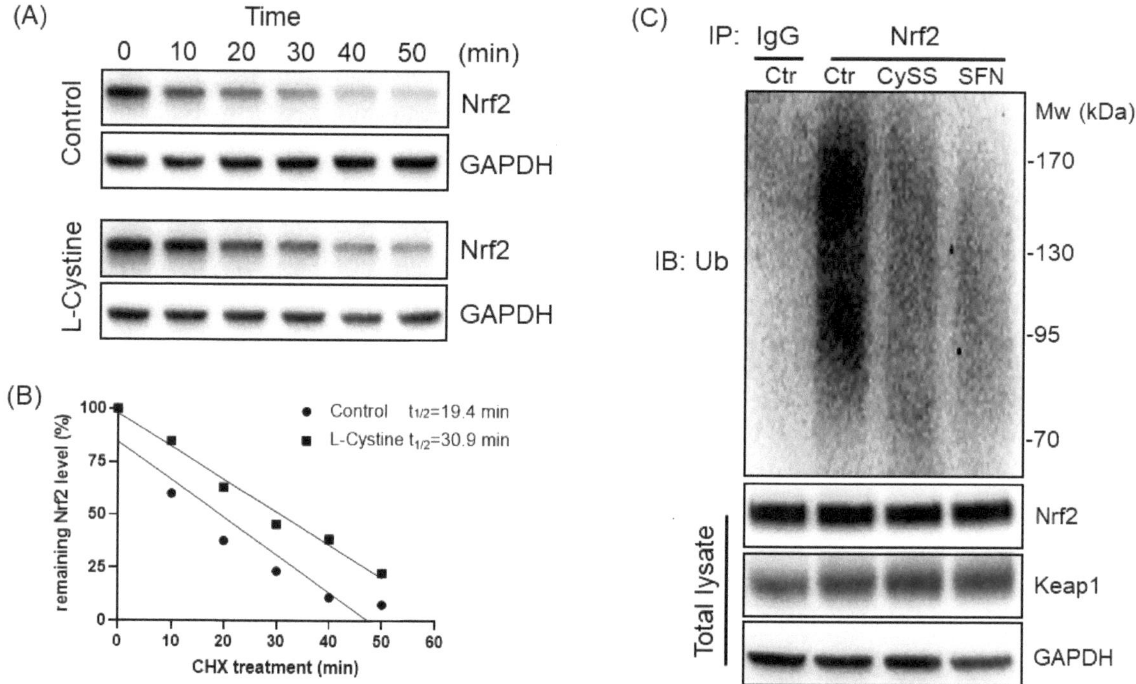

Figure 10. L-Cystine Extends the Half-Life of the Nrf2 Protein and decreases Nrf2 ubiquitylation. HeLa cells were treated with 0 (control) or 0.8 mM L-Cystine for 4 h followed by treatment of 100 µg/mL cycloheximide (CHX) for the indicated time. Cells were harvested at the end of the time points for detection of Nrf2 protein by Western Blot, using GAPDH as a loading control (**A**). The relative Nrf2 expression was plotted over time after adding CHX for the calculation of the half-life by linear regression (**B**). HeLa cells were treated with 0.8 mM L-Cystine (CySS) or 5 µM SFN in the presence of 10 µM MG132 for 4 h. Nrf2 protein was immunoprecipitated from cell lysate with an anti-Nrf2 antibody and resolved in 5% SDS-page gel for detection of ubiquitin by immunoblotting (**C**). Endogenous Nrf2, Keap1 and GAPDH from MG132-treated cells were included as a loading control (**C**).

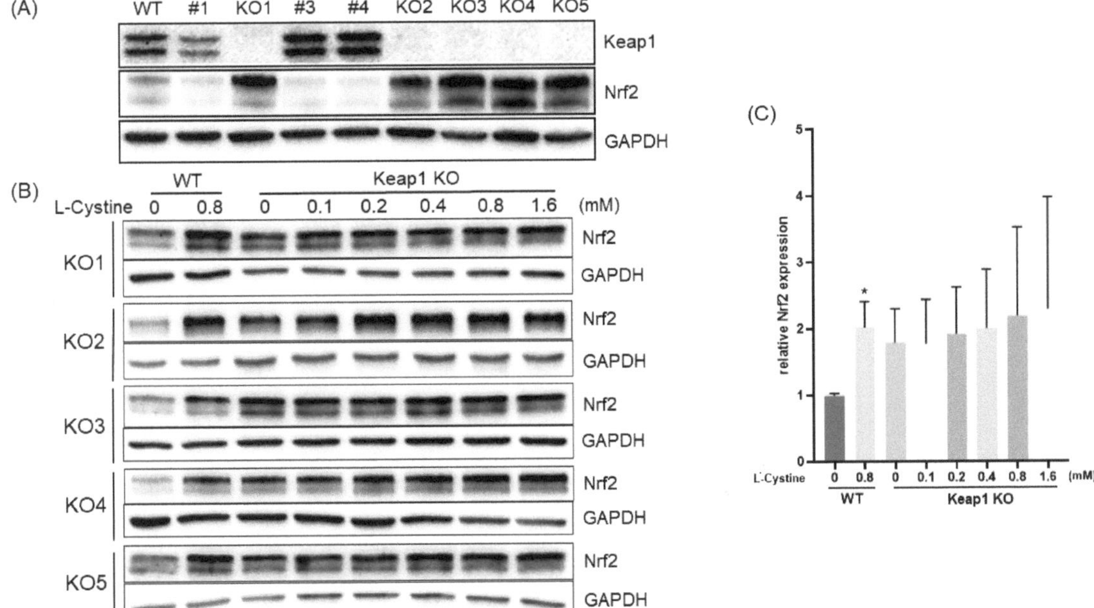

Figure 11. Keap1 Dependent Nrf2 Protein Induction by L-Cystine. HeLa cells were transfected with Keap1 CRISPR/Cas9 KO Plasmid for clonal selection. The clones with Keap1 knockout display a loss of Keap1 protein and elevated Nrf2 protein (**A**). The parent wild-type or Keap1 KO clones were treated with indicated doses of L-Cystine for 4 h before harvesting to measure the induction of Nrf2 protein by Western Blot, using GAPDH as a loading control (**B**). The intensities of the bands from wild-type or five clones of Keap1 KO cells were quantified using NIH ImageJ (**C**). * indicates statistically significant compared to 0 mM L-Cystine in WT cells by one-way ANOVA, corrected by Dunnett's multiple comparisons test (**C**).

3.3. L-Cystine Protects against ROS Generation and Cell Death

To demonstrate that induction of Nrf2 by L-Cystine serves as a cytoprotective mechanism, we tested the inhibition of ROS generation and the anti-apoptotic capacity of L-Cystine. We found that L-Cystine reduced ROS generation by H_2O_2 (Figure 12A) and suppressed H_2O_2 from inducting caspase cleavage (Figure 12B,C). Doxorubicin is a chemotherapeutic drug that induces apoptosis [30]. HeLa cells lost approximately 60% of viability as measured by MTT assay following 24 h of exposure to 2 μM doxorubicin (Figure 13A). L-Cystine at 0.4–1.6 mM showed some protection against a loss of cell viability (Figure 13A). When caspase activity was assayed as a quantitative measure of apoptosis, inhibition of caspase activation was observed with increasing doses of L-Cystine, starting at 0.1 mM (Figure 13B). Measurements of caspase-3 cleavage confirmed the protective effect of L-Cystine (Figure 13C,D). The dependency of Nrf2 for L-Cystine-induced cytoprotection was confirmed using a siRNA to knock down Nrf2 (Figure 13E). These lines of data support that L-Cystine induces cytoprotection via Nrf2.

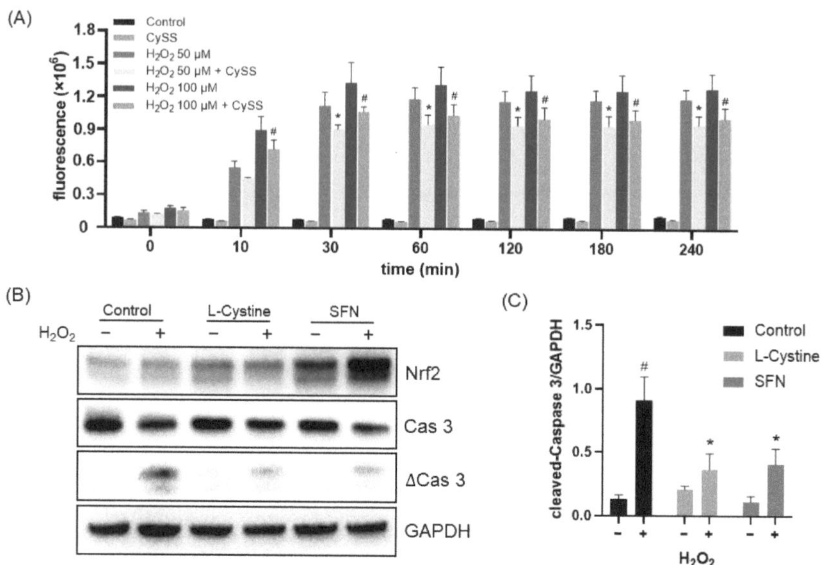

Figure 12. L-Cystine Reduces ROS Generation and Protects Against Oxidant-Induced Cell Death. HeLa cells were pre-treated with 0.8 mM L-Cystine (CySS) for 16 h before 30-min incubation with 2′,7′-dichlorofluorescein diacetate (DCFDA, **A**). H_2O_2 (50, 100 μM) was added to cells to induce hydroxyl radical formation (**A**). The fluorescence was recorded by a microplate reader at the time point indicated (**A**). To induce apoptosis, following 16 h of pretreatment with 0.8 mM L-Cystine or 5 μM SFN, the cells were treated with 1 mM H_2O_2 for 6 h before harvesting for Western blot (**B,C**). The data are shown as means ± SD from triplicates (**A**) or from three independent experiments by quantification of band intensities with NIH ImageJ (**C**). * or # indicates statistically significant compared to 50 or 100 μM H_2O_2 treatment alone, respectively, at the same time point (**A**), or compared to H_2O_2 treatment alone or to control without H_2O_2 treatment, respectively (**C**), by two-way ANOVA, corrected by Turkey's multiple comparisons test (**A,C**).

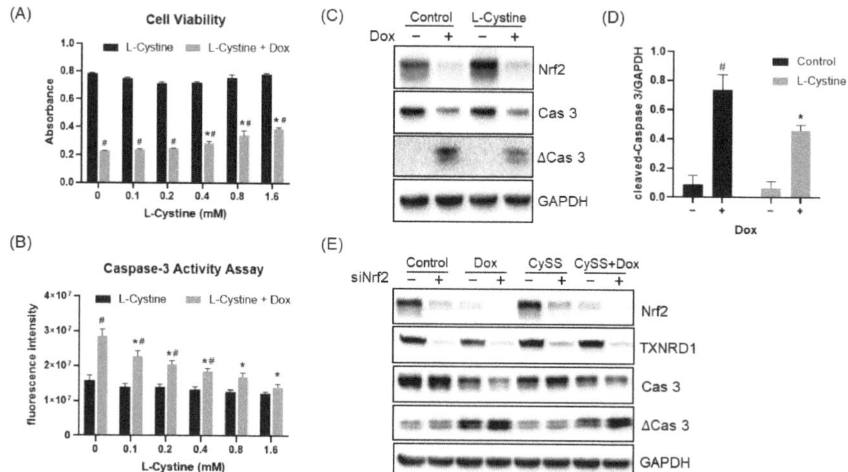

Figure 13. L-Cystine Protects Against Apoptosis. HeLa cells without transfection (**A–D**) or with transfection of control or Nrf2 siRNA for 48 h (**E**) were pre-treated with indicated doses of L-Cystine (**A,B**) or 0.8 mM L-Cystine (**C–E**) for 8 h followed by 2 μM doxorubicin (Dox) treatment for 24 h. Cells

were collected for measurements of cell viability using MTT assay (**A**), apoptosis using caspase assay with Ac-DEVD-AMC as the substrate (**B**), caspase-3 cleavage by Western blot, using GAPDH as a loading control (**C–E**). Nrf2 protein and its downstream gene TXNRD1-encoded protein were measured to demonstrate the efficacy of siRNA (**D,E**). The data are presented as means ± SD from triplicates. * or # indicates statistically significant compared to Dox treatment alone, or the same dose of L-Cystine without Dox treatment, respectively, as the adjusted *p* value < 0.05 by two-way ANOVA, corrected by Turkey's multiple comparisons test (**A,B,D**).

4. Discussion

This study has demonstrated that L-Cystine induces the elevation of Nrf2 protein and activates the Nrf2 transcription factor at the concentration of 0.2 mM as in the culture medium. The concentration of L-Cystine higher than that in the culture medium, e.g., 0.8–1.6 mM, induces a profound Nrf2 activation. L-Cystine-induced Nrf2 activation has been validated by RNA-seq, which demonstrated the oxidation-reduction process as a main event of L-Cystine exposure. Activation of the Nrf2 transcriptional cascade appears to be Keap1-dependent, as the Keap1 gene knockout clones failed to respond to L-Cystine for elevating Nrf2 protein. Since L-Cystine is considered a non-toxic amino acid derivative, its activation of Nrf2 provides a means for cytoprotection without concern for toxicity.

L-Cystine is a disulfide amino acid derivative, a dimer of L-Cysteine, that benefits the survival and proliferation of many cell types in cultures [34]. The reduced form of L-Cystine, i.e., L-Cysteine, is an essential amino acid capable of undergoing reversible oxidation under biological conditions, therefore serving for redox signaling and cellular control [35]. L-Cysteine enters into cells via an amino acid transporter [36]. In contrast, L-Cystine enters cells by a cystine/glutamate antiporter. Upon entering into cells, L-Cystine is reduced to L-Cysteine by cystine reductase or glutathione and thioredoxin, thereby providing the amino acid L-Cysteine essential for protein synthesis and glutathione production. The redox cycling of L-Cystine to L-Cysteine can result in protein oxidation without additional oxidants [35,37]. Because Keap1 is rich in cysteine residues, Keap1 dependency on Nrf2 induction by L-Cystine points to Keap1 oxidation as the main mechanism.

Keap1 oxidation or alkylation serves as a common molecular event leading to Nrf2 activation by small molecule Nrf2 inducers [3,5,9–11]. A well-established Nrf2 inducer, SFN, modifies Keap1 at cysteine residue 151 by thiocarbamylation [38,39]. The nature of SFN as an electrophile contributes to its toxicity and eliminates its clinical application, despite SFN's popularity in cell culture and animal experiments. Two categories of Nrf2 inducers that currently have ongoing clinical trials are dimethyl fumarate and bardoxolone methyl (CDDO-Me, bardoxolone) or the related compound omaveloxolone [4,19,20]. These Nrf2 inducers are electrophiles that alkylate Keap1 at Cys 151/273/288, disabling the Keap1 checkpoint on Nrf2 [40,41]. This feature poses the possibility of off-target effects or toxicity. A Phase III clinical trial of bardoxolone for the treatment of late-stage renal failure was terminated early due to bardoxolone toxicity and increased mortality [42]. This latter result points to the urgency of developing Nrf2 inducers without safety concerns for humans.

L-Cystine fits well into the criteria for a non-toxic Nrf2 inducer. The potential limitation of utilizing L-Cystine for cytoprotection is its dependency on the presence of the cystine/glutamate antiporter for cellular uptake of L-Cystine. Many cell types contain the cystine/glutamate antiporter, a heterodimer composed of the xCT subunit of the transporter encoded by the SLC7A11 gene and ancillary 4F2hc subunit for plasma membrane localization encoded by the SLC3A2 gene [34]. Many types of cancer cells have an elevated expression of SLC7A11 and SLC3A2, a feature benefiting L-Cystine uptake and high metabolic activity for tumor growth [32,43,44]. However, not all cell types express the cystine/glutamate antiporter [34]. With cells in culture, fibroblasts, epithelial cells and macrophages are known to be efficient at L-Cystine uptake, whereas lymphocytes and stem cells are incapable of L-Cystine uptake and are therefore dependent on feeder cells to provide a source of L-Cysteine [34]. Transcriptomic profiling of gene expression among different human tissues showed higher SLC7A11 expression in the brain to a level comparable to that of cultured fibroblasts [45] (https://www.gtexportal.org/home/gene/SLC7A11 (accessed on 8 December 2022)). Proteomic profiling has revealed high levels of the cystine/glutamate transporter protein in adrenal glands, prefrontal cortex, brain, lung, heart and spinal cord [46,47]. These lines of evidence support the potential use of L-Cystine as a cytoprotective agent in those tissues.

Author Contributions: W.D. performed all experiments, generated the data and wrote the manuscript. Q.M.C. initiated the concept of a fresh culture medium for Nrf2 induction, supervised the work, wrote and edited the manuscript. The authors consent to the authorship and statements in the manuscript. All authors have read and agreed to the published version of the manuscript.

Funding: The research programs under the direction of Qin M. Chen have been supported by National Institutes of Health (NIH) grants (R01 GM125212, R01 GM126165), Holsclaw Endowment and the University of Arizona, College of Pharmacy start-up fund.

Institutional Review Board Statement: Not applicable.

Informed Consent Statement: Not applicable.

Data Availability Statement: The data supporting reported results are included in the article and the raw RNA-seq data are available upon request.

Acknowledgments: The authors would like to thank Benjamin N. Yuchen and Joseph S. Alpert for editing the manuscript. The graphical abstract was generated in part using Servier Medical Art under a Creative Commons Attribution 3.0 unported license.

Conflicts of Interest: The authors declare no conflict of interest.

Abbreviations

AKR1C1	Aldo-keto Reductase 1 family member C1
CySS	L-Cystine
GAPDH	Glyceraldehyde-3-Phosphate Dehydrogenase
GCLC	Glutamate-cysteine Ligase catalytic subunit
GCLM	Glutamate-cysteine Ligase regulatory subunit
HMOX1	Heme Oxygenase 1
NQO1	NAD(P)H Quinone Oxidoreductase 1
OSGIN1	Oxidative Stress-Induced Growth Inhibitor 1
PVDF	polyvinylidene fluoride membrane
SRXN1	Sulfiredoxin 1
b-TrCP	b-transducin repeat-containing protein
TXNRD1	thioredoxin reductase 1

References

1. Moi, P.; Chan, K.; Asunis, I.; Cao, A.; Kan, Y.W. Isolation of NF-E2-related factor 2 (Nrf2), a NF-E2-like basic leucine zipper transcriptional activator that binds to the tandem NF-E2/AP1 repeat of the beta-globin locus control region. *Proc. Natl. Acad. Sci. USA* **1994**, *91*, 9926–9930. [CrossRef] [PubMed]
2. Hayes, J.D.; Dinkova-Kostova, A.T. The Nrf2 regulatory network provides an interface between redox and intermediary metabolism. *Trends Biochem. Sci.* **2014**, *39*, 199–218. [CrossRef] [PubMed]
3. Yamamoto, M.; Kensler, T.W.; Motohashi, H. The KEAP1-NRF2 System: A Thiol-Based Sensor-Effector Apparatus for Maintaining Redox Homeostasis. *Physiol. Rev.* **2018**, *98*, 1169–1203. [CrossRef]
4. Cuadrado, A.; Rojo, A.I.; Wells, G.; Hayes, J.D.; Cousin, S.P.; Rumsey, W.L.; Attucks, O.C.; Franklin, S.; Levonen, A.L.; Kensler, T.W.; et al. Therapeutic targeting of the NRF2 and KEAP1 partnership in chronic diseases. *Nat. Rev. Drug Discov.* **2019**, *18*, 295–317. [CrossRef] [PubMed]
5. Chen, Q.M. Nrf2 for cardiac protection: Pharmacological options against oxidative stress. *Trends Pharm. Sci.* **2021**, *42*, 729–744. [CrossRef] [PubMed]
6. Chen, Q.M. Nrf2 for protection against oxidant generation and mitochondrial damage in cardiac injury. *Free Radic. Biol. Med.* **2022**, *179*, 133–143. [CrossRef] [PubMed]

7.   Nguyen, T.; Nioi, P.; Pickett, C.B. The Nrf2-antioxidant response element signaling pathway and its activation by oxidative stress. *J. Biol. Chem.* **2009**, *284*, 13291–13295. [CrossRef]
8.   Raghunath, A.; Sundarraj, K.; Nagarajan, R.; Arfuso, F.; Bian, J.; Kumar, A.P.; Sethi, G.; Perumal, E. Antioxidant response elements: Discovery, classes, regulation and potential applications. *Redox Biol.* **2018**, *17*, 297–314. [CrossRef]
9.   Ma, Q.; He, X. Molecular basis of electrophilic and oxidative defense: Promises and perils of Nrf2. *Pharmacol. Rev.* **2012**, *64*, 1055–1081. [CrossRef]
10. Rojo de la Vega, M.; Chapman, E.; Zhang, D.D. NRF2 and the Hallmarks of Cancer. *Cancer Cell* **2018**, *34*, 21–43. [CrossRef]
11. Baird, L.; Yamamoto, M. The Molecular Mechanisms Regulating the KEAP1-NRF2 Pathway. *Mol. Cell. Biol.* **2020**, *40*, e00099-20. [CrossRef] [PubMed]
12. Wu, T.; Zhao, F.; Gao, B.; Tan, C.; Yagishita, N.; Nakajima, T.; Wong, P.K.; Chapman, E.; Fang, D.; Zhang, D.D. Hrd1 suppresses Nrf2-mediated cellular protection during liver cirrhosis. *Genes Dev.* **2014**, *28*, 708–722. [CrossRef] [PubMed]
13. Purdom-Dickinson, S.E.; Sheveleva, E.V.; Sun, H.; Chen, Q.M. Translational Control of Nrf2 Protein in Activation of Antioxidant Response Element by Oxidants. *Mol. Pharm.* **2007**, *72*, 1074–1081. [CrossRef] [PubMed]
14. Zhang, J.; Dinh, T.N.; Kappeler, K.; Tsaprailis, G.; Chen, Q.M. La autoantigen mediates oxidant induced de novo Nrf2 protein translation. *Mol. Cell. Proteom.* **2012**, *11*, M111.015032. [CrossRef]
15. Xu, B.; Zhang, J.; Strom, J.; Lee, S.; Chen, Q.M. Myocardial ischemic reperfusion induces de novo Nrf2 protein translation. *Biochim. Biophys. Acta* **2014**, *1842*, 1638–1647. [CrossRef]
16. Lee, S.C.; Zhang, J.; Strom, J.; Yang, D.; Dinh, T.N.; Kappeler, K.; Chen, Q.M. G-Quadruplex in the NRF2 mRNA 5′ Untranslated Region Regulates De Novo NRF2 Protein Translation under Oxidative Stress. *Mol. Cell. Biol.* **2017**, *37*, e00122-16. [CrossRef]
17. Dai, W.; Qu, H.; Zhang, J.; Thongkum, A.; Dinh, T.N.; Kappeler, K.V.; Chen, Q.M. Far Upstream Binding Protein 1 (FUBP1) participates in translational regulation of Nrf2 protein under oxidative stress. *Redox Biol.* **2021**, *41*, 101906. [CrossRef]
18. Li, W.; Thakor, N.; Xu, E.Y.; Huang, Y.; Chen, C.; Yu, R.; Holcik, M.; Kong, A.N. An internal ribosomal entry site mediates redox-sensitive translation of Nrf2. *Nucleic Acids Res.* **2010**, *38*, 778–788. [CrossRef]
19. Robledinos-Anton, N.; Fernandez-Gines, R.; Manda, G.; Cuadrado, A. Activators and Inhibitors of NRF2: A Review of Their Potential for Clinical Development. *Oxid. Med. Cell. Longev.* **2019**, *2019*, 9372182. [CrossRef]
20. Staurengo-Ferrari, L.; Badaro-Garcia, S.; Hohmann, M.S.N.; Manchope, M.F.; Zaninelli, T.H.; Casagrande, R.; Verri, W.A., Jr. Contribution of Nrf2 Modulation to the Mechanism of Action of Analgesic and Anti-inflammatory Drugs in Pre-clinical and Clinical Stages. *Front. Pharmacol.* **2018**, *9*, 1536. [CrossRef]
21. Jeong, W.S.; Jun, M.; Kong, A.N. Nrf2: A potential molecular target for cancer chemoprevention by natural compounds. *Antioxid. Redox Signal.* **2006**, *8*, 99–106. [CrossRef] [PubMed]
22. Chen, Q.M.; Maltagliati, A.J. Nrf2 at the heart of oxidative stress and cardiac protection. *Physiol. Genom.* **2018**, *50*, 77–97. [CrossRef]
23. Ilatovskaya, D.V.; Pavlov, T.S.; Levchenko, V.; Staruschenko, A. ROS production as a common mechanism of ENaC regulation by EGF, insulin, and IGF-1. *Am. J. Physiol. Cell Physiol.* **2013**, *304*, C102–C111. [CrossRef] [PubMed]
24. Yazaki, K.; Matsuno, Y.; Yoshida, K.; Sherpa, M.; Nakajima, M.; Matsuyama, M.; Kiwamoto, T.; Morishima, Y.; Ishii, Y.; Hizawa, N. ROS-Nrf2 pathway mediates the development of TGF-beta1-induced epithelial-mesenchymal transition through the activation of Notch signaling. *Eur. J. Cell Biol.* **2021**, *100*, 151181. [CrossRef]
25. Zhuang, J.; Jiang, T.; Lu, D.; Luo, Y.; Zheng, C.; Feng, J.; Yang, D.; Chen, C.; Yan, X. NADPH oxidase 4 mediates reactive oxygen species induction of CD146 dimerization in VEGF signal transduction. *Free Radic. Biol. Med.* **2010**, *49*, 227–236. [CrossRef]
26. Cock, P.J.; Fields, C.J.; Goto, N.; Heuer, M.L.; Rice, P.M. The Sanger FASTQ file format for sequences with quality scores, and the Solexa/Illumina FASTQ variants. *Nucleic Acids Res.* **2010**, *38*, 1767–1771. [CrossRef]
27. Kim, D.; Langmead, B.; Salzberg, S.L. HISAT: A fast spliced aligner with low memory requirements. *Nat. Methods* **2015**, *12*, 357–360. [CrossRef] [PubMed]
28. Langmead, B.; Salzberg, S.L. Fast gapped-read alignment with Bowtie 2. *Nat. Methods* **2012**, *9*, 357–359. [CrossRef] [PubMed]
29. Li, B.; Dewey, C.N. RSEM: Accurate transcript quantification from RNA-Seq data with or without a reference genome. *BMC Bioinform.* **2011**, *12*, 323. [CrossRef]
30. Chen, Q.M.; Alexander, D.; Sun, H.; Xie, L.; Lin, Y.; Terrand, J.; Morrissy, S.; Purdom, S. Corticosteroids inhibit cell death induced by doxorubicin in cardiomyocytes: Induction of antiapoptosis, antioxidant, and detoxification genes. *Mol. Pharmacol.* **2005**, *67*, 1861–1873. [CrossRef]
31. Brennan, M.S.; Matos, M.F.; Richter, K.E.; Li, B.; Scannevin, R.H. The NRF2 transcriptional target, OSGIN1, contributes to monomethyl fumarate-mediated cytoprotection in human astrocytes. *Sci. Rep.* **2017**, *7*, 42054. [CrossRef] [PubMed]
32. Lewerenz, J.; Hewett, S.J.; Huang, Y.; Lambros, M.; Gout, P.W.; Kalivas, P.W.; Massie, A.; Smolders, I.; Methner, A.; Pergande, M.; et al. The cystine/glutamate antiporter system x(c)(-) in health and disease: From molecular mechanisms to novel therapeutic opportunities. *Antioxid. Redox Signal.* **2013**, *18*, 522–555. [CrossRef] [PubMed]
33. Patel, D.; Kharkar, P.S.; Gandhi, N.S.; Kaur, E.; Dutt, S.; Nandave, M. Novel analogs of sulfasalazine as system xc (-) antiporter inhibitors: Insights from the molecular modeling studies. *Drug Dev. Res.* **2019**, *80*, 758–777. [CrossRef] [PubMed]
34. Ishii, T.; Mann, G.E. Redox status in mammalian cells and stem cells during culture in vitro: Critical roles of Nrf2 and cystine transporter activity in the maintenance of redox balance. *Redox Biol.* **2014**, *2*, 786–794. [CrossRef]

35. Jones, D.P.; Go, Y.M.; Anderson, C.L.; Ziegler, T.R.; Kinkade, J.M., Jr.; Kirlin, W.G. Cysteine/cystine couple is a newly recognized node in the circuitry for biologic redox signaling and control. *FASEB J.* **2004**, *18*, 1246–1248. [CrossRef]
36. Knickelbein, R.G.; Seres, T.; Lam, G.; Johnston, R.B., Jr.; Warshaw, J.B. Characterization of multiple cysteine and cystine transporters in rat alveolar type II cells. *Am. J. Physiol.* **1997**, *273*, L1147–L1155. [CrossRef]
37. Ghezzi, P. Oxidoreduction of protein thiols in redox regulation. *Biochem. Soc. Trans.* **2005**, *33*, 1378–1381. [CrossRef]
38. Zhang, D.D.; Lo, S.C.; Cross, J.V.; Templeton, D.J.; Hannink, M. Keap1 is a redox-regulated substrate adaptor protein for a Cul3-dependent ubiquitin ligase complex. *Mol. Cell. Biol.* **2004**, *24*, 10941–10953. [CrossRef]
39. Hu, C.; Eggler, A.L.; Mesecar, A.D.; van Breemen, R.B. Modification of keap1 cysteine residues by sulforaphane. *Chem. Res. Toxicol.* **2011**, *24*, 515–521. [CrossRef]
40. Liby, K.; Hock, T.; Yore, M.M.; Suh, N.; Place, A.E.; Risingsong, R.; Williams, C.R.; Royce, D.B.; Honda, T.; Honda, Y.; et al. The synthetic triterpenoids, CDDO and CDDO-imidazolide, are potent inducers of heme oxygenase-1 and Nrf2/ARE signaling. *Cancer Res.* **2005**, *65*, 4789–4798. [CrossRef]
41. Suzuki, T.; Muramatsu, A.; Saito, R.; Iso, T.; Shibata, T.; Kuwata, K.; Kawaguchi, S.I.; Iwawaki, T.; Adachi, S.; Suda, H.; et al. Molecular Mechanism of Cellular Oxidative Stress Sensing by Keap1. *Cell Rep.* **2019**, *28*, 746–758. [CrossRef] [PubMed]
42. de Zeeuw, D.; Akizawa, T.; Audhya, P.; Bakris, G.L.; Chin, M.; Christ-Schmidt, H.; Goldsberry, A.; Houser, M.; Krauth, M.; Lambers Heerspink, H.J.; et al. Bardoxolone methyl in type 2 diabetes and stage 4 chronic kidney disease. *N. Engl. J. Med.* **2013**, *369*, 2492–2503. [CrossRef]
43. Koppula, P.; Zhuang, L.; Gan, B. Cystine transporter SLC7A11/xCT in cancer: Ferroptosis, nutrient dependency, and cancer therapy. *Protein Cell* **2021**, *12*, 599–620. [CrossRef] [PubMed]
44. Liang, J.; Sun, Z. Overexpression of membranal SLC3A2 regulates the proliferation of oral squamous cancer cells and affects the prognosis of oral cancer patients. *J. Oral Pathol. Med.* **2021**, *50*, 371–377. [CrossRef] [PubMed]
45. GTEx Consortium. The GTEx Consortium atlas of genetic regulatory effects across human tissues. *Science* **2020**, *369*, 1318–1330. [CrossRef]
46. Lautenbacher, L.; Samaras, P.; Muller, J.; Grafberger, A.; Shraideh, M.; Rank, J.; Fuchs, S.T.; Schmidt, T.K.; The, M.; Dallago, C.; et al. ProteomicsDB: Toward a FAIR open-source resource for life-science research. *Nucleic Acids Res.* **2022**, *50*, D1541–D1552. [CrossRef]
47. Samaras, P.; Schmidt, T.; Frejno, M.; Gessulat, S.; Reinecke, M.; Jarzab, A.; Zecha, J.; Mergner, J.; Giansanti, P.; Ehrlich, H.C.; et al. ProteomicsDB: A multi-omics and multi-organism resource for life science research. *Nucleic Acids Res.* **2020**, *48*, D1153–D1163. [CrossRef]

Article

Evaluation of Malondialdehyde Levels, Oxidative Stress and Host–Bacteria Interactions: *Escherichia coli* and *Salmonella* Derby

Vardan Tsaturyan [1], Armen Poghosyan [2], Michał Toczyłowski [3] and Astghik Pepoyan [4],*

1   Military Therapy Department, Yerevan State Medical University, Yerevan 0025, Armenia
2   The International Scientific-Educational Center of NAS RA, Yerevan 0019, Armenia
3   Food Safety and Biotechnology Department, Scientific Research Institute of Food Science and Biotechnology, Armenian National Agrarian University, Yerevan 0009, Armenia
4   Scientific Research Institute of Food Science and Biotechnology, Armenian National Agrarian University, Yerevan 0009, Armenia
*   Correspondence: apepoyan@gmail.com; Tel.: +374-9-432490

Abstract: Either extracts, cell-free suspensions or bacterial suspensions are used to study bacterial lipid peroxidation processes. Along with gas chromatography-mass spectrometry, liquid chromatography-mass spectrometry, and several other strategies, the thiobarbituric acid test is used for the determination of malondialdehyde (MDA) as the basis for the commercial test kits and the colorimetric detection of lipid peroxidation. The aim of the current study was to evaluate lipid peroxidation processes levels in the suspensions, extracts and culture supernatants of *Escherichia coli* and *Salmonella* Derby strains. The dependence of the formation of thiobarbituric acid-reactive substances levels in the cell extracts, the suspensions and cell-free supernatants on bacterial species, and their concentration and growth phase were revealed. The effect of bacterial concentrations on MDA formation was also found to be more pronounced in bacterial suspensions than in extracts, probably due to the dynamics of MDA release into the intercellular space. This study highlights the possible importance of MDA determination in both cell-free suspensions and extracts, as well as in bacterial suspensions to elucidate the role of lipid peroxidation processes in bacterial physiology, bacteria–host interactions, as well as in host physiology.

Keywords: oxidative stress; malondialdehyde; intercellular interactions; probiotic; commensal *E. coli*; host–bacteria interaction

Citation: Tsaturyan, V.; Poghosyan, A.; Toczyłowski, M.; Pepoyan, A. Evaluation of Malondialdehyde Levels, Oxidative Stress and Host–Bacteria Interactions: *Escherichia coli* and *Salmonella* Derby. Cells **2022**, *11*, 2989. https://doi.org/10.3390/cells11192989

Academic Editors: Alessia Remigante and Rossana Morabito

Received: 10 September 2022
Accepted: 23 September 2022
Published: 26 September 2022

Publisher's Note: MDPI stays neutral with regard to jurisdictional claims in published maps and institutional affiliations.

1. Introduction

The interdependence between the lipid peroxidation processes (LPP) and the composition of lipids can be considered both as a physicochemical system of regulation and as one of the normal forms of renewal of the composition of lipids of bacterial cell membranes [1,2]. However, prolonged non-enzymatic free radical oxidation of lipids leads to a sharp disruption of the physicochemical structure of membranes. This, in particular, relates to permeability, stability of lipid–protein complexes, as well as inactivation of lipid-dependent membrane-bound enzymes [3–5]. Detecting the dynamic variations of oxidative stress is informative for the clarification of its impact in the basic cellular processes and for its regulation [6]. One of the most popular markers for the assessment of oxidative stress is malondialdehyde (MDA) [7], the endogenous genotoxic product for both enzymatic and non-enzymatic LPP [8,9]. The different approaches are used for the sample preparation to monitor the concentrations of MDA for bacteria, including homogenates of fresh [8] and freeze-dried bacterial cultures [10] and bacterial culture broths [6]. Different mechanical pressure, acoustic, temperature and chemical methods for the extract preparation for bacteria are also used [11], and the mechanical methods are evaluated as one of the appropriate methods for oxidative stress monitoring among these.

A diverse community of large numbers of commensal bacteria from the human and animals' mucosal and epidermal surfaces plays a crucial role in host life [12–16]. To attempt to understand the immunologic impact of individual commensal species within the microbiota, Brown and co-authors systematically profiled the immunologic fingerprint of commensals from the human intestinal major phyla, showing that *Bacteroidetes* and *Firmicutes* have distinct effects on intestinal immunity by differentially inducing genes' primary and secondary responses [17]. Parallel to the species from *Actinobacteria*, *Bacteroidetes* and *Firmicutes*, the importance of gut commensal *Escherichia coli* (*E. coli*) from the Proteobacteria for the human host is described in the literature [18–20]. A series of our publications, mainly relating to the changes in growth of gut commensal *E. coli* in diseased people or animals, also emphasize the importance of gut commensal *E. coli* for the host's physiology [21–26]. DNA synthesis is sensitive to changes of phospholipids, fatty acids and lipid peroxidation in the bacterial membranes [1,27].

Previously, we hypothesized that membrane interactions influenced the assessment of LPP intensity in the bacterial extracts and suspensions of *Salmonella* Derby strains K89 and K82 [8,28]. Taking into account this and all the above-mentioned information, the aim of the current study was LPP evaluation of *E. coli* G35 strains from healthy and diseased human gut microbiota, with attention on the changes in thiobarbituric acid-reactive substances (*TBARs*) levels in the bacterial extracts, suspensions and in the cell-free supernatants. In addition to our previous investigations [28], the TBARs levels of the *S.* Derby strain K134, the ultraviolet-sensitive (UV) mutant strain, were also evaluated during the current study. The main point of the research was whether the determination of malonic dialdehyde/TBARs levels in bacterial suspensions, extracts and cell-free supernatants could be important in elucidating the role of bacterial lipid peroxidation processes in bacterial physiology, bacteria–host interactions and host physiology.

2. Materials and Methods

2.1. Bacterial Strains

E. coli strain G35 N49, from the feces of a breast cancer patient [8,21], probioic *E. coli* strain G35 N61 from a healthy person (Passport N 01-07/89, State Research Institute of Standardization and Control of Medical Biological Preparations, Moscow, Russia) [29] and UV-mutant strain *Salmonella enterica* subspecies *enterica* serotype Derby (*S.* Derby) K134 and K95 from the microbial strain collection of the Armenian National Agrarian University were used during these investigations.

2.2. Bacterial Suspension/Extract Preparation

A colony of *E. coli/S.* Derby cells were inoculated in 12 mL of Luria–Bertani broth (Sigma-Aldrich, UK) and were grown at 37 °C for 2.5–3.0 h (logarithmic phase culture) and for 24 h (stationary phase culture). The bacterial suspensions with different optical densities (OD), measured spectrophotometrically (spectrophotometer SF-46 LOMO, Saint Petersburg, Russia), were prepared from the logarithmic phase/overnight cultures at OD_{600}. For this, after centrifugation ($5000\times g$) for 5 min of the exponential or stationary phase bacterial culture, the different optical densities ($OD_{600} = 0.2$–0.7) were adjusted with distilled water using the spectrophotometer.

For the preparation of bacterial extracts from the suspensions, the suspensions were vortexed with the 0.5 mm glass beads (Cat No./ID: 13116-400, QIAGEN, Milano, QIAGEN) (eight times for 20 s, under cold conditions), and the content of MDA was determined in the cell-free extracts.

2.3. MDA Determination

The LPP in bacterial suspensions, extracts and cell-free supernatants was assumed by thiobarbituric acid (TBA) reaction according to Vladimirov and Archakov [30]. The incubation mixture, at a volume of 1 mL, consisted of 40 mM Tris hydrochloride buffer (pH 7.4), ammonium iron(II) sulfate hexahydrate (Sigma Aldrich, Glasgow, UK) (12 µM) and ascorbate (Sigma Aldrich, UK) (0.8 mM). The incubation was carried out at 37 °C for 0.5–2.0 h. To investigate the impact of the growth phase on the formation of TBARs levels in cell suspensions of *E. coli,* the duration of incubation was 2 h. The "TBA"-reaction was stopped by adding trichloroacetic acid (Sigma Aldrich, UK). The residue was removed by centrifugation at $3000 \times g$ for 5 min. The concentration of MDA was calculated per mg of protein. The protein concentration in the cell suspension was calculated by the use of 2 N NaOH with a modification of Yakovleva and co-authors [31]. For this, 1 mL of 2 N NaOH was added to 80 mg of cells; after incubation at 37 °C for 18 h and centrifugation at $5000 \times g$ for 5 min, the concentration of protein was determined in the supernatant according to a Lowry assay [32]. The protein concentration in the extracts was also determined according to Lowry [32].

After the centrifugation of the bacterial suspension at $5000 \times g$ for 5 min, 0.2 mL of the supernatant was used for the determination of TBARs levels in cell-free supernatants [8,27].

MDA in *E. coli* G35 bacterial suspensions was determined both by the above mentioned method and by the instruction of the commercial kit (Lipid Peroxidation (MDA) Assay Kit (Colorimetric) (ab233471)), making it possible to compare the results.

2.4. Statistical Analysis

The Mann–Whitney and Student's t tests (Excel 16) were used for statistical analyses. A probability of $p < 0.05$ was considered significant.

3. Results

3.1. TBARs Levels in the Bacterial Suspensions

The results for *TBARs* levels in bacterial suspensions are shown in Figure 1. The results indicate that the maximum amounts of MDA for bacteria are observed after 1.5 h. The formation of malondialdehyde is observed depending on the reaction time (0.5 h–2.0 h) at $OD_{600} = 0.4$ and $OD_{600} = 0.7$ concentrations of bacteria for *E. coli* G35 N61 cells: 95.34–120.91 µg/mg protein ($OD_{600} = 0.4$) and 31.1–55.1 µg/mg protein ($OD_{600} = 0.7$) (Figure 1). In the case of *E. coli* G35 N49 with a concentration of $OD_{600} = 0.4$, 118.47–150.26 µg/mg protein of MDA is formed within 0.5 h–2.0 h; at a concentration of $OD_{600} = 0.7$, 38.65–68.47 µg/mg protein of MDA was formed. At a concentration of $OD_{600} = 0.4$ of *S.* Derby K89 within 0.5 h–2.0 h, the MDA is formed with 93.47–118.54 µg/mg protein concentrations, and at a concentration of $OD_{600} = 0.7$, the MDA is formed in 30.39–54.02 µg/mg protein concentrations. For *S.* Derby K82, *S.* Derby K134 and *S.* Derby K95 strains, at $OD_{600} = 0.4$, an increase of levels of MDA (30.58 µg/mg protein to 36.95 µg/mg protein, 23.9 µg/mg protein to 47.44 µg/mg protein, and 8 µg/mg protein to 35.4 µg/mg protein) is recorded. Respectively, at $OD_{600} = 0.7$, an increase of MDA concentrations from 19.81 µg/mg protein to 26 µg/mg protein, 15 µg/mg protein to 26.1 µg/mg protein and 21.2 µg/mg protein to 30.7 µg/mg protein is recorded. Moreover, within 2 hours, at $OD_{600} = 0.4$ concentration of *E. coli* G35 N49 strains, MDA is formed at 120.91 µg/mg of protein, and at $OD_{600} = 0.7$, 55.1 µg/mg of protein (Figure 1). A similar trend is observed for *E. coli* G35 N49 and *S.* Derby K134, and for *S.* Derby K95 at a concentration of $OD_{600} = 0.7$, an increase in the amount of malondialdehyde formed is observed, starting from 8 µg/mg of protein and increasing to 21.2 µg/ mg of protein (Figure 1).

3.2. TBARs Levels in the Bacterial Extracts

The results on TBARs levels in the bacterial extracts are presented in Figure 2. According to the results of Figure 2, the levels of formed MDA, depending on the species/strain of bacteria, remained relatively constant or increased in bacterial extracts of $OD_{600} = 0.4$ and $OD_{600} = 0.7$ concentrations (Figure 2).

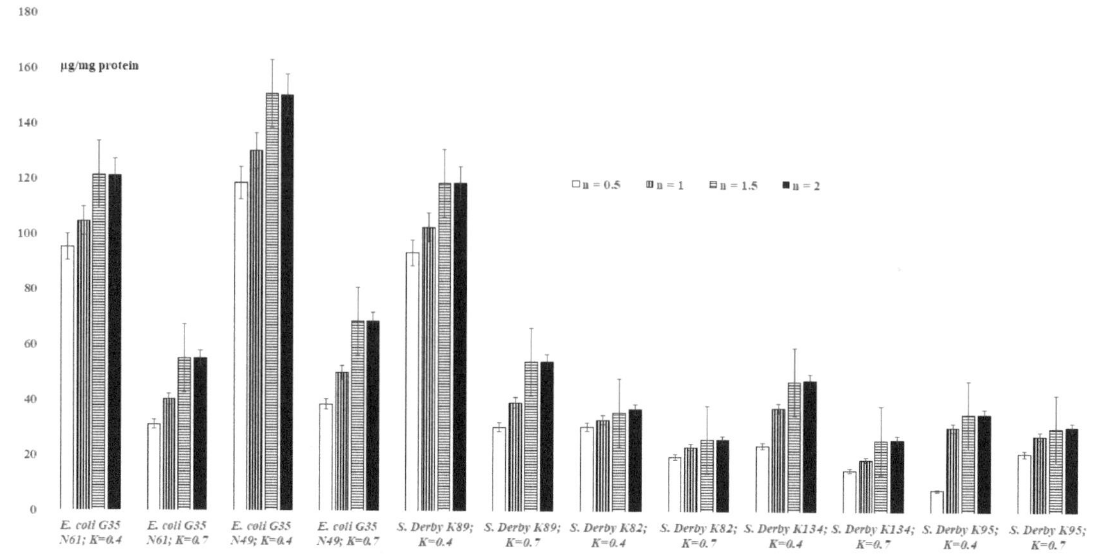

Figure 1. Levels of TBARs formed during lipid peroxidation processes in bacterial cells (suspensions; overage ± standard error) depending on the incubation time (n, hour); K—"OD_{600}".

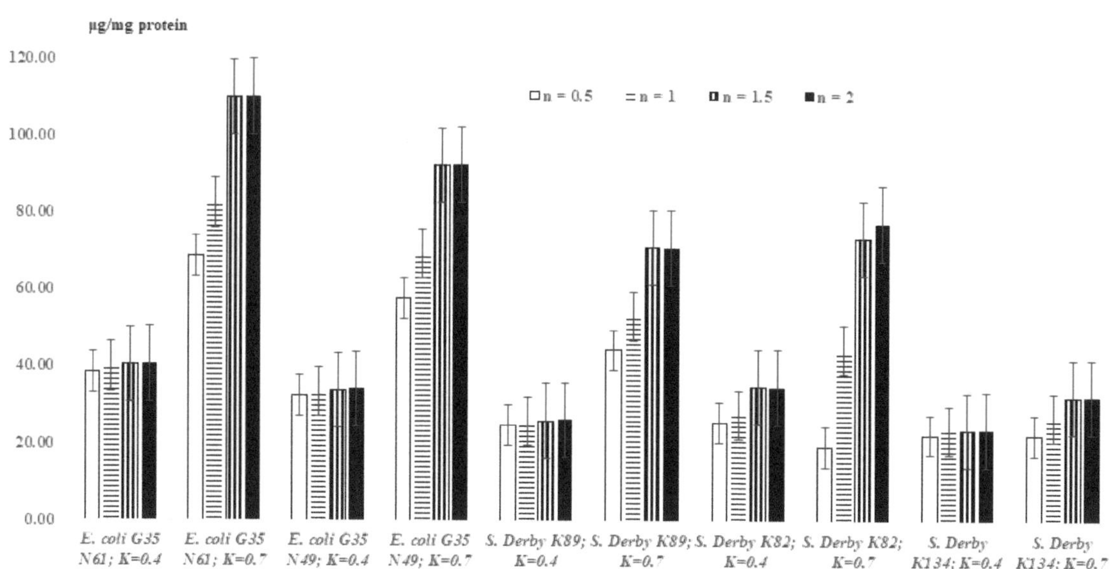

Figure 2. Levels of malondialdehyde formed during lipid peroxidation processes in bacterial extracts (overage ± standard error) depending on the incubation time (n, hour); K—"OD_{600}".

In addition, the levels of formed MDA increased or stayed relatively constant in the bacterial extracts depending on the duration of the TBA reaction (Figure 2). Thus, after 0.5 h of incubation, the content of MDA for the bacterial extracts (OD_{600} = 0.4) was 38.7 ± 1.93 µg/mg protein. There were no significant statistical changes in this "value" during the following incubation periods: 38.7 ± 1.93 vs. 40.05 ± 2.0, 40.41 ± 2.02 and

40.77 ± 2.04; 32.47 ± 1.62 vs. 33.64 ± 1.68, 33.94 ± 1.7 and 34.24 ± 1.71 (for the *E. coli* cells); and 22.40 ± 1.12 vs. 23.52 ± 1.18, 23.71 ± 1.19 and 23.80 ± 1.19 (for the *S.* Derby K89 cells) ($p > 0.05$) (Figure 2). In the case of *S.* Derby K82 cells ($OD_{600} = 0.4$), the increase of the content of MDA was detected after 1.5 hour of incubation, in comparison with that after $0.5/1$ hour of incubation (Figure 2); the level remained unchanged in 2.0 h of incubation (Figure 2). In comparison with the concentration $OD_{600} = 0.4$, the levels of MDA increased for $OD_{600} = 0.7$ (all the investigated bacterial cells) (Figure 2).

3.3. TBARs *Levels in the Logarithmic and Stationary Phase E. coli Cells*

The results of TBARs levels for the logarithmic and stationary phase *E. coli* cells are presented in Table 1. According to these results, the concentrations of MDA formed for the bacterial suspensions of *E. coli* received from the logarithmic and stationary growth phases were different (Table 1).

Table 1. Impact of growth phase on thiobarbituric acid-reactive substances (TBARs) levels * in cell suspension of *Escherichia coli*, µg/mg protein (overage $\pm$ standard error, incubation time: 2 h).

Strains	Logarithmic Phase of Growth		Stationary Phase of Growth		
	$OD_{600} = 0.1$	$OD_{600} = 0.2$	$OD_{600} = 0.2$	$OD_{600} = 0.4$	$OD_{600} = 0.7$
E. coli G35 N61	143.36	55.58 ± 2.78	37.06 ± 1.85 $p * < 0.05$	121.3 ± 2.4	55.02
E. coli G35 N49	178.15	69.07 ± 3.45	46.05 ± 2.3 $p * < 0.05$	150.74 ± 1.75	68.37

* NADPH-dependent lipid peroxidation. $p < 0.05$ was considered significant (comparison of the levels of MDA in $OD_{600} = 0.2$).

3.4. TBARs *Levels for the* E. coli *G35 Cells*

The results on comparative analysis of TBARs levels for the stationary phase *E. coli* G35 cells are presented in Table 2. According to these results, the concentrations of MDA formed in the bacterial suspensions, extracts and cell-free supernatant of *E. coli* G35 N61 (probiotic strain) differed from that of the *E. coli* G35 N49 strain (Table 2).

Table 2. Shifts in thiobarbituric acid-reactive substances (*TBARs*) levels (µg/mg protein) of *Escherichia coli* G35 cells (overage $\pm$ standard error, incubation time: 2 h).

TBARs Levels	Strains	
	E. coli G35 N61	*E. coli* G35 N49
Suspension	121.3 ± 2.4 (75.55) *	150.74 ± 1.75 (93.4) * $p < 0.05$
Extract	40.72 ± 1.05	34.2 ± 1.23 $p < 0.05$
Cell-free supernatant	0.625 ± 0.08	0.171 ± 0.06 $p < 0.05$

$p < 0.05$ was considered significant (comparison of the strains ($OD_{600} = 0.4$)). * The data received by the use of 'ab233471' (the average of three experiments).

4. Discussion

4.1. TBARs *Levels in the Bacterial Suspensions and Extracts*

As expected (according to the instructions of commercial kits), the formation of maximum amounts of MDA in bacterial suspensions is observed after 2 h for all *E. coli* and *S.* Derby cells. This proves that 2 h is sufficient for a correct quantitative assessment of MDA in bacterial suspensions. According to the present study, the amount of malondialdehyde formed in bacterial suspensions, in addition to the species and strain of bacteria, as well as the duration of the reaction, also depends on the concentration of bacteria in suspensions. It is interesting to note that sometimes in the case of bacterial suspensions and more rarely for the extracts, higher concentrations of "$OD_{600} = 0.7$" show a lower level of MDA compared to lower concentrations of "$OD_{600} = 0.4$" (Figures 1 and 2).

These results are in accordance with our preliminary investigations, where the possible influence of membrane interactions on assessment of LPP intensity in the bacterial extracts and suspensions of *S.* Derby were discussed [8,28]. The levels of MDA for both *E. coli* and UV sensitive mutant cells of *S.* Derby K134, detected during the current study, refer to the formula (1) for the bacterial suspensions, and formula (2) for the bacterial extracts, as previously were supposed by Pepoyan and co-authors [8,28]:

$$d[\text{МДА}]_c/dt = K_o N\,[\text{МДА}]_o\,(S_o - S_k) \tag{1}$$

$$d[\text{МДА}]_c/dt = 2\,K_o N\,[\text{МДА}]_o\,(S_o - S_k{}^{\ni}) \tag{2}$$

where K_o—the constant of the process of MDA release from the membrane into the intercellular space, N—the concentration of bacterial cells in the system, $[\text{MDA}]_o$—the concentration of MDA, S_o—surface of cell membrane, S_k and $S_k{}^{\ni}$—surface of cell membrane engaged in intercellular contacts for the "intact" and "destroyed" cells, and $(S_o - S_k)$—the surface of the membrane free from intercellular contacts.

Integrating the Equation (1)/(2), for the relatively constant bacterial concentrations (S_k = constant and $S_k{}^{\ni}$ = constant), the Formula (3) can be formed:

$$[\text{МДА}]_c = S_o - S_k/2(S_o - S_k{}^{\ni})[\text{МДА}]_{\ni} \tag{3}$$

Formula (3) provides evidence of the existence of dependence between the formation of MDA in the system, and S_k and $S_k{}^{\ni}$ both for the "intact" and "destroyed" cells of *E. coli* and *S.* Derby. Consequently, at S_o = constant, the ratio $[\text{МДА}]_c/[\text{МДА}]_{\ni}$, determined experimentally, will be determined by S_k and $S_k{}^{\ni}$, depending on the concentrations of bacteria. In the case of extracts, one should expect a rapid increase in $S_k{}^{\ni}$ with a concentration and a steady-state value being reached even at lower cell concentrations than in the case of intact cells. This, perhaps, is due to the complementarity of the membranes of the intact cells, which have a spherical or ellipsoid shape. According to our previous studies, *E. coli* and *S.* Derby cells differ from each other in their morphological and physiological properties [21,25], which could be the reason for the change in the S_k and/or $S_k{}^{\ni-}$ surfaces. Bacteria in different logarithmic and stationary phases of growth also differ in their membrane and metabolic properties, which probably causes changes in the S_k and $S_k{}^{e}$ surfaces and, therefore, affects the formation of MDA, the end-product of oxidative stress of the same bacteria (Table 1).

Based on the above-mentioned data/discussions, the actual amount of MDA formed by the studied bacteria is observed at an $OD_{600} = 0.4$, when, according to the results of this study, the concentration of bacteria (intercellular interactions) does not significantly affect the release of MDA into the intercellular space (Formula (3), Figures 1 and 2).

4.1.1. Comparative Evaluation of *TBARs* Levels in the Bacterial Cells

Oxidative stress in the cells is determined by the predominance of levels of reactive oxygen species over antioxidant levels [33]; for example, the intracellular concentration of hydrogen peroxide under aerobic conditions increases in *E. coli* by $\sim$0.2 nM $O_2{}^-$ and $\sim$50 nM H_2O_2 [34], which can change under the influence of exogenous factors [35]. To mitigate damage caused by oxidative stress, bacteria activate various regulatory responses to stress, depending on the stressor's nature [5]. In addition, the bacterial membrane is semi-permeable to H_2O_2, and H_2O_2 produced by one bacterium can enter and potentially harm other bacteria in the host microbiome [35]. The association of the gut microbiota with altered oxidative stress is now well established for neurodegenerative diseases [33].

As far back as the 1980s, several scientific publications indicated the association between the intestinal non-pathogenic *E. coli* cells and hosts with colorectal cancer [36,37]. Karapetyan, one of the authors of these publications, claimed that the *E. coli* G35 N49 strain predominates in the intestines of tumor patients, especially in the intestines of colorectal cancer patients [36]. The dominant *E. coli* G35 N49 strain isolated by Karapetyan from the patient's fecal microbiota was described by him and later by his colleagues as a non-pathogenic, non-lactose-fermenting strain dominant in the intestinal microbiota of patients with colon cancer. Comparative properties of *E. coli* G35 N49 and the strain isolated from healthy human microbiota (*E. coli* G35 N61 strain) have been actively investigated, considering growth and proliferation, membrane properties [21,22,27,29], and interaction with tumor cells [38], and studies have demonstrated the differences in characteristics of predominant commensal *E. coli* strains isolated from the gut microbiota of different tumor patients [39]. In 2020, Tang and colleagues showed that, compared to healthy individuals, colorectal cancer patients harbored a lower diversity of intestinal *E. coli* isolates [40]. The authors hypothesized that "diseased" isolates suppressed the growth of healthy isolates under the nutrient-limited culture conditions [40]. Nowadays, not only an immense antioxidative and anti-inflammatory role of healthy gut microbiota is well known, but it is also known that altered gut-microbiota-mediated oxidative stress is associated with different diseases, even including neurodegenerative [41] and several skin diseases [42].

4.1.2. Comparative Evaluation of TBARs Levels in the Cells of *E. coli* G35

The anti-cancer properties of *E. coli* probiotics [29,36,43,44], and the probiotics' effectiveness in association with the pre- and post-radiation nutrition, are known [45,46]. Regarding the latter, Pepoyan and co-authors tried to explain the probiotic's participation in the host's free radical metabolism [45], which is possible also in the case of *E. coli* probiotics [3]. Previously, Mirzoyan and co-authors reported the differences in physico-chemical and physiological properties of E. coli G35 cells, related to the growth and cells' membrane functions [21]. The strains of *E. coli* G35 N49 (prevailed *E. coli* strain from the cancer patient) and *E. coli* G35 N61, despite the same membrane potentials, were basically different in total and N,N′-dicyclohexylcarbodiimide-sensitive rates of energy-dependent transmembrane H^+ and K^+ transport, demonstrating a low level of H_2 production [22]. According to current investigations, the *E. coli* G35 strains also differ significantly from each other by the LPP intensity (Table 2). Probably, the changes in MDA concentrations in the bacterial suspensions, extracts and cell-free supernatant of the *E. coli* G35 N49 affect not only the bacterial cells, but also the metabolic processes of the host organism.

4.1.3. Comparative Evaluation of TBARs Levels in the UV-Sensitive Cells of *S.* Derby K134

According to Figures 1 and 2, the content of MDA was different in the wild *S.* Derby strain K89, and in its UV-sensitive mutant strain *S.* Derby K134, indicating that the mutation has influence not only on the bacterial UV-resistance and on the bacterial growth [25], but also on the dynamics of LPP of the cells (118.54 ± 5.93 μg/mg protein vs. 47.44 ± 2.37 μg/mg protein (for suspension), and 26.30 ± 1.32 μg/mg protein vs. 23.80 ± 1.19 μg/mg protein (for extract)). The differences in LPP formed in *S.* Derby K134 and *S.* Derby K95 (plasmid-free derivative of K134 [8]) might be explained by the effect of the R-plasmid on host membrane integrity [8]. As in the case of *E. coli* G35, we hypothesize that the differences in TBARs levels in *S.* Derby suspensions may play an "environmentally" important role for the potential bacterial host.

Thus, despite the commercial kits for the assessment of malondialdehyde formation, mainly determining the amounts of malondialdehyde formed in human (also rat and mouse) serum and plasma, as well as the amounts of malondialdehyde formed in various tissue homogenates, these kits can also be easily used to assess malondialdehyde formation in bacterial extracts, suspensions, and supernatants. However, the dependence of TBARs levels in the cell extracts, the suspensions and cell-free supernatants on bacterial species, their concentration and the growth phase were revealed during the current investigations. The effect of bacterial concentrations on MDA formation was also found to be more pronounced in bacterial suspensions than in extracts, probably due to the dynamics of MDA release into the intercellular space.

On the other hand, interestingly, the comparative assessment of TBARs levels in the cell extracts, the suspensions and cell-free supernatants in one of the important representatives of intestinal microbiota, *E. coli*, revealed differences between the probiotic and the "diseased" strains (Table 2); meanwhile, usually, either extracts, cell-free suspensions or bacterial suspensions are used to evaluate bacterial lipid peroxidation processes, which may lead to incorrect discussions and conclusions.

5. Conclusions

Nowadays, it is known that, in addition to commensal bacteria [47], "non-pleasant"/ pathogenic microbes [21,48] can also live inside hosts without causing noticeable diseases. Along with adhesion processes, which are mainly determined by the bacterial membrane structures and are important in bacteria–host interactions, reactive oxygen species and nitrogen species are also actively discussed by researchers as a defensive tool against pathogens [49,50]. The bacterial membrane is semipermeable to H_2O_2, and H_2O_2 formed by a microbe can be destructive to the host microbiome [35]. The different approaches are used for sample preparation to monitor the concentrations of MDA for bacteria, including homogenates of fresh [8] and freeze-dried bacterial cultures [10] and bacterial culture broths [6]. Current investigations revealed levels of thiobarbituric acid-reactive substances in cell extracts, suspensions and cell-free supernatants for *S.* Derby and *E. coli* strains. This study highlights the importance of simultaneous assessment of oxidative processes in bacterial extracts, suspensions and culture liquids to elucidate the role of lipid peroxidation processes in bacterial physiology, bacteria–host interactions, as well as in host physiology. Logarithmic/stationary growth phases of bacteria might also be important in such evaluation processes. However, further investigations are needed to specifically define the role of the formed MDA in cell-free supernatants or suspensions or extracts in the bacteria–host interaction.

Author Contributions: Conceptualization, A.P. (Astghik Pepoyan); data curation, A.P. (Armen Poghosyan) and M.T.; formal analysis, A.P. (Armen Poghosyan); Funding acquisition, M.T.; Investigation, V.T. and M.T.; Methodology, A.P. (Astghik Pepoyan); Project administration, A.P. (Astghik Pepoyan); Resources, V.T.; Supervision, A.P. (Astghik Pepoyan); Validation, V.T.; Visualization, A.P. (Armen Poghosyan); Writing—original draft, A.P. (Astghik Pepoyan); Writing—review & editing, A.P. (Astghik Pepoyan). All authors have read and agreed to the published version of the manuscript.

Funding: This work was supported by the State Committee of Science Armenia (projects 21AG-4D065, 10-15/I-5 and EAPI2020-065).

Informed Consent Statement: Not applicable.

Data Availability Statement: The data that support the findings of this study are available on request from the author.

Conflicts of Interest: The authors declare no conflict of interest.

References

1. Zhang, Y.M.; Rock, C. Membrane Lipid Homeostasis in Bacteria. *Nat. Rev. Microbiol.* **2008**, *6*, 222–233. [CrossRef] [PubMed]
2. Araújo, J.R.; Tazi, A.; Burlen-Defranoux, O.; Vichier-Guerre, S.; Nigro, G.; Licandro, H.; Demignot, S.; Sansonetti, P.J. Fermentation Products of Commensal Bacteria Alter Enterocyte Lipid Metabolism. *Cell Host Microbe* **2020**, *27*, 358–375. [CrossRef] [PubMed]
3. Ayala, A.; Munoz, M.F.; Arguelles, S. Lipid Peroxidation: Production, Metabolism, and Signaling Mechanisms of Malondialdehyde and 4-hydroxy-2-nonenal. *Oxid. Med. Cell. Longev.* **2014**, *2014*, 360438. [CrossRef] [PubMed]
4. Mih, N.; Monk, J.M.; Fang, X.; Catoiu, E.; Heckmann, D.; Yang, L.; Palsson, B. Adaptations of *Escherichia coli* Strains to Oxidative Stress are Reflected in Properties of their Structural Proteomes. *BMC Bioinform.* **2020**, *21*, 162. [CrossRef]
5. Imlay, J.A. Where in the World do Bacteria Experience Oxidative Stress? *Environ. Microbiol.* **2019**, *21*, 521–530. [CrossRef]
6. Hsu, K.C.; Hsu, P.F.; Chen, Y.C.; Lin, H.C.; Hung, C.C.; Chen, P.C.; Huang, Y.L. Oxidative Stress During Bacterial Growth Characterized through Microdialysis Sampling Coupled with HPLC/fluorescence Detection of Malondialdehyde. *J. Chromatogr. B Analyt. Technol. Biomed. Life Sci.* **2016**, *1019*, 112–116. [CrossRef] [PubMed]

7. Toto, A.; Wild, P.; Graille, M.; Turcu, V.; Crézé, C.; Hemmendinger, M.; Sauvain, J.J.; Bergamaschi, E.; GusevaCanu, I.; Hopf, N.B. Urinary Malondialdehyde (MDA) Concentrations in the General Population-A Systematic Literature Review and Meta-Analysis. *Toxics* **2022**, *10*, 160. [CrossRef] [PubMed]

8. Pepoyan, A.Z. Role of Membranes in UV Resistance of *Enterobacteriaceae* Cells. Ph.D. Thesis, Institute of Molecular Biology, NAS RA, Yerevan, Armenia, 2002; p. 222.

9. Zhuravlev, A.V.; Vetrovoy, O.V.; Savvateeva-Popova, E.V. Enzymatic and Non-enzymatic Pathways of Kynurenines' Dimerization: The Molecular Factors for Oxidative Stress Development. *PLoS Comput. Biol.* **2018**, *14*, e1006672. [CrossRef] [PubMed]

10. Hassan, A.H.A.; Hozzein, W.N.; Mousa, A.S.M.; Rabie, W.; Alkhalifah, D.H.M.; Selim, S.; AbdElgawad, H. Heat Stress as an Innovative Approach to Enhance the Antioxidant Production in Pseudooceanicola and Bacillus Isolates. *Sci. Rep.* **2020**, *10*, 15076. [CrossRef] [PubMed]

11. Cole, S.D.; Miklos, A.E.; Chiao, A.C.; Sun, Z.Z.; Lux, M.W. Methodologies for Preparation of Prokaryotic Extracts for Cell-free Expression Systems. *Synth. Syst. Biotechnol.* **2020**, *5*, 252–267. [CrossRef] [PubMed]

12. Balayan, M.H.; Manvelyan, A.M.; Marutyan, S.; Isajanyan, M.; Tsaturyan, V.V.; Pepoyan, A.Z.; Marotta, F.; Torok, T. Impact of *Lactobacillus acidophilus* INMIA 9602 Er-2 and *Escherichia coli* M-17 on Some Clinical Blood Characteristics of Familial Mediterranean Fever Disease Patients from the Armenian Cohort. *Int. J. Probiotics Prebiotics* **2015**, *10*, 91–95.

13. Khan, R.; Petersen, F.C.; Shekhar, S. Commensal Bacteria: An Emerging Player in Defense Against Respiratory Pathogens. *Front. Immunol.* **2019**, *10*, 1203. [CrossRef] [PubMed]

14. Ragonnaud, E.; Biragyn, A. Gut Microbiota as the Key Controllers of "Healthy" Aging of Elderly People. *Immun. Ageing.* **2021**, *18*, 2. [CrossRef]

15. Pepoyan, A.Z.; Tsaturyan, V.V.; Badalyan, M.; Weeks, R.; Kamiya, S.; Chikindas, M.L. Blood Protein Polymorphisms and the Gut Bacteria: Impact of Probiotic *Lactobacillus acidophilus* Narine on *Salmonella* Carriage in Sheep. *Benef. Microbes* **2020**, *7*, 1–8. [CrossRef] [PubMed]

16. Pepoyan, A.Z.; Balayan, M.H.; Manvelyan, A.M.; Galstyan, L.; Pepoyan, S.; Petrosyan, S.D.; Tsaturyan, V.V.; Kamiya, S.; Torok, T.; Chikindas, M. Probiotic *Lactobacillus acidophilus* strain INMIA 9602 Er317/402 Administration Reduces the Numbers of *Candida albicans* and Abundance of enterobacteria in the Gut Microbiota of Familial Mediterranean Fever Patients. *Front. Immunol.* **2018**, *9*, 1426. [CrossRef] [PubMed]

17. Brown, R.L.; Larkinson, M.L.Y.; Clarke, T.B. Immunological Design of Commensal Communities to Treat Intestinal Infection and Inflammation. *PLoS Pathog.* **2021**, *17*, e1009191. [CrossRef] [PubMed]

18. Bok, E.; Mazurek, J.; Myc, A.; Stosik, M.; Wojciech, M.; Baldy-Chudzik, K. Comparison of Commensal *Escherichia coli* Isolates from Adults and Young Children in Lubuskie Province, Poland: Virulence Potential, Phylogeny and Antimicrobial Resistance. *Int. J. Environ. Res. Public Health* **2018**, *15*, 617. [CrossRef]

19. Mazurek-Popczyk, J.; Pisarska, J.; Bok, E.; Baldy-Chudzik, K. Antibacterial Activity of Bacteriocinogenic Commensal *Escherichia coli* Against Zoonotic Strains Resistant and Sensitive to Antibiotics. *Antibiotics* **2020**, *9*, 411. [CrossRef] [PubMed]

20. Escribano-Vazquez, U.; Verstraeten, S.; Martin, R.; Chain, F.; Langella, P.; Thomas, M.; Cherbuy, C. The Commensal *Escherichia coli* CEC15 Reinforces Intestinal Defences in Gnotobiotic Mice and is Protective in a Chronic Colitis Mouse Model. *Sci. Rep.* **2019**, *9*, 11431. [CrossRef] [PubMed]

21. Mirzoyan, N.S.; Pepoyan, A.Z.; Trchounian, A.H. Modification of the Biophysical Characteristics of Membranes Incommensal *Escherichia coli* Strains from Breast Cancer Patients. *FEMS Microbiol. Lett.* **2006**, *254*, 81–86. [CrossRef] [PubMed]

22. Stepanyan, K.; Balayan, M.H.; Vassilian, A.; Pepoyan, A.Z.; Trchounian, A.H. Growth Peculiarities and Some Characteristics of Membrane for Probiotic Strain of *Escherichia coli*. *Memb. Cell Biol.* **2007**, *1*, 333–335. [CrossRef]

23. Balayan, M.H.; Mirzabekyan, S.S.; Pepoyan, A.Z.; Trchounian, A.H. Some Peculiarities of Growth and Functional Activity of *Escherichia coli* Strain from Probiotic Formula "ASAP". *World Acad. Sci. Eng. Technol.* **2010**, *44*, 858–862.

24. Gasparyan, G.; Balayan, M.H.; Grigoryan, A.; Hakopyan, A.; Manvelyan, A.M.; Mirzabekyan, S.S.; Pepoyan, A.Z. Growth Peculiarities of Commensal *Escherichia coli* Isolates from the Gut Microflora of Crohn's Disease Patients. *Biofizika* **2013**, *58*, 690–696. (In Russian) [CrossRef]

25. Pepoyan, A.Z.; Balayan, M.H.; Manvelyan, A.M.; Tsaturyan, V.V. Growth and Motility of Gut Commensal *Escherichia coli* in Health and Disease. *Biophys. J.* **2014**, *106*, 726. [CrossRef]

26. Pepoyan, A.Z.; Balayan, M.H.; Manvelyan, A.M.; Mamikonyan, V.; Isajanyan, M.; Tsaturyan, V.V.; Kamiya, S.; Netrebov, V.; Chikindas, M. *Lactobacillus acidophilus* INMIA 9602 Er-2 strain 317/402 Probiotic Regulates Growth of Commensal *Escherichia coli* in Gut Microbiota of Familial Mediterranean Fever Disease Subjects. *Lett. Appl. Microbiol.* **2017**, *64*, 254–260. [CrossRef] [PubMed]

27. Pepoyan, A.Z.; Mirzoyan, N.S.; Saakyan, M.O.; Kirakosyan, L.A.; КарарезяН, K.G. Some Peculiarities of Antioxidative System of *E. coli* G35 Bacterial Strains. *Rep. Acad. Sci.* **2002**, *102*, 70–75.

28. Pepoyan, A.Z.; Ktsoyan, Z.A.; Shaginyan, A.A.; Osipyan, L.M.; Karagezyan, K.G. The Role of Membrane Interactions in Lipid-peroxidation Processes in Plasmid and Unplasmid Cells of *Salmonella derby*. *Biophysika.* **1991**, *36*, 475–479. (In Russian)

29. Shahinyan, A.; Garibyan, J.; Pepoyan, A.; Karapetyan, O. Cancerolitic Action of *E. coli*. *J. Nat. Sci.* **2003**, *1*, 53–58.

30. Lowry, O.H.; Rosebrouh, N.J.; Farr, A.L.; Randall, R.J. Protein Measurement with the Folin Phenol Reagent. *J. Biol. Chem.* **1951**, *193*, 265–275. [CrossRef]

31. Vladimirov, Y.A.; Archakov, A.I. *Lipid Peroxidation in Biological Membranes*; Nauka: Moscow, Russia, 1972; p. 252. (In Russian)

32. Yakovleva, V.I.; Maloreeva, I.V.; Zueva, M.N.; Andreeva, V.P.; Gubnetskiy, L.S.; Shcherbakova, V.N.; Berezin, I.V. Synthesis of L-aspartic Acid from Ammonium Fumarate by Free and Immobilized Cells of *Escherichia Coli*. *Prikl. Biokhim. Mikrobiol.* **1979**, *15*, 328–336.

33. Leyane, T.S.; Jere, S.W.; Houreld, N.N. Oxidative Stress in Ageing and Chronic Degenerative Pathologies: Molecular Mechanisms Involved in Counteracting Oxidative Stress and Chronic Inflammation. *Int. J. Mol. Sci.* **2022**, *23*, 7273. [CrossRef]

34. Imlay, J.A. The Molecular Mechanisms and Physiological Consequences of Oxidative Stress: Lessons from a Model Bacterium. *Nat. Publ. Gr.* **2013**, *11*, 443–454. [CrossRef]

35. Fasnacht, M.; Polacek, N. Oxidative Stress in Bacteria and the Central Dogma of Molecular Biology. *Front. Mol. Biosci.* **2021**, *8*, 671037. [CrossRef]

36. Karapetyan, A.O. Biological Antagonism of Some Representatives of the Intestinal Microflora and Tumor Cells. Ph.D. Thesis, Moscow State University, Moscow, Russia, 1987; p. 101. (In Russian).

37. Shachlamov, W.A.; Karapetyan, A.O.; Karapetyan, O.G. *Materials of Actual Quest. of Modern Histology*; Nauka: Moscow, Russia, 1988; p. 101. (In Russian)

38. Nersesyan, L.E.; Garibyan, D.V.; Garibjanyan, B.T.; Shahinyan, A.A. The Change in the Structure of Tumor DNA under the Combined Action of Bacterial Culture, E. coli and Doxorubicin A In Vivo. *Electron. J. Nat. Sci.* **2004**, *3*, 36–39.

39. Manvelyan, A.; Balayan, M.; Pepoyan, E.; Aloyan, S.; Pepoyan, A. Laboratory Assessment of Gut Dysbiosis. *Biol. J. RA* **2007**, *12*, 35–39. (In Russian)

40. Tang, L.; Zhou, Y.J.; Zhu, S.; Liang, G.D.; Zhuang, H.; Zhao, M.; Chang, X.Y.; Li, H.N.; Liu, Z.; Guo, Z.R.; et al. *E. coli* Diversity: Low in Colorectal Cancer. *BMC Med. Genom.* **2020**, *13*, 59. [CrossRef]

41. Shandilya, S.; Kumar, S.; Kumar Jha, N.; Kumar Kesari, K.; Ruokolainen, J. Interplay of Gut Microbiota and Oxidative Stress: Perspective on Neurodegeneration and Neuroprotection. *J. Adv. Res.* **2021**, *38*, 223–244. [CrossRef]

42. Ni, Q.; Zhang, P.; Li, Q.; Han, Z. Oxidative Stress and Gut Microbiome in Inflammatory Skin Diseases. *Front. Cell Dev. Biol.* **2022**, *10*, 849985. [CrossRef]

43. Kulchitskaya, M. Development of a Probiotic Preparation Ocarin for Medicine and Veterinary Medicine. Ph.D. Thesis, Nizhny Novgorod State Agricultural Academy, Nizhny Novgorod, Russia, 2002; p. 193.

44. Alizadeh, S.; Esmaeili, A.; Omidi, Y. Anti-cancer Properties of *Escherichia coli* Nissle 1917 against HT-29 Colon Cancer Cells through Regulation of Bax/Bcl-xL and AKT/PTEN Signaling Pathways. *Iran. J. Basic Med. Sci.* **2020**, *23*, 886–893. [CrossRef]

45. Pepoyan, A.Z.; Manvelyan, A.M.; Balayan, M.H.; McCabe, G.; Tsaturyan, V.V.; Melnikov, V.G.; Chikindas, M.L.; Weeks, R.; Karlyshev, A.V. The Effectiveness of Potential Probiotics *Lactobacillus rhamnosus* Vahe and *Lactobacillus delbrueckii* IAHAHI in Irradiated Rats Depends on the Nutritional Stage of the Host. *Probiotics Antimicro. Prot.* **2020**, *12*, 1439–1450. [CrossRef]

46. Harutyunyan, N.; Kushugulova, A.; Hovhannisyan, N.; Pepoyan, A. One Health Probiotics as Biocontrol Agents: One Health Tomato Probiotics. *Plants* **2022**, *11*, 1334. [CrossRef]

47. Stones, D.H.; Krachler, A.M. Against the Tide: The Role of Bacterial Adhesion in Host Colonization. *Biochem. Soc. Trans.* **2016**, *44*, 1571–1580. [CrossRef]

48. Casadevall, A.; Pirofski, L.A. Host-pathogen Interactions: Basic Concepts of Microbial Commensalism, Colonization, Infection, and Disease. *Infect. Immun.* **2000**, *68*, 6511–6518. [CrossRef]

49. Chautrand, T.; Souak, D.; Chevalier, S.; Duclairoir-Poc, C. Gram-Negative Bacterial Envelope Homeostasis under Oxidative and Nitrosative Stress. *Microorganisms* **2022**, *10*, 924. [CrossRef]

50. Knaus, U.G.; Hertzberger, R.; Pircalabioru, G.G.; Yousefi, S.P.; Branco Dos Santos, F. Pathogen Control at the Intestinal Mucosa-H_2O_2 to the Rescue. *Gut Microbes* **2017**, *8*, 67–74. [CrossRef]

Article

Açai Berry Mitigates Vascular Dementia-Induced Neuropathological Alterations Modulating Nrf-2/Beclin1 Pathways

Daniela Impellizzeri [1,†], Ramona D'Amico [1,†], Roberta Fusco [1], Tiziana Genovese [1], Alessio Filippo Peritore [1], Enrico Gugliandolo [2], Rosalia Crupi [2], Livia Interdonato [1], Davide Di Paola [1], Rosanna Di Paola [2,*], Salvatore Cuzzocrea [1,3,*], Rosalba Siracusa [1,‡] and Marika Cordaro [4,‡]

1 Department of Chemical, Biological, Pharmaceutical and Environmental Sciences, University of Messina, Viale Ferdinando Stagno D'Alcontres 31, 98166 Messina, Italy
2 Department of Veterinary Sciences, University of Messina, 98168 Messina, Italy
3 Department of Pharmacological and Physiological Science, Saint Louis University School of Medicine, Saint Louis, MO 63104, USA
4 Department of Biomedical, Dental and Morphological and Functional Imaging, University of Messina, Via Consolare Valeria, 98125 Messina, Italy
* Correspondence: dipaolar@unime.it (R.D.P.); salvator@unime.it (S.C.)
† These authors contributed equally to this work.
‡ These authors shared senior authorship.

Citation: Impellizzeri, D.; D'Amico, R.; Fusco, R.; Genovese, T.; Peritore, A.F.; Gugliandolo, E.; Crupi, R.; Interdonato, L.; Di Paola, D.; Di Paola, R.; et al. Açai Berry Mitigates Vascular Dementia-Induced Neuropathological Alterations Modulating Nrf-2/Beclin1 Pathways. *Cells* **2022**, *11*, 2616. https://doi.org/10.3390/cells11162616

Academic Editors: Alessia Remigante and Rossana Morabito

Received: 28 July 2022
Accepted: 20 August 2022
Published: 22 August 2022

Publisher's Note: MDPI stays neutral with regard to jurisdictional claims in published maps and institutional affiliations.

Abstract: The second-most common cause of dementia is vascular dementia (VaD). The majority of VaD patients experience cognitive impairment, which is brought on by oxidative stress and changes in autophagic function, which ultimately result in neuronal impairment and death. In this study, we examine a novel method for reversing VaD-induced changes brought on by açai berry supplementation in a VaD mouse model. The purpose of this study was to examine the impact of açai berries on the molecular mechanisms underlying VaD in a mouse model of the disease that was created by repeated ischemia–reperfusion (IR) of the whole bilateral carotid artery. Here, we found that açai berry was able to reduce VaD-induced behavioral alteration, as well as hippocampal death, in CA1 and CA3 regions. These effects are probably due to the modulation of nuclear factor erythroid 2-related factor 2 (Nrf-2) and Beclin-1, suggesting a possible crosstalk between these molecular pathways. In conclusion, the protective effects of açai berry could be a good supplementation in the future for the management of vascular dementia.

Keywords: vascular dementia; oxidative stress; autophagy; açai berry

1. Introduction

In the Western world, vascular dementia (VaD) is the second-most common cause of organic acquired cognitive dysfunction; in several Asian nations, it is likely the main cause [1].

There are no viable pharmaceuticals approved for the treatment of VaD in any country despite the recent investments made in experimental and clinical neuroscience [2]. Smoking, not exercising, eating poorly, and being exposed to pollutants are all known to worsen the neuronal environment and increase oxidative stress, which can lead to the cognitive loss associated with aging and neurodegenerative disorders. Important brain areas like the hippocampus should be more vulnerable to oxidative and inflammatory stresses since these conditions might impair synaptic plasticity and memory by causing dendritic alteration and neuronal death. Furthermore, the brain is also subjected to an altered expression of two important neuronal markers of wellbeing: MAP-2 and β-Tubulin.

As it promotes microtubule stiffness, MAP2—which is found in cell bodies and dendrites—is regarded as a microtubule stabilizer. When expression is changed, MAP2 builds up into granules, causing neurotoxicity and neuronal degeneration. During cortical development, the tubulin family primarily expresses itself in post-mitotic neurons with a particular geographic and temporal expression pattern. Microtubules' primary building blocks, tubulin, are essential for the mechanisms of the central nervous system's (CNS) development, including neuronal migration and axonal guidance. The altered expression of these proteins may result in the death of neurons. The loss of MAP2 and β-Tubulin have been strongly associated with VaD [3,4].

Additionally, defective autophagy has been linked to increased oxidative stress in the brain, altering protein "quality control," accumulating undesired proteins and organelles in brain cells [5].

A fundamental leucine zipper redox-sensitive transcription factor called nuclear factor erythroid 2-related factor 2 (Nrf2) regulates the redox status of the cell under damaging stressors. Under normal circumstances, Kelch-like ECH-associated protein 1(Keap1), Nrf2's inhibitor, anchors Nrf2 in the cytoplasm. However, during oxidative stress, Nrf2 separates from Keap1, moves into the nucleus, and interacts with the antioxidant response element (ARE) to control the production of antioxidant genes such heme oxygenase-1 (HO-1). Additionally, recent studies have shown that Nrf2 could control the initiation of autophagy, furthering its protective benefits [6–8]. In fact, inclusion bodies, which are pathogenic markers of many neurodegenerative disorders, are known to accumulate excessively as a result of dysfunctional autophagy. Numerous studies have demonstrated that entire fruits and dietary polyphenols such as anthocyanins, stilbenes, melatonin, kaemferol, quercetin, and resveratrol protect neurons by a modulation of autophagy process [9–20].

Scientists have recently developed an interest in açai seeds. The berry known as the açai berry has a variety of beneficial nutritional properties as well as some possible medicinal uses. This fruit has a tart flavor and is produced by the *Euterpe oleracea* palm, which can only be found in the Amazon. For millennia, Amazonian Indians have employed the açai fruit, which is regarded as a high-energy meal, as a food source and a natural remedy for a number of illnesses [21–29].

The pulp of the açai fruit has been extensively researched due to its high bioactive nutritional and phytochemical content. Açai berry pulp composition contains a wide range of physiologically active phytochemicals as well as significant amounts of mono- and polyunsaturated fatty acids, which are uncommon in most fruits and other berries. Phytochemicals such proanthocyanidins, anthocyanins, and other flavonoids are present in açai pulp. Additionally, phytochemical analyses revealed that the açai berry contains a significant amount of luteolin, quercetin, dihydrokaempferol, and chrysoerial, among other polyphenolics, as well as a variety of anthocyanins, including cyanidin, delphinidin, malvidin, pelargonidin, and peonidin. Açai fruit pulp included five different types of carotenoids, including carotene, lycopene, astaxanthin, lutein, and zeaxanthin [30].

Numerous pharmacological benefits of açai berry extract and its bioactive components include anti-inflammatory and anti-anxiety properties through modulation of oxidative stress, inflammation, autophagy, and Nrf2 expression in the hippocampus and frontal cortex [31–39]. However, more data are needed to confirm the neuroprotective, anti-inflammatory and antioxidant effect. For this reason, we used a consolidated murine model of VaD to investigate the potential beneficial effects of açai supplementation and the molecular way by which it acts.

2. Materials and Methods

2.1. Animals

CD1 male mice (8-week-old, 18–24 g) were acquired from Envigo (Milan, Italy) and located in a controlled environment. All animal experiments complied with the new Italian regulations (D. Lgs 2014/26), EU regulations (EU Directive 2010/63) and the ARRIVE guidelines.

2.2. Experimental Design and Groups

Mice were subjected to temporary bilateral carotid occlusion as previously described by Cordaro et al. [40]. Briefly, following anesthesia, the bilateral carotid arteries underwent three cycles of ligation and release lasting a total of 10 min each. After the threading was taken out, the incision was stitched up. After 15 days of induction, the mice were sacrificed, and their brains were removed and processed for histological and biochemical analysis [40,41].

Mice were arbitrarily divided into groups:

Group 1: Sham = mice were subjected to the surgery without carotid arteries ligation and treated daily orally with saline (vehicle) for 15 days.

Group 2: Sham + açai berry = same as above, but açai berry at a dose of 500 mg/kg, dissolved in saline, was administered orally once a day for 15 days (data not shown).

Group 3: VaD = mice were subjected to the VaD surgery described above and treated with vehicle.

Group 4: VaD + açai berry = same as above, but açai berry at a dose of 500 mg/kg was administered daily orally for 15 days.

A total of six samples was used for each technique. Freeze-dried açai extract was dissolved in distilled water. This substance (Cas Number: 879496-95, 906351-38-0) was purchased from Farmalabor Srl (Canosa di Puglia, Barletta, Italy) (see Supplementary File for technical datasheet). The dose and the route of administration of açai berry were chosen based on our previous study [42]. No significant difference was found between the sham and sham + açai berry, so only data regarding the sham groups are shown.

2.3. Behavioral Tests

2.3.1. Open Field Test (OFT)

Locomotor activity and anxiety-like behavior were monitored by the OFT. Each mouse was carefully placed in the center of the box after training, and any action was counted as a line crossing if a mouse moved all four paws out of one square and into another [43,44].

2.3.2. Novel Object Recognition (NOR) Test

The spontaneous inclination of mice to spend time investigating a novel object or a familiar one was examined with the NOR test. After a training session, mice were switched out in the box for a five-minute exam, during which the examiner, at random, switched out one of the familiar objects for a new one. The amount of time the mouse spent investigating each item was measured [41,45].

2.3.3. Social Interaction Test (SIT)

A three-chambered device was used to create the social interaction test as previously described. The number of active interactions and the duration of contacts were measured [41].

2.4. Histopathological Evaluation

Brains were stained with hematoxylin/eosin (H/E) or cresyl-violet and blindly analyzed using light microscopy LeicaDM6 connected to an imaging system (LasX Navigator)(Leica Microsystem, Buccinasco, Italy) by two investigators without knowledge of the experimental groups [46–48]. The severity of brain injury and numbers of dead cells was assessed as previously described [40,41].

2.5. Immunohistochemical Evaluation

Brain sections were incubated with anti-NRF2 antibody (1:250, Santa Cruz Biotechnology) as previously described by Cordaro et al. [49,50]. At the end of the protocol, the digital photos were analyzed by two observers blinded to the treatment as previously made in our laboratory [51–55].

2.6. Terminal Deoxynucleotidyl Nick-End Labeling (TUNEL) Assay

TUNEL staining for apoptotic cell nuclei and DAPI staining for all cell nuclei were performed in brain sections as described previously [56–59].

2.7. Immunofluorescence Evaluation

Brain sections were incubated with the following primary antibodies: HO-1 (Santa Cruz Biotechnology; 1:50 in PBS, *v/v*) or β-Tubulin (Santa Cruz Biotechnology, Heidelberg, Germany; 1:50) or MAP-2 (Santa Cruz Biotechnology; 1:50), as previously described [60–65]. After the incubation, sections were washed with PBS and incubated with secondary antibody FITC-conjugated anti-mouse Alexa Fluor-488 (1:2000 *v/v* Molecular Probes, UK) and 40,60-diamidino-2-phenylindole (DAPI; Hoechst, Frankfurt; Germany). Sections were observed and photographed using a Leica DM6 microscope (Leica Microsystems SpA, Milan, Italy) [66–70].

2.8. RT-qPCR

Total RNA was isolated from the hippocampus using TRIzol Reagent (Invitrogen, USA) following the manufacturer's recommendations. Reverse transcription of cDNA was performed according to the manufacturer's instructions as previously described [71]. The following primer sequences (5′–3′) were used: Beclin-1 (F) GCTGTAGCCAGCCTCTGAAA (R) AATGGCTCCTGTGAGTTCCTG; LC3B (F) GGGACCCTAACCCCATAGGA (R) TCTC-CCCCTTGTATCGCTCT; P62 (F) ACTGCTCAGGAGGAGACGAT (R) CCGGGGATCAGC-CTCTGTAG; Bcl-2 (F) GCGTCAACAGGGAGATGTCA (R) GCATGCTGGGGCCATATAGT; Bax (F) CTGGATCCAAGACCAGGGTG (R) GTGAGGACTCCAGCCACAAA β-Actin (F) ACACTCTCCCAGAAGGAGGG (R) TTTATAGGACGCCACAGCGG. Relative expression of mRNA was calculated by the delta–delta C_T method [72,73]. All values for the mRNA species were normalized to β-actin.

2.9. Western Blot Analysis of Cytosolic and Nuclear Extracts

Extracts from cytosol and nucleus were prepared as previously described and incubated with the antibodies anti-NRF-2 (1–500, SCB, Heidelberg, Germany, #sc-365949), anti-heme oxygenase 1 (HO-1; 1-500, SCB, Heidelberg, Germany, #sc-136960), anti-β-tubulin (1–500, SCB, Heidelberg, Germany, #sc-166729), and anti-MAP2 (1–500, SCB, Heidelberg, Germany, #sc-74421) in $1\times$ PBS, 5% *w/v* non-fat dried milk and 0.1% Tween-20 at 4 °C overnight [46–48,74–83]. For the cytosolic and nuclear fraction, blots were also probed with β-actin and lamin A/C protein to ensure that they were filled with equivalent amounts of proteins (1:500; Santa Cruz Biotechnology). Signals were detected as previously described in our works [49,84–86].

2.10. Materials

All compounds were purchased from Sigma-Aldrich (Milan, Italy). All solutions used for in vivo infusions were prepared using nonpyrogenic saline (0.9% NaCl; Baxter Healthcare Ltd., Thetford, Norfolk, UK).

2.11. Statistical Evaluation

In this study, the data are expressed as the average $\pm$ SEM and represent at least three experiments carried out on different days. For in vivo studies, N represents the number of animals used. The number of animals used for in vivo studies was determined by G*Power 3.1 software (Die Heinrich-Heine-Universität Düsseldorf, Düsseldorf, Germany). The images used in the histology, immunofluorescence, and immunohistochemistry came from at least three independent investigations. For multiple comparisons, a one-way ANOVA was employed, then a Bonferroni post hoc analysis; 0.05 was a significant *p*-value.

3. Results

3.1. Açai Berry Improve Behavioral Changes Vascular Dementia-Induced

We observed that after the unfamiliar mouse was placed in the three-chamber test arena, the number of contacts (Figure 1A) considerably increased while the total duration of contacts (Figure 1B) decreased compared to the control group. On the other hand, after oral açai administration for 15 days at the dose of 500 mg/kg, we noticed that the behavior was more similar to the control group than the vehicle group. Additionally, we assessed alterations in cognitive function using the novel object recognition test (Figure 1C,D). Controls and VaD animals did not significantly differ in their exploration of items during training; however, 15 days after carotid artery ligation, VaD animals dramatically decreased their preference for the novel object, indicating compromised cognitive function. However, after taking açai orally for 15 days at a dose of 500 mg/kg, we found that the behavior was more like the control group than the vehicle group.

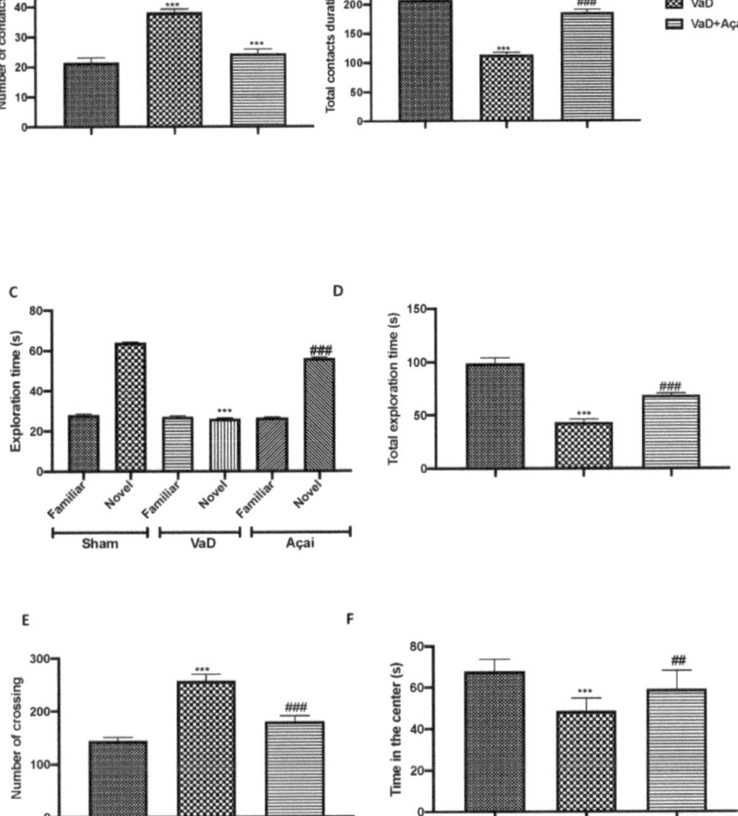

Figure 1. Açai berry improves vascular-dementia-induced behavioral changes. (**A**) Number of contacts and (**B**) total contact duration measured during social interaction test; (**C**) time spent to explore familiar or novel object, and (**D**) total exploration time measured during novel object recognition test; (**E**) number of crossing and (**F**) time in the center measured during open field test. Values are means ± SEM of six mice for all groups. See manuscript for further details. *** $p < 0.001$ vs. sham; ### $p < 0.001$ vs. VaD; ## $p < 0.01$ vs. VaD.

3.2. Açai Berry Limits Histological Changes Vascular Dementia-Induced

At 15 days after the induction of vascular dementia, all slices were stained with H&E to observe the severity of the hippocampal areas' damage. As showed in Figure 2B (for CA1 region) and 2E (for CA3 region), the hippocampal regions of the animals that had suffered VaD damage exhibited disorderly and rigidly labeled neurons. After daily oral administration of açai, brain sections showed a clear reorganization of the hippocampal CA1 and CA3 regions with an increased number of hippocampal neurons (Figure 2C for CA1 region, 2F for CA3 region). No alteration was observed in sham animals (Figure 2A for CA1 region, 2D for CA3 region; see histological score G).

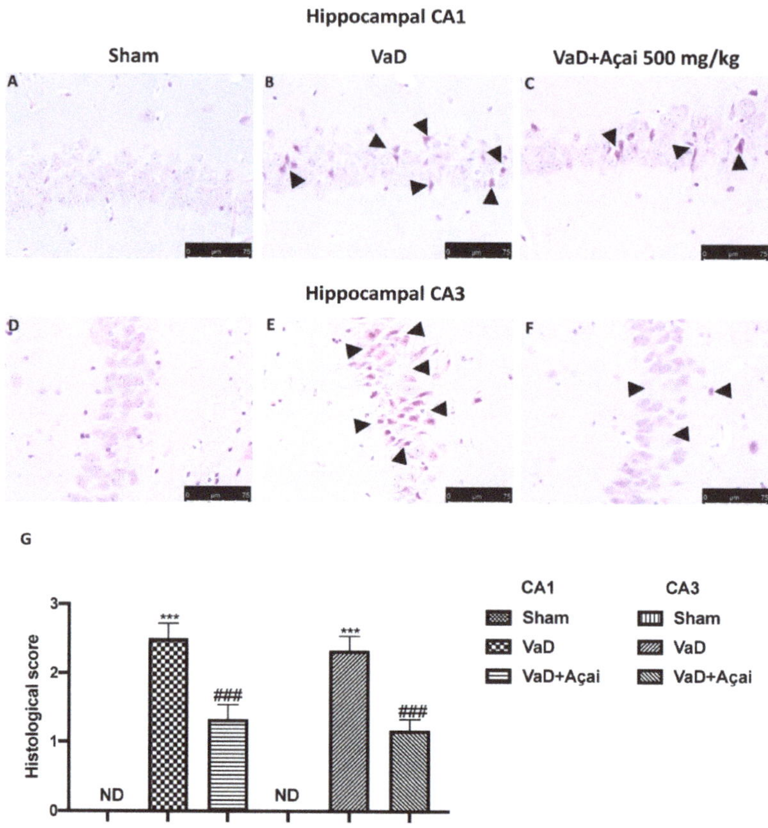

Figure 2. Açai berry limits vascular-dementia-induced histological changes. Hippocampal CA1 figure of Sham (**A**), VaD (**B**), Açai Berry (**C**); Hippocampal CA3 figure of Sham (**D**), VaD (**E**), Açai Berry (**F**); Histological score (**G**). Arrows indicate altered neurons and loss of hippocampal architecture. Figures are representative from least three experiments. Values are means ± SEM of six mice for all groups. See manuscript for further details. *** $p < 0.001$ vs. sham; ### $p < 0.001$ vs. VaD.

3.3. Açai Berry Limits Neuronal Death in the Hippocampus

Next, we investigated whether açai berry could attenuate neuronal death in the hippocampal structure, which is an essential feature in VaD patients [87]. For this, we used cresil violet and TUNEL staining to assess the effects of açai berry administration on the loss of pyramidal neurons in the hippocampus, especially in the CA1 and CA3 subregion. As showed in Figure 3B,E and in Figure 4B,E, the vehicle group was characterized by an intense neuronal loss due to the death of pyramidal neurons in both CA1 and CA3 regions compared to the control group, Figure 3A,D and in Figure 4A,D. On the other hand, daily administration of açai berry at the dose of 500 mg/kg was able to significantly reduce neuronal death in both regions (Figures 3C,F and 4C,F; see relative score, Figures 3G and 4G).

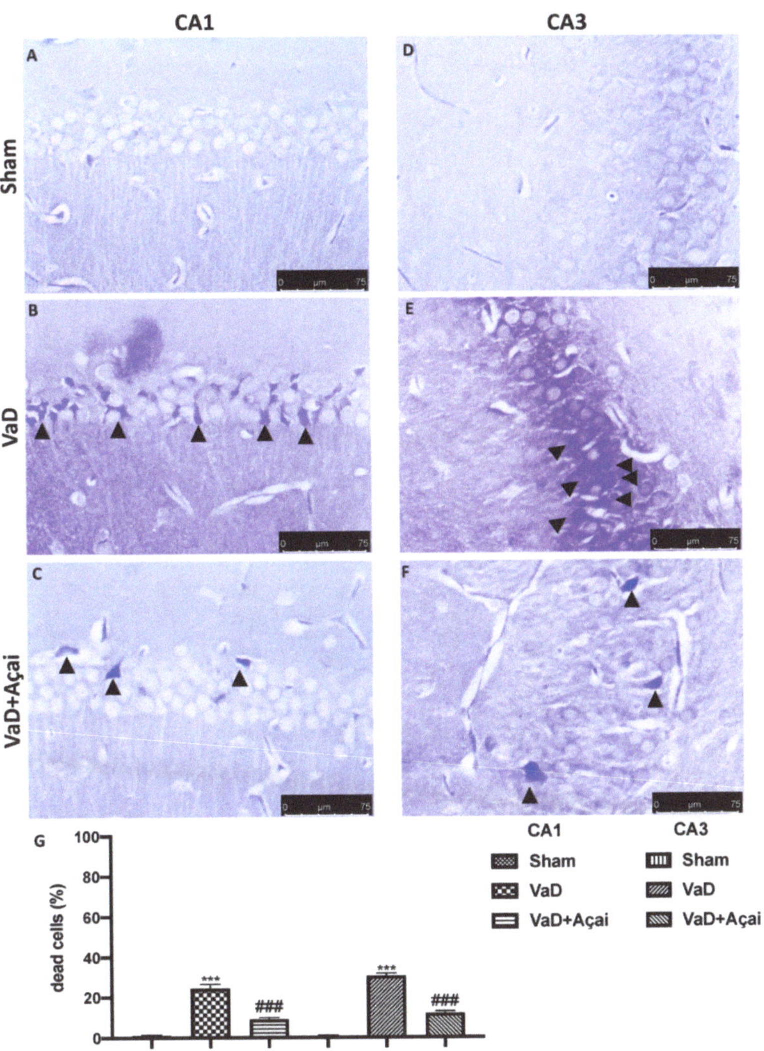

Figure 3. Açai berry limits neuronal hippocampal deaths. Hippocampal CA1 figure of Sham (**A**), VaD (**B**), Açai Berry (**C**); Hippocampal CA3 figure of Sham (**D**), VaD (**E**), Açai Berry (**F**); count of dead cells (**G**). Arrows indicate death neurons in both hippocampal sections. Figures are representative from at least three experiments. Values are means ± SEM of six mice for all groups. See manuscript for further details. *** $p < 0.001$ vs. sham; ### $p < 0.001$ vs. VaD.

3.4. Açai Berry Modulates Nrf-2 Pathways

In our work, by immunohistochemistry staining and western blot analysis, we found an increase in NRF-2 expression after VaD induction (see Figure 5B for CA1, Figure 5E for CA3, Figure 5G for quantification, and Figure 5H,I for western blot) compared to the control group (see Figure 5A for CA1, Figure 5D for CA3, Figure 5G for quantification, and Figure 5H,I for western blot). As expected, after açai berry administration, the physiological response was significant (see Figure 5C for CA1, Figure 5F for CA3, Figure 5G for quantification, and Figure 5H,I for western blot). The same trend was also observed in the expression of HO-1, the most important enzyme regulated by Nrf-2 (see Figure 6A,A1,A2 for CA1 region and Figure 6D,D1,D2 for CA3 region of Sham group, Figure 6B,B1,B2 for CA1 region and Figure 6E,E1,E2 for CA3 region of VaD group, Figure 6C,C1,C2 for CA1 region and Figure 6F,F1,F2 for CA3 region of Açai Berry group, and Figure 6H,I for western blot).

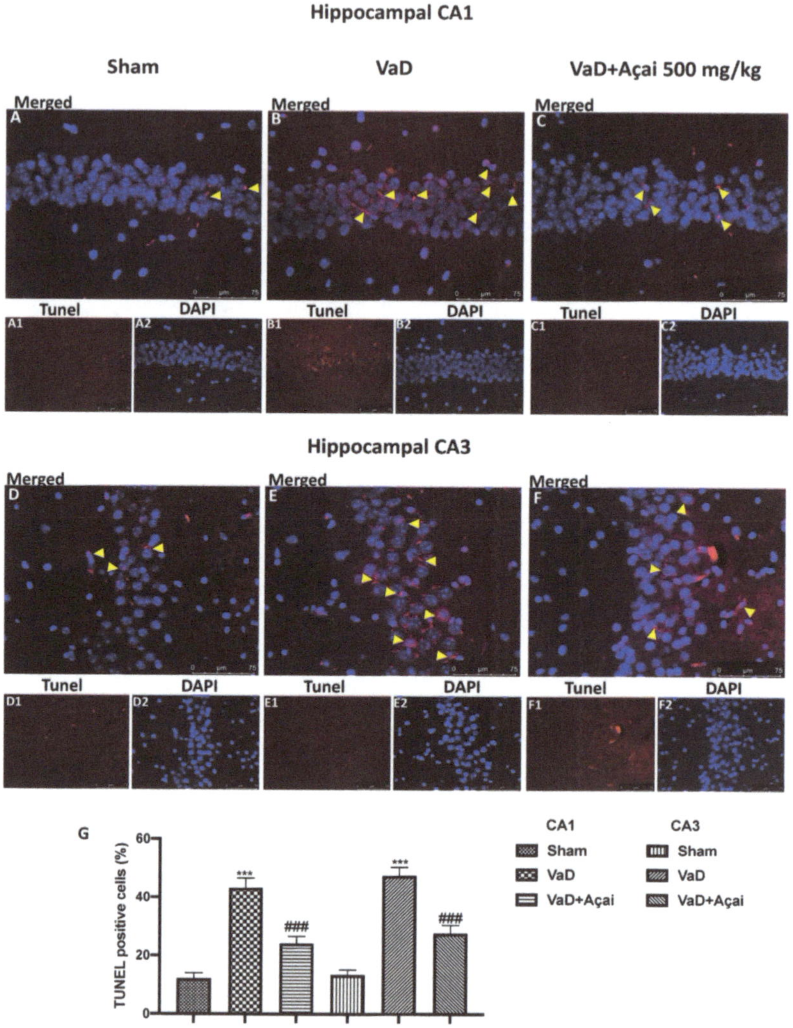

Figure 4. Açai berry reduces hippocampal TUNEL positive staining. Hippocampal CA1 figure of Sham (**A,A1,A2**), VaD (**B,B1,B2**), Açai Berry (**C,C1,C2**); Hippocampal CA3 figure of Sham (**D,D1,D2**), VaD (**E,E1,E2**), Açai Berry (**F,F1,F2**); count of dead cells (**G**). Arrows indicate TUNEL positive cells, merged with DAPI. Figures are representative from at least three experiments. Values are means $\pm$ SEM of six mice for all groups. See manuscript for further details. *** $p < 0.001$ vs. sham; ### $p < 0.001$ vs. VaD.

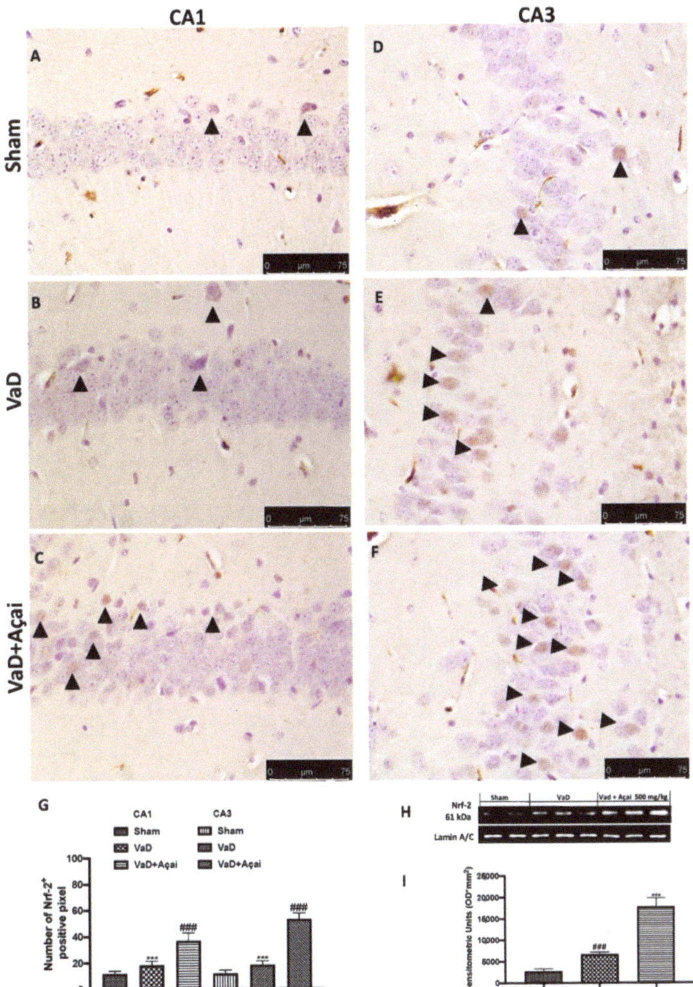

Figure 5. Açai berry improve NRF-2 expression. Hippocampal CA1 figure of Sham (**A**), VaD (**B**), Açai Berry (**C**); Hippocampal CA3 figure of Sham (**D**), VaD (**E**), Açai Berry (**F**); Number of Nrf-2 positive pixel (**G**); (**H**) western blot of NRF-2 expression and (**I**) relative densitometric analysis. Arrows indicate Nrf-2 positive cells. Figures are representative from at least three experiments. Values are means ± SEM of six mice for all groups. See manuscript for further details. *** $p < 0.001$ vs. sham; ### $p < 0.001$ vs. VaD.

3.5. Açai Berry Modulates Apoptotic and Autophagic Pathways

To further investigated neuronal death, we made RT-qPCR for Bax, and Bcl-2 and we found that after VaD induction there was a significant increase in Bax expression and a significant decrease in Bcl-2 expression compared to sham group (Figure 7A,B). Vice versa, the daily oral administration of açai berry at a dose of 500 mg/kg was able to restore both apoptotic markers to almost physiological values (Figure 7A,B). Autophagy and apoptosis interact in a complex way because, depending on the biological context, it can either promote cell survival or cell death [88]. In our study, açai berry supplementation increased autophagy in the hippocampus by activating Beclin1 (Figure 7C) decreasing LC3B turnover (Figure 7D), which led also to a decrease in p62 accumulation (Figure 7E).

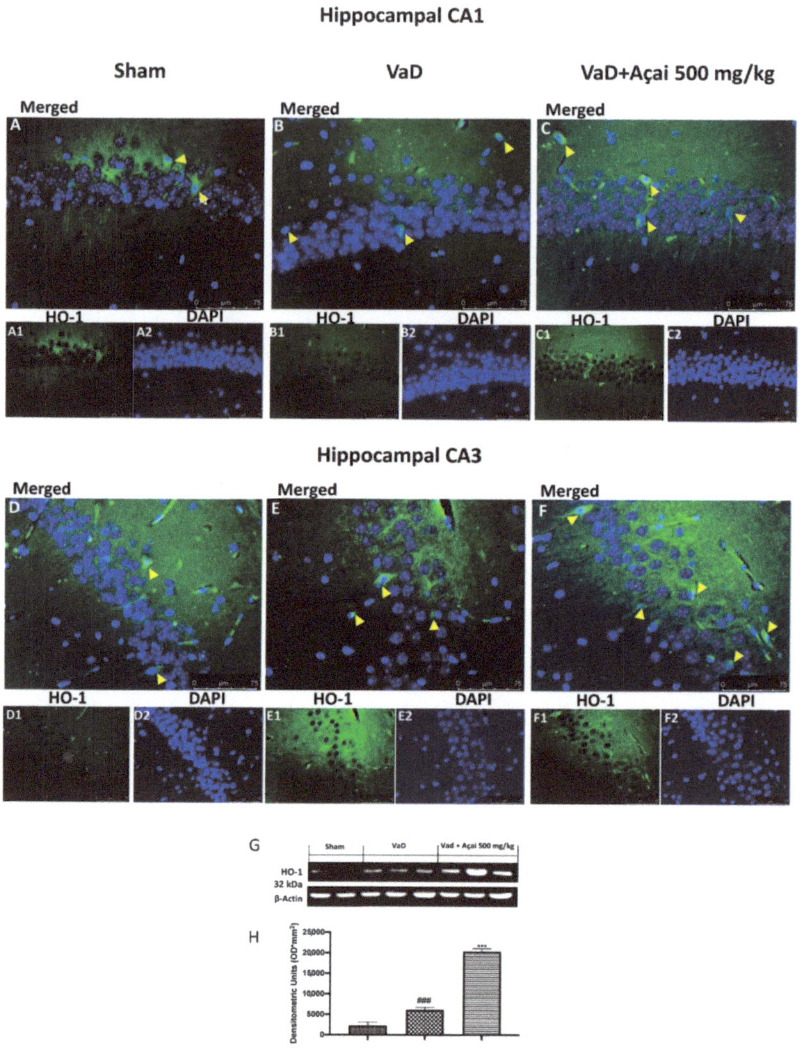

Figure 6. Açai berry increase HO-1 expression. Hippocampal CA1 figure of Sham (**A,A1,A2**), VaD (**B,B1,B2**), Açai Berry (**C,C1,C2**); Hippocampal CA3 figure of Sham (**D,D1,D2**), VaD (**E,E1,E2**), Açai Berry (**F,F1,F2**); (**G**) western blot of HO-1 and (**H**) relative densitometric analysis. Arrows indicate HO-1-DAPI merged cells in both hippocampal sections. Values are means $\pm$ SEM of six mice for all groups. Figures are representative from at least three experiments. See manuscript for further details. *** $p < 0.001$ vs. sham; ### $p < 0.001$ vs. VaD.

3.6. Açai Berry Mitigates Brain Structure Change VaD-Induced

It is well-known that reduced levels of tubulin were found in patients with VaD [3]. In our study, we detected by immunofluorescence a significant decrease of b-tubulin in animals subjected to the operation compared to Sham group in both hippocampal regions. This decrease was significantly reduced after daily administration of açai berry at the dose of 500 mg/kg (see Figure 8A,A1,A2 for CA1 region and Figure 8D,D1,D2 for CA3 region of Sham group, Figure 8B,B1,B2 for CA1 region and Figure 8E,E1,E2 for CA3 region of VaD group, Figure 8C,C1,C2 for CA1 region and Figure 8F,F1,F2 for CA3 region of Açai

Berry group). The same result was also observed with the immunofluorescence analysis of MAP-2. In this case as well, we observed a significant reduction of MAP-2 expression after VaD induction compared to the control group, which was significant restored after açai berry administration (see Figure 9A,A1,A2 for CA1 region and Figure 9D,D1,D2 for CA3 region of sham group, Figure 9B,B1,B2 for CA1 region and Figure 9E,E1,E2 for CA3 region of VaD group, Figure 9C,C1,C2 for CA1 region and Figure 9F,F1,F2 for CA3 region of Açai Berry group). These results were also confirmed by western blots analysis of β-tubulin and MAP-2 (see Figure 8G,H and Figure 9G,H, respectively).

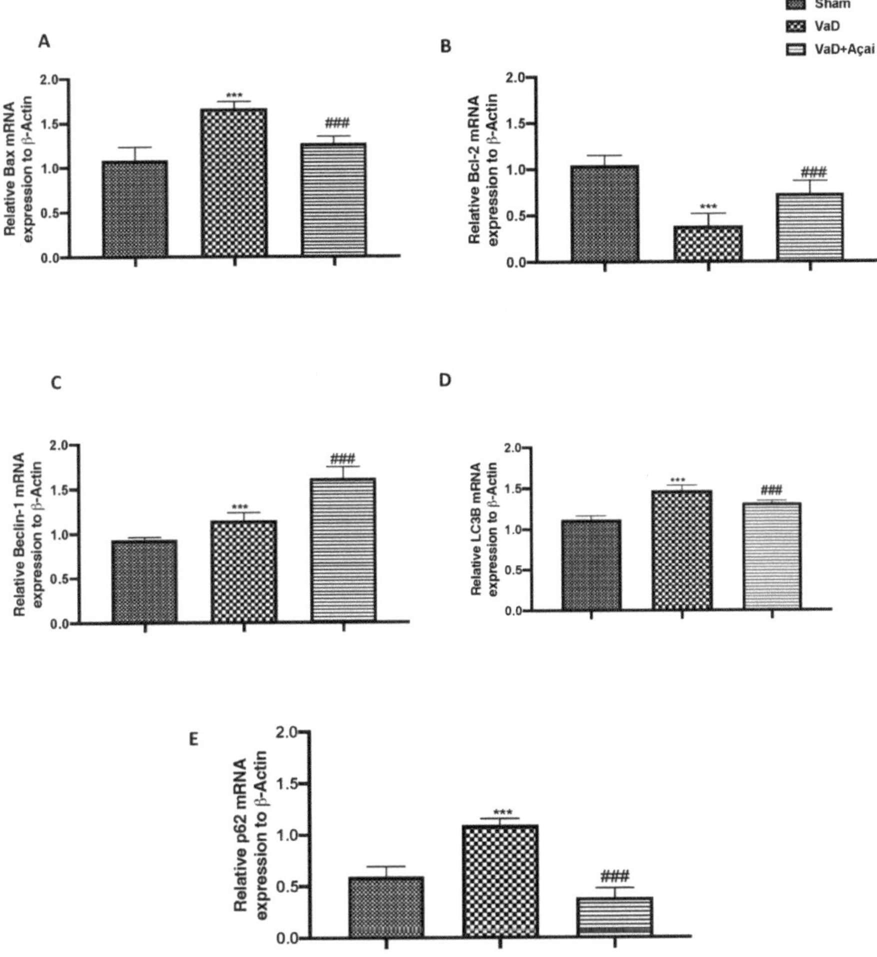

Figure 7. Açai berry modulates apoptotic and autophagic pathways. (**A**) Bax, (**B**) Bcl-2, (**C**) Beclin-1, (**D**) LC3B, (**E**) p62, mRNA levels in the hippocampus. Figures are representative from at least three experiments. Values are means ± SEM of six mice for all groups. See manuscript for further details. *** $p < 0.001$ vs. sham; ### $p < 0.001$ vs. VaD.

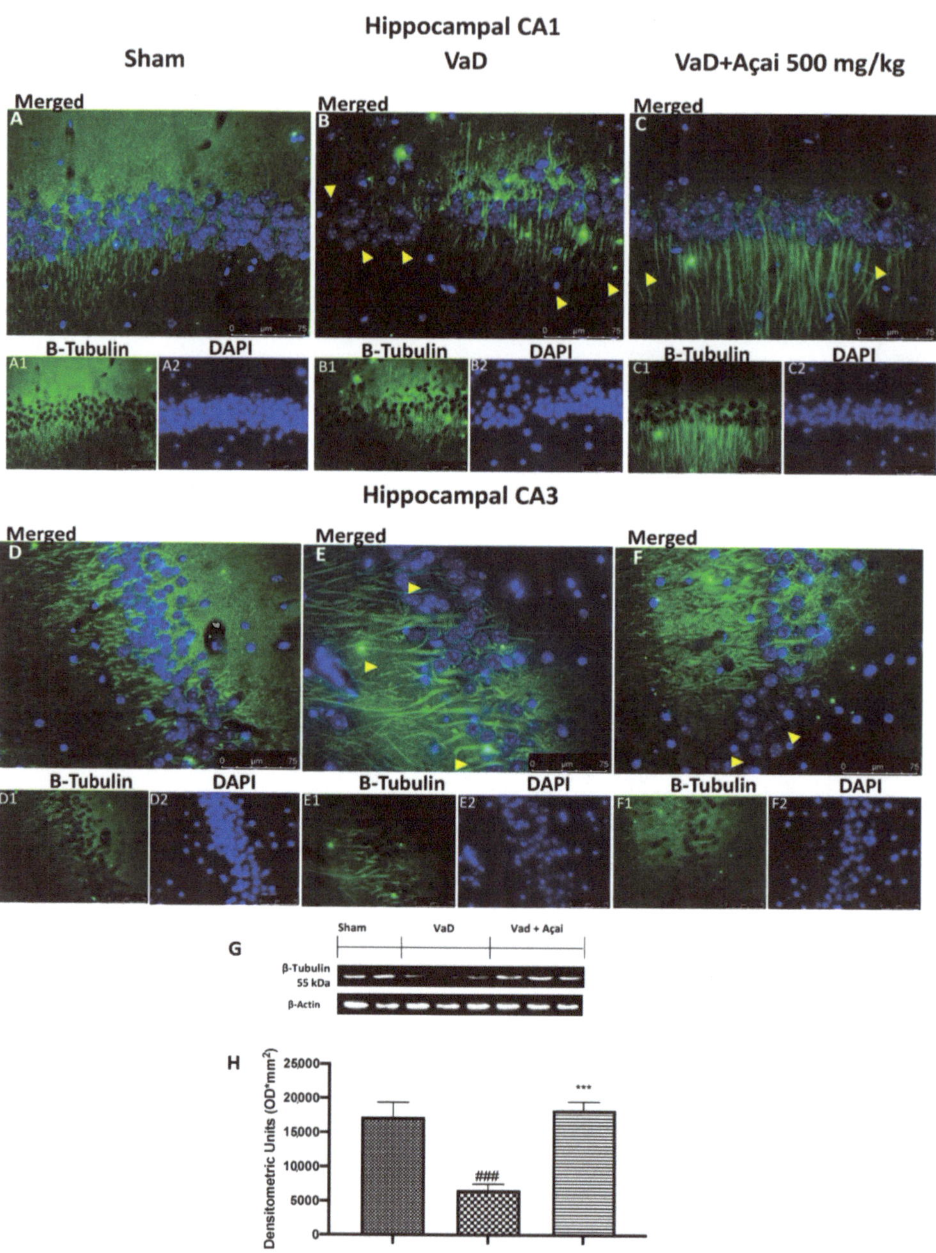

Figure 8. Açai berry mitigates β-tubulin alteration. Hippocampal CA1 figure of Sham (**A**,**A1**,**A2**), VaD (**B**,**B1**,**B2**), Açai Berry (**C**,**C1**,**C2**); Hippocampal CA3 figure of Sham (**D**,**D1**,**D2**), VaD (**E**,**E1**,**E2**), Açai Berry (**F**,**F1**,**F2**); β-tubulin western blot (**G**) and relative densitometric analysis (**H**). Arrows indicate the alteration or absence of β-tubulin. Values are means ± SEM of six mice for all groups. Figures are representative from at least three experiments. See manuscript for further details. *** $p < 0.001$ vs. sham; ### $p < 0.001$ vs. VaD.

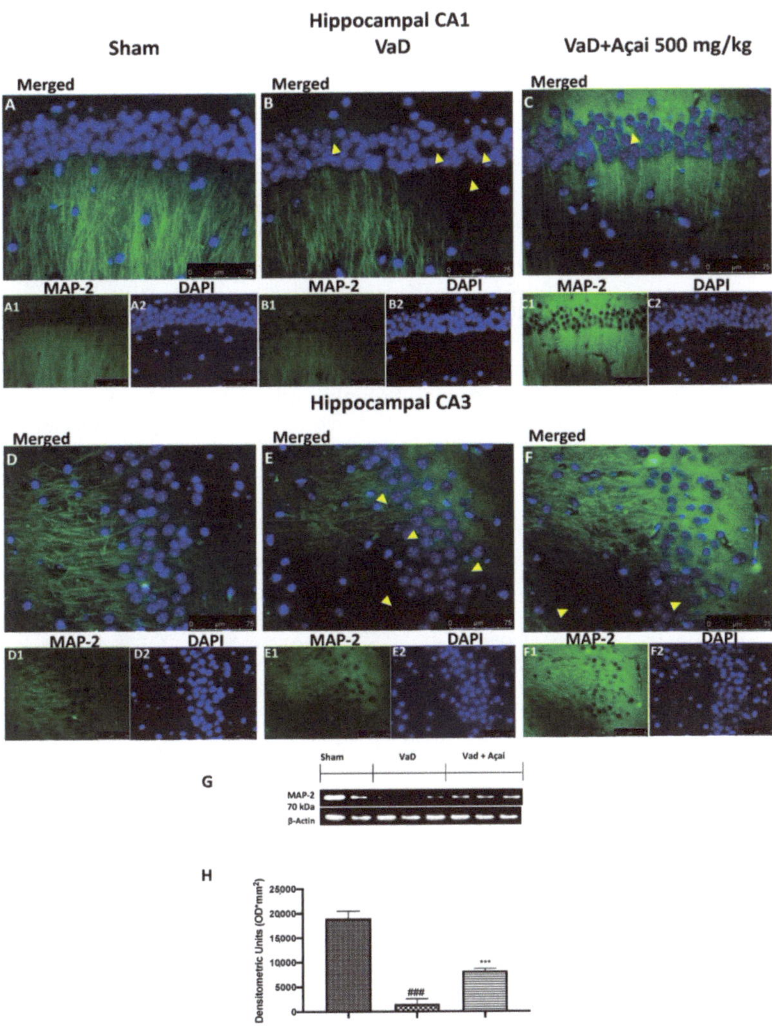

Figure 9. Açai berry mitigates MAP-2 alteration. Hippocampal CA1 figure of Sham (**A,A1,A2**), VaD (**B,B1,B2**), Açai Berry (**C,C1,C2**); Hippocampal CA3 figure of Sham (**D,D1,D2**), VaD (**E,E1,E2**), Açai Berry (**F,F1,F2**); MAP-2 western blot (**G**), and relative densitometric analysis (**H**) Arrows indicate the alteration or absence of MAP-2. Values are means ± SEM of 6 mice for all groups. Figures are representative from at least three experiments. See manuscript for further details. *** $p < 0.001$ vs. sham; ### $p < 0.001$ vs. VaD.

4. Discussion

The pathologic disease known as VaD, which affects the elderly, is characterized by disturbed cerebral blood flow. Loss of judgment, reason, and especially cognitive and memory abilities are its symptoms [89]. Advanced age, previous stroke episodes, hypertension, higher blood pressure, diabetes mellitus, smoking, high hematocrit, changes in hemostasis, dyslipoproteinemia, high alcohol consumption, aspirin use, psychological stress in early life, occupational exposure to pesticides, herbicides, liquid plastic or rubber, and premorbid personality traits made the subjects more susceptible to developing VaD [90]. Neuronal degeneration, axonal injury, and white matter injury in VaD are caused by a number of processes. They frequently share vascular dysfunction-induced common molecular and regulatory pathways of neuronal survival and death. Of these, oxidative stress, apoptosis, and autophagy pathways typically demonstrate different processes of their own. Along with age and other risk factors acting in concert, a crosstalk between these signaling pathways contributes to myelinated axonal damage and cerebrovascular diseases. The diseased characteristics cause changes in the health of the neuronal population as well as a series of harmful events and white matter destruction. Cognitive impairment and brain atrophy are the results of these diseased traits.

Although numerous papers describe the pathologic mechanisms behind vascular dementia, it is uncertain how the signaling pathways involved in vascular neuropathologies and cerebrovascular dysfunction interact with one another [91]. Accumulated data from numerous studies show that supplementing the diet with fruits, nuts, and vegetables of vibrant colors enhances the motor, memory, and cognitive abilities of both people and animals [5,66–70]. The antioxidant defense system may be compromised by increased oxidative stress brought on by a poor diet. On the other hand, a nutritious diet that is well-balanced and rich in a variety of nutrients, such as numerous servings of fruits and vegetables, low dosages of omega-3 fatty acids, tea, coffee, and wine, may provide neuroprotection [92,93].

The new food that is discussed in this paper is a fruit from the Euterpe genus of tropical palm trees that is native to South America and is commonly referred to as "Açai." Researchers have been investigating *Euterpe oleracea* due to its high antioxidant content when compared to other fruits and berries. The composition of açai pulp has also been studied, and it contains a number of phytochemicals that have physiological activity. Numerous studies have demonstrated the neuroprotective properties of açai berries [30]. The causes of many of these diseases are multifactorial, including oxidative stress, chronic neuroinflammation, excitotoxicity, mitochondrial dysfunction, alteration in autophagy, abnormal protein accumulation in brain tissues, and other cellular aetiologias. These diseases are brought on by a combination of aging, genetic disorders, and exposure to one or more environmental factors. Açai berry extracts have been shown in studies to have neuroprotective effects by, among other things, restoring calcium homeostasis and mitochondrial function, preventing the formation of toxic protein aggregates, and demonstrating antioxidant and anti-inflammatory activities. Additionally, açai fruit has anticonvulsant and antidepressant effects that may be beneficial for those with certain neurological conditions [94–101].

Keeping this aim in our mind, we decided to explore the neuroprotective effects of açai berry administration in a murine model of VaD. Depressive, anxious, or apathetic mood swings are common in people with vascular dementia. A person with vascular dementia may start acting strangely as their condition worsens. For instance, they might become angrier, hostile, or perplexed [102,103]. In our study, using different behavioral tests, we found that açai supplementation was able to reduce VaD-induced behavioral decline. It is well-known that a strong relationship exists between cognitive impairment and neuronal loss in the hippocampus. For this, we investigated by different staining what happen in CA1 and CA3. In particular, we found that both hippocampal sections showed, after VaD induction, a significant increase in altered and dead neurons, compared to Sham animals. After açai berry supplementation at the dose of 500 mg/kg, we notice an improvement in histological alteration as well as a decrease in dead neurons in both hippocampal regions.

Nrf2 is the master regulator of redox status and controls the transcription of a panel of antioxidative and anti-inflammatory genes. By cooperating with NF-κB (nuclear factor κB), Nrf2 also coordinates cellular oxidative and inflammatory balance. Nrf2 deficiency is linked to aging, and mounting evidence supports Nrf2's role in slowing the VaD process. By reducing neuroinflammation and oxidative stress, targeting Nrf2 has become a popular strategy for treating neurodegenerative diseases [104–106]. In our study, we found that açai berry supplementation was able to improve physiological anti-oxidant defense, improving Nrf-2 expression as well as HO-1, reducing oxidative stress. Additionally, recent studies suggest that autophagy and the Kelch-like ECH-associated protein

1 (Keap1-Nrf2) signaling pathway cooperate to avoid oxidative injury in the brain, and that p62 phosphorylation activates Nrf2 in response to oxidative stress. This is particularly intriguing because recent articles have demonstrated dynamic Beclin-1 alterations in the hippocampus of male mice with VaD. Beclin-1 is a crucial regulator of several trafficking pathways, including autophagy and receptor recycling in microglia. [71,107]. Additionally, it has also been demonstrated that Beclin 1 regulates signaling through the neuroprotective TGF-β pathway by decreasing TGF-β1 receptor II (TBRII). TGF-β1 plays a key function in hippocampus synaptic plasticity, memory, and neuronal survival. It also regulates brain homeostasis [108–110]. In our study, we found that açai berry restored autophagic flux at physiological levels. Interesting studies have shown that the altered expression of MAP-2 and β-tubulin leads to neuronal death; such results could be used in the future target for neuroprotective therapy targeting [111–114]. In our work, we found that açai berry stops the process of cell death and neurodegeneration and restores the microtubule network.

5. Conclusions

In conclusion, we demonstrate for the first time that açai berry supplementation at a dose of 500 mg/kg was able to enhance the physiological anti-oxidant defense, suggesting a possible protective role during VaD events.

Supplementary Materials: The following are available online at https://www.mdpi.com/article/10.3390/cells11162616/s1, Supplementary File: Technical Datasheet of EUTERPE OLERACEA P.E. 10% POLYPHENOLS.

Author Contributions: Conceptualization, R.S. and M.C.; Formal analysis, L.I. and D.D.P.; Funding acquisition, R.D.P. and S.C.; Methodology, D.I., R.D. and E.G.; Software, R.F. and T.G.; Supervision, R.D.P.; Validation, A.F.P. and R.C.; Writing—original draft, R.S.; Writing—review & editing, M.C. All authors have read and agreed to the published version of the manuscript.

Funding: This research received no external funding.

Institutional Review Board Statement: The animal study protocol was approved by the Institutional Review Board of University of Messina.

Informed Consent Statement: Not applicable.

Data Availability Statement: The data used to support the findings of this study are available from the corresponding author upon request.

Conflicts of Interest: The authors declare no conflict of interest.

References

1. Orgogozo, J.M.; Abadie, E. Vascular dementia: European perspectives. *Alzheimer Dis. Assoc. Disord.* **1999**, *13* (Suppl. 3), S192–S200. [CrossRef] [PubMed]
2. Baskys, A.; Hou, A.C. Vascular dementia: Pharmacological treatment approaches and perspectives. *Clin. Interv. Aging* **2007**, *2*, 327. [CrossRef] [PubMed]
3. Santiago-Mujika, E.; Luthi-Carter, R.; Giorgini, F.; Kalaria, R.N.; Mukaetova-Ladinska, E.B. Tubulin and Tubulin Posttranslational Modifications in Alzheimer's Disease and Vascular Dementia. *Front. Aging Neurosci.* **2021**, *13*, 730107. [CrossRef] [PubMed]
4. Sabbatini, M.; Catalani, A.; Consoli, C.; Marletta, N.; Tomassoni, D.; Avola, R. The hippocampus in spontaneously hypertensive rats: An animal model of vascular dementia? *Mech. Ageing Dev.* **2002**, *123*, 547–559. [CrossRef]
5. Poulose, S.M.; Bielinski, D.F.; Carey, A.; Schauss, A.G.; Shukitt-Hale, B. Modulation of oxidative stress, inflammation, autophagy and expression of Nrf2 in hippocampus and frontal cortex of rats fed with acai-enriched diets. *Nutr. Neurosci.* **2017**, *20*, 305–315. [CrossRef]
6. Jiang, T.; Harder, B.; Rojo de la Vega, M.; Wong, P.K.; Chapman, E.; Zhang, D.D. p62 links autophagy and Nrf2 signaling. *Free Radic. Biol. Med.* **2015**, *88*, 199–204. [CrossRef]
7. Jo, C.; Gundemir, S.; Pritchard, S.; Jin, Y.N.; Rahman, I.; Johnson, G.V. Nrf2 reduces levels of phosphorylated tau protein by inducing autophagy adaptor protein NDP52. *Nat. Commun.* **2014**, *5*, 3496. [CrossRef]
8. Shah, S.Z.A.; Zhao, D.; Hussain, T.; Sabir, N.; Mangi, M.H.; Yang, L. p62-Keap1-NRF2-ARE Pathway: A Contentious Player for Selective Targeting of Autophagy, Oxidative Stress and Mitochondrial Dysfunction in Prion Diseases. *Front. Mol. Neurosci.* **2018**, *11*, 310. [CrossRef]

9. Zhang, L.; Wang, H.; Fan, Y.; Gao, Y.; Li, X.; Hu, Z.; Ding, K.; Wang, Y.; Wang, X. Fucoxanthin provides neuroprotection in models of traumatic brain injury via the Nrf2-ARE and Nrf2-autophagy pathways. *Sci. Rep.* **2017**, *7*, 46763. [CrossRef]

10. Xiao, X.T.; He, S.Q.; Wu, N.N.; Lin, X.C.; Zhao, J.; Tian, C. Green Tea Polyphenols Prevent Early Vascular Aging Induced by High-Fat Diet via Promoting Autophagy in Young Adult Rats. *Curr. Med. Sci.* **2022**. *Epub ahead of print.* [CrossRef]

11. Arias, C.; Salazar, L.A. Autophagy and Polyphenols in Osteoarthritis: A Focus on Epigenetic Regulation. *Int. J. Mol. Sci.* **2021**, *23*, 421. [CrossRef]

12. Brimson, J.M.; Prasanth, M.I.; Malar, D.S.; Thitilertdecha, P.; Kabra, A.; Tencomnao, T.; Prasansuklab, A. Plant Polyphenols for Aging Health: Implication from Their Autophagy Modulating Properties in Age-Associated Diseases. *Pharmaceuticals* **2021**, *14*, 982. [CrossRef]

13. Garcia-Aguilar, A.; Palomino, O.; Benito, M.; Guillen, C. Dietary Polyphenols in Metabolic and Neurodegenerative Diseases: Molecular Targets in Autophagy and Biological Effects. *Antioxidants* **2021**, *10*, 142. [CrossRef]

14. Musial, C.; Siedlecka-Kroplewska, K.; Kmiec, Z.; Gorska-Ponikowska, M. Modulation of Autophagy in Cancer Cells by Dietary Polyphenols. *Antioxidants* **2021**, *10*, 123. [CrossRef]

15. Abbaszadeh, F.; Fakhri, S.; Khan, H. Targeting apoptosis and autophagy following spinal cord injury: Therapeutic approaches to polyphenols and candidate phytochemicals. *Pharmacol. Res.* **2020**, *160*, 105069. [CrossRef]

16. Wu, M.; Luo, Q.; Nie, R.; Yang, X.; Tang, Z.; Chen, H. Potential implications of polyphenols on aging considering oxidative stress, inflammation, autophagy, and gut microbiota. *Crit. Rev. Food Sci. Nutr.* **2021**, *61*, 2175–2193. [CrossRef]

17. Zhou, H.; Chen, Y.; Huang, S.W.; Hu, P.F.; Tang, L.J. Regulation of autophagy by tea polyphenols in diabetic cardiomyopathy. *J. Zhejiang Univ. Sci. B* **2018**, *19*, 333–341. [CrossRef]

18. Shabalala, S.; Muller, C.J.F.; Louw, J.; Johnson, R. Polyphenols, autophagy and doxorubicin-induced cardiotoxicity. *Life Sci.* **2017**, *180*, 160–170. [CrossRef]

19. Zhang, P.W.; Tian, C.; Xu, F.Y.; Chen, Z.; Burnside, R.; Yi, W.J.; Xiang, S.Y.; Xie, X.; Wu, N.N.; Yang, H.; et al. Green Tea Polyphenols Alleviate Autophagy Inhibition Induced by High Glucose in Endothelial Cells. *Biomed. Environ. Sci.* **2016**, *29*, 524–528. [CrossRef]

20. Pallauf, K.; Rimbach, G. Autophagy, polyphenols and healthy ageing. *Ageing Res. Rev.* **2013**, *12*, 237–252. [CrossRef]

21. Maciel-Silva, F.W.; Buller, L.S.; MBB Gonçalves, M.L.; Rostagno, M.A.; Forster-Carneiro, T. Sustainable development in the Legal Amazon: Energy recovery from açaí seeds. *Biofuels Bioprod. Biorefin.* **2021**, *15*, 1174–1189. [CrossRef]

22. Melo, P.S.; Massarioli, A.P.; Lazarini, J.G.; Soares, J.C.; Franchin, M.; Rosalen, P.L.; Alencar, S.M. Simulated gastrointestinal digestion of Brazilian acai seeds affects the content of flavan-3-ol derivatives, and their antioxidant and anti-inflammatory activities. *Heliyon* **2020**, *6*, e05214. [CrossRef]

23. Melo, P.S.; Selani, M.M.; Gonçalves, R.H.; De Oliveira Paulino, J.; Massarioli, A.P.; De Alencar, S.M. Açaí seeds: An unexplored agro-industrial residue as a potential source of lipids, fibers, and antioxidant phenolic compounds. *Ind. Crops Prod.* **2021**, *161*, 113204. [CrossRef]

24. Rodrigues, R.B.; Lichtenthaler, R.; Zimmermann, B.F.; Papagiannopoulos, M.; Fabricius, H.; Marx, F.; Maia, J.G.; Almeida, O. Total oxidant scavenging capacity of *Euterpe oleracea* Mart. (acai) seeds and identification of their polyphenolic compounds. *J. Agric. Food Chem.* **2006**, *54*, 4162–4167. [CrossRef] [PubMed]

25. De Moura, R.S.; Pires, K.M.; Santos Ferreira, T.; Lopes, A.A.; Nesi, R.T.; Resende, A.C.; Sousa, P.J.; Da Silva, A.J.; Porto, L.C.; Valenca, S.S. Addition of acai (*Euterpe oleracea*) to cigarettes has a protective effect against emphysema in mice. *Food Chem. Toxicol.* **2011**, *49*, 855–863. [CrossRef]

26. Lee, J.Y.; Kim, N.; Choi, Y.J.; Nam, R.H.; Lee, S.; Ham, M.H.; Suh, J.H.; Choi, Y.J.; Lee, H.S.; Lee, D.H. Anti-inflammatory and Anti-tumorigenic Effects of Acai Berry in Helicobacter felis-infected mice. *J. Cancer Prev.* **2016**, *21*, 48–54. [CrossRef]

27. Moura, R.S.; Ferreira, T.S.; Lopes, A.A.; Pires, K.M.; Nesi, R.T.; Resende, A.C.; Souza, P.J.; Silva, A.J.; Borges, R.M.; Porto, L.C.; et al. Effects of *Euterpe oleracea* Mart. (ACAI) extract in acute lung inflammation induced by cigarette smoke in the mouse. *Phytomedicine* **2012**, *19*, 262–269. [CrossRef]

28. Poulose, S.M.; Fisher, D.R.; Larson, J.; Bielinski, D.F.; Rimando, A.M.; Carey, A.N.; Schauss, A.G.; Shukitt-Hale, B. Anthocyanin-rich acai (*Euterpe oleracea* Mart.) fruit pulp fractions attenuate inflammatory stress signaling in mouse brain BV-2 microglial cells. *J. Agric. Food Chem.* **2012**, *60*, 1084–1093. [CrossRef]

29. Santos, I.B.; De Bem, G.F.; Da Costa, C.A.; De Carvalho, L.; De Medeiros, A.F.; Silva, D.L.B.; Romao, M.H.; De Andrade Soares, R.; Ognibene, D.T.; De Moura, R.S.; et al. Acai seed extract prevents the renin-angiotensin system activation, oxidative stress and inflammation in white adipose tissue of high-fat diet-fed mice. *Nutr. Res.* **2020**, *79*, 35–49. [CrossRef]

30. ALNasser, M.N.; Mellor, I.R. Neuroprotective activities of acai berries (*Euterpe* sp.): A review. *J. Herbmed Pharmacol.* **2022**, *11*, 166–181. [CrossRef]

31. Kang, J.; Li, Z.; Wu, T.; Jensen, G.S.; Schauss, A.G.; Wu, X. Anti-oxidant capacities of flavonoid compounds isolated from acai pulp (*Euterpe oleracea* Mart.). *Food Chem.* **2010**, *122*, 610–617. [CrossRef]

32. De Moraes Arnoso, B.J.; Magliaccio, F.M.; De Araujo, C.A.; De Andrade Soares, R.; Santos, I.B.; De Bem, G.F.; Fernandes-Santos, C.; Ognibene, D.T.; De Moura, R.S.; Resende, A.C.; et al. Acai seed extract (ASE) rich in proanthocyanidins improves cardiovascular remodeling by increasing antioxidant response in obese high-fat diet-fed mice. *Chem. Biol. Interact.* **2022**, *351*, 109721. [CrossRef] [PubMed]

33. Bellucci, E.R.B.; Dos Santos, J.M.; Carvalho, L.T.; Borgonovi, T.F.; Lorenzo, J.M.; Silva-Barretto, A.C.D. Acai extract powder as natural antioxidant on pork patties during the refrigerated storage. *Meat Sci.* **2022**, *184*, 108667. [CrossRef] [PubMed]

34. Da Silva, T.V.N.; Torres, M.F.; Sampaio, L.A.; Hamoy, M.; Monserrat, J.M.; Barbas, L.A.L. Dietary *Euterpe oleracea* Mart. attenuates seizures and damage to lipids in the brain of Colossoma macropomum. *Fish Physiol. Biochem.* **2021**, *47*, 1851–1864. [CrossRef]

35. Kim, K.J.; Kim, Y.; Jin, S.G.; Kim, J.Y. Acai berry extract as a regulator of intestinal inflammation pathways in a Caco-2 and RAW 264.7 co-culture model. *J. Food Biochem.* **2021**, *45*, e13848. [CrossRef]

36. Silva, M.; Costa, J.H.; Pacheco-Fill, T.; Ruiz, A.; Vidal, F.C.B.; Borges, K.R.A.; Guimaraes, S.J.A.; Azevedo-Santos, A.P.S.; Buglio, K.E.; Foglio, M.A.; et al. Acai (*Euterpe oleracea* Mart.) Seed Extract Induces ROS Production and Cell Death in MCF-7 Breast Cancer Cell Line. *Molecules* **2021**, *26*, 3546. [CrossRef]

37. Martins, G.R.; Guedes, D.; Marques de Paula, U.L.; De Oliveira, M.; Lutterbach, M.T.S.; Reznik, L.Y.; Servulo, E.F.C.; Alviano, C.S.; Ribeiro da Silva, A.J.; Alviano, D.S. Acai (*Euterpe oleracea* Mart.) Seed Extracts from Different Varieties: A Source of Proanthocyanidins and Eco-Friendly Corrosion Inhibition Activity. *Molecules* **2021**, *26*, 3433. [CrossRef]

38. De Bem, G.F.; Okinga, A.; Ognibene, D.T.; Da Costa, C.A.; Santos, I.B.; Soares, R.A.; Silva, D.L.B.; Da Rocha, A.P.M.; Isnardo Fernandes, J.; Fraga, M.C.; et al. Anxiolytic and antioxidant effects of *Euterpe oleracea* Mart. (acai) seed extract in adult rat offspring submitted to periodic maternal separation. *Appl. Physiol. Nutr. Metab.* **2020**, *45*, 1277–1286. [CrossRef]

39. Remigante, A.; Spinelli, S.; Straface, E.; Gambardella, L.; Caruso, D.; Falliti, G.; Dossena, S.; Marino, A.; Morabito, R. Acai (*Euterpe oleracea*) Extract Protects Human Erythrocytes from Age-Related Oxidative Stress. *Cells* **2022**, *11*, 2391. [CrossRef]

40. Cordaro, M.; D'Amico, R.; Fusco, R.; Peritore, A.F.; Genovese, T.; Interdonato, L.; Franco, G.; Arangia, A.; Gugliandolo, E.; Crupi, R.; et al. Discovering the Effects of Fisetin on NF-kappaB/NLRP-3/NRF-2 Molecular Pathways in a Mouse Model of Vascular Dementia Induced by Repeated Bilateral Carotid Occlusion. *Biomedicines* **2022**, *10*, 1448. [CrossRef]

41. Siracusa, R.; Impellizzeri, D.; Cordaro, M.; Crupi, R.; Esposito, E.; Petrosino, S.; Cuzzocrea, S. Anti-Inflammatory and Neuroprotective Effects of Co-UltraPEALut in a Mouse Model of Vascular Dementia. *Front. Neurol.* **2017**, *8*, 233. [CrossRef]

42. Genovese, T.; D'Amico, R.; Fusco, R.; Impellizzeri, D.; Peritore, A.F.; Crupi, R.; Interdonato, L.; Gugliandolo, E.; Cuzzocrea, S.; Paola, R.D.; et al. Acai (*Euterpe oleraceae* Mart.) Seeds Regulate NF-kappaB and Nrf2/ARE Pathways Protecting Lung against Acute and Chronic Inflammation. *Cell Physiol. Biochem.* **2022**, *56*, 1–20. [CrossRef]

43. Prut, L.; Belzung, C. The open field as a paradigm to measure the effects of drugs on anxiety-like behaviors: A review. *Eur. J. Pharmacol.* **2003**, *463*, 3–33. [CrossRef]

44. Crupi, R.; Cambiaghi, M.; Spatz, L.; Hen, R.; Thorn, M.; Friedman, E.; Vita, G.; Battaglia, F. Reduced adult neurogenesis and altered emotional behaviors in autoimmune-prone B-cell activating factor transgenic mice. *Biol. Psychiatry* **2010**, *67*, 558–566. [CrossRef]

45. Pan, Z.; Cui, M.; Dai, G.; Yuan, T.; Li, Y.; Ji, T.; Pan, Y. Protective Effect of Anthocyanin on Neurovascular Unit in Cerebral Ischemia/Reperfusion Injury in Rats. *Front. Neurosci.* **2018**, *12*, 947. [CrossRef]

46. Di Paola, D.; Iaria, C.; Capparucci, F.; Cordaro, M.; Crupi, R.; Siracusa, R.; D'Amico, R.; Fusco, R.; Impellizzeri, D.; Cuzzocrea, S.; et al. Aflatoxin B1 Toxicity in Zebrafish Larva (*Danio rerio*): Protective Role of *Hericium erinaceus*. *Toxins* **2021**, *13*, 710. [CrossRef]

47. Di Paola, D.; Natale, S.; Gugliandolo, E.; Cordaro, M.; Crupi, R.; Siracusa, R.; D'Amico, R.; Fusco, R.; Impellizzeri, D.; Cuzzocrea, S.; et al. Assessment of 2-Pentadecyl-2-oxazoline Role on Lipopolysaccharide-Induced Inflammation on Early Stage Development of Zebrafish (*Danio rerio*). *Life* **2022**, *12*, 128. [CrossRef]

48. Di Paola, D.; Capparucci, F.; Abbate, J.M.; Cordaro, M.; Crupi, R.; Siracusa, R.; D'Amico, R.; Fusco, R.; Genovese, T.; Impellizzeri, D.; et al. Environmental Risk Assessment of Oxaliplatin Exposure on Early Life Stages of Zebrafish (*Danio rerio*). *Toxics* **2022**, *10*, 81. [CrossRef]

49. Cordaro, M.; Impellizzeri, D.; Gugliandolo, E.; Siracusa, R.; Crupi, R.; Esposito, E.; Cuzzocrea, S. Adelmidrol, a Palmitoylethanolamide Analogue, as a New Pharmacological Treatment for the Management of Inflammatory Bowel Disease. *Mol. Pharmacol.* **2016**, *90*, 549–561. [CrossRef]

50. Fusco, R.; Siracusa, R.; D'Amico, R.; Cordaro, M.; Genovese, T.; Gugliandolo, E.; Peritore, A.F.; Crupi, R.; Di Paola, R.; Cuzzocrea, S.; et al. Mucosa-Associated Lymphoid Tissue Lymphoma Translocation 1 Inhibitor as a Novel Therapeutic Tool for Lung Injury. *Int. J. Mol. Sci.* **2020**, *21*, 7761. [CrossRef]

51. Fusco, R.; Siracusa, R.; Peritore, A.F.; Gugliandolo, E.; Genovese, T.; D'Amico, R.; Cordaro, M.; Crupi, R.; Mandalari, G.; Impellizzeri, D.; et al. The Role of Cashew (*Anacardium occidentale* L.) Nuts on an Experimental Model of Painful Degenerative Joint Disease. *Antioxidants* **2020**, *9*, 511. [CrossRef]

52. Peritore, A.F.; Siracusa, R.; Fusco, R.; Gugliandolo, E.; D'Amico, R.; Cordaro, M.; Crupi, R.; Genovese, T.; Impellizzeri, D.; Cuzzocrea, S.; et al. Ultramicronized Palmitoylethanolamide and Paracetamol, a New Association to Relieve Hyperalgesia and Pain in a Sciatic Nerve Injury Model in Rat. *Int. J. Mol. Sci.* **2020**, *21*, 3509. [CrossRef]

53. Fusco, R.; Cordaro, M.; Siracusa, R.; D'Amico, R.; Genovese, T.; Gugliandolo, E.; Peritore, A.F.; Crupi, R.; Impellizzeri, D.; Cuzzocrea, S.; et al. Biochemical Evaluation of the Antioxidant Effects of Hydroxytyrosol on Pancreatitis-Associated Gut Injury. *Antioxidants* **2020**, *9*, 781. [CrossRef]

54. Sawant, S.; Gokulan, R.; Dongre, H.; Vaidya, M.; Chaukar, D.; Prabhash, K.; Ingle, A.; Joshi, S.; Dange, P.; Joshi, S.; et al. Prognostic role of Oct4, CD44 and c-Myc in radio-chemo-resistant oral cancer patients and their tumourigenic potential in immunodeficient mice. *Clin. Oral. Investig.* **2016**, *20*, 43–56. [CrossRef]

55. Siracusa, R.; Paterniti, I.; Cordaro, M.; Crupi, R.; Bruschetta, G.; Campolo, M.; Cuzzocrea, S.; Esposito, E. Neuroprotective Effects of Temsirolimus in Animal Models of Parkinson's Disease. *Mol. Neurobiol.* **2018**, *55*, 2403–2419. [CrossRef]

56. Gugliandolo, E.; D'Amico, R.; Cordaro, M.; Fusco, R.; Siracusa, R.; Crupi, R.; Impellizzeri, D.; Cuzzocrea, S.; Di Paola, R. Effect of PEA-OXA on neuropathic pain and functional recovery after sciatic nerve crush. *J. Neuroinflamm.* **2018**, *15*, 264. [CrossRef]

57. Genovese, T.; Siracusa, R.; D'Amico, R.; Cordaro, M.; Peritore, A.F.; Gugliandolo, E.; Crupi, R.; Trovato Salinaro, A.; Raffone, E.; Impellizzeri, D.; et al. Regulation of Inflammatory and Proliferative Pathways by Fotemustine and Dexamethasone in Endometriosis. *Int. J. Mol. Sci.* **2021**, *22*, 5998. [CrossRef] [PubMed]

58. Peritore, A.F.; Crupi, R.; Scuto, M.; Gugliandolo, E.; Siracusa, R.; Impellizzeri, D.; Cordaro, M.; D'Amico, R.; Fusco, R.; Di Paola, R.; et al. The Role of Annexin A1 and Formyl Peptide Receptor 2/3 Signaling in Chronic Corticosterone-Induced Depression-Like behaviors and Impairment in Hippocampal-Dependent Memory. *CNS Neurol. Disord. Drug Targets* **2020**, *19*, 27–43. [CrossRef]

59. Fusco, R.; Cordaro, M.; Siracusa, R.; Peritore, A.F.; Gugliandolo, E.; Genovese, T.; D'Amico, R.; Crupi, R.; Smeriglio, A.; Mandalari, G.; et al. Consumption of *Anacardium occidentale* L. (Cashew Nuts) Inhibits Oxidative Stress through Modulation of the Nrf2/HO-1 and NF-kB Pathways. *Molecules* **2020**, *25*, 4426. [CrossRef] [PubMed]

60. Cordaro, M.; Siracusa, R.; Crupi, R.; Impellizzeri, D.; Peritore, A.F.; D'Amico, R.; Gugliandolo, E.; Di Paola, R.; Cuzzocrea, S. 2-Pentadecyl-2-Oxazoline Reduces Neuroinflammatory Environment in the MPTP Model of Parkinson Disease. *Mol. Neurobiol.* **2018**, *55*, 9251–9266. [CrossRef] [PubMed]

61. Impellizzeri, D.; Cordaro, M.; Bruschetta, G.; Crupi, R.; Pascali, J.; Alfonsi, D.; Marcolongo, G.; Cuzzocrea, S. 2-pentadecyl-2-oxazoline: Identification in coffee, synthesis and activity in a rat model of carrageenan-induced hindpaw inflammation. *Pharmacol. Res.* **2016**, *108*, 23–30. [CrossRef]

62. Cordaro, M.; Siracusa, R.; Fusco, R.; D'Amico, R.; Peritore, A.F.; Gugliandolo, E.; Genovese, T.; Scuto, M.; Crupi, R.; Mandalari, G.; et al. Cashew (*Anacardium occidentale* L.) Nuts Counteract Oxidative Stress and Inflammation in an Acute Experimental Model of Carrageenan-Induced Paw Edema. *Antioxidants* **2020**, *9*, 660. [CrossRef]

63. D'Amico, R.; Fusco, R.; Cordaro, M.; Siracusa, R.; Peritore, A.F.; Gugliandolo, E.; Crupi, R.; Scuto, M.; Cuzzocrea, S.; Di Paola, R.; et al. Modulation of NLRP3 Inflammasome through Formyl Peptide Receptor 1 (Fpr-1) Pathway as a New Therapeutic Target in Bronchiolitis Obliterans Syndrome. *Int. J. Mol. Sci.* **2020**, *21*, 2144. [CrossRef]

64. Esposito, E.; Impellizzeri, D.; Bruschetta, G.; Cordaro, M.; Siracusa, R.; Gugliandolo, E.; Crupi, R.; Cuzzocrea, S. A new co-micronized composite containing palmitoylethanolamide and polydatin shows superior oral efficacy compared to their association in a rat paw model of carrageenan-induced inflammation. *Eur. J. Pharmacol.* **2016**, *782*, 107–118. [CrossRef]

65. Impellizzeri, D.; Siracusa, R.; Cordaro, M.; Peritore, A.F.; Gugliandolo, E.; Mancuso, G.; Midiri, A.; Di Paola, R.; Cuzzocrea, S. Therapeutic potential of dinitrobenzene sulfonic acid (DNBS)-induced colitis in mice by targeting IL-1beta and IL-18. *Biochem. Pharmacol.* **2018**, *155*, 150–161. [CrossRef]

66. Di Paola, R.; Fusco, R.; Impellizzeri, D.; Cordaro, M.; Britti, D.; Morittu, V.M.; Evangelista, M.; Cuzzocrea, S. Adelmidrol, in combination with hyaluronic acid, displays increased anti-inflammatory and analgesic effects against monosodium iodoacetate-induced osteoarthritis in rats. *Arthritis Res. Ther.* **2016**, *18*, 291. [CrossRef]

67. Fusco, R.; Cordaro, M.; Genovese, T.; Impellizzeri, D.; Siracusa, R.; Gugliandolo, E.; Peritore, A.F.; D'Amico, R.; Crupi, R.; Cuzzocrea, S.; et al. Adelmidrol: A New Promising Antioxidant and Anti-Inflammatory Therapeutic Tool in Pulmonary Fibrosis. *Antioxidants* **2020**, *9*, 601. [CrossRef]

68. Fusco, R.; Gugliandolo, E.; Siracusa, R.; Scuto, M.; Cordaro, M.; D'Amico, R.; Evangelista, M.; Peli, A.; Peritore, A.F.; Impellizzeri, D.; et al. Formyl Peptide Receptor 1 Signaling in Acute Inflammation and Neural Differentiation Induced by Traumatic Brain Injury. *Biology* **2020**, *9*, 238. [CrossRef]

69. Impellizzeri, D.; Siracusa, R.; Cordaro, M.; Crupi, R.; Peritore, A.F.; Gugliandolo, E.; D'Amico, R.; Petrosino, S.; Evangelista, M.; Di Paola, R.; et al. N-Palmitoylethanolamine-oxazoline (PEA-OXA): A new therapeutic strategy to reduce neuroinflammation, oxidative stress associated to vascular dementia in an experimental model of repeated bilateral common carotid arteries occlusion. *Neurobiol. Dis.* **2019**, *125*, 77–91. [CrossRef]

70. Cordaro, M.; Cuzzocrea, S.; Crupi, R. An Update of Palmitoylethanolamide and Luteolin Effects in Preclinical and Clinical Studies of Neuroinflammatory Events. *Antioxidants* **2020**, *9*, 216. [CrossRef]

71. Deng, M.; Huang, L.; Zhong, X.; Huang, M. Dynamic Changes of Beclin-1 in the Hippocampus of Male Mice with Vascular Dementia at Different Time Points. *J. Mol. Neurosci.* **2020**, *70*, 1611–1618. [CrossRef] [PubMed]

72. Livak, K.J.; Schmittgen, T.D. Analysis of relative gene expression data using real-time quantitative PCR and the 2(-Delta Delta C(T)) Method. *Methods* **2001**, *25*, 402–408. [CrossRef] [PubMed]

73. Lacruz, R.S.; Smith, C.E.; Smith, S.M.; Hu, P.; Bringas, P., Jr.; Sahin-Toth, M.; Moradian-Oldak, J.; Paine, M.L. Chymotrypsin C (caldecrin) is associated with enamel development. *J. Dent. Res.* **2011**, *90*, 1228–1233. [CrossRef] [PubMed]

74. Paola, D.D.; Capparucci, F.; Natale, S.; Crupi, R.; Cuzzocrea, S.; Spano, N.; Gugliandolo, E.; Peritore, A.F. Combined Effects of Potassium Perchlorate and a Neonicotinoid on Zebrafish Larvae (*Danio rerio*). *Toxics* **2022**, *10*, 203. [CrossRef]

75. Di Paola, D.; Capparucci, F.; Lanteri, G.; Cordaro, M.; Crupi, R.; Siracusa, R.; D'Amico, R.; Fusco, R.; Impellizzeri, D.; Cuzzocrea, S.; et al. Combined Toxicity of Xenobiotics Bisphenol A and Heavy Metals on Zebrafish Embryos (*Danio rerio*). *Toxics* **2021**, *9*, 344. [CrossRef]

76. Di Paola, D.; Natale, S.; Iaria, C.; Crupi, R.; Cuzzocrea, S.; Spano, N.; Gugliandolo, E.; Peritore, A.F. Environmental Co-Exposure to Potassium Perchlorate and Cd Caused Toxicity and Thyroid Endocrine Disruption in Zebrafish Embryos and Larvae (*Danio rerio*). *Toxics* **2022**, *10*, 198. [CrossRef]

77. Di Paola, D.; Abbate, J.M.; Iaria, C.; Cordaro, M.; Crupi, R.; Siracusa, R.; D'Amico, R.; Fusco, R.; Impellizzeri, D.; Cuzzocrea, S.; et al. Environmental Risk Assessment of Dexamethasone Sodium Phosphate and Tocilizumab Mixture in Zebrafish Early Life Stage (*Danio rerio*). *Toxics* **2022**, *10*, 279. [CrossRef]

78. Di Paola, D.; Capparucci, F.; Lanteri, G.; Crupi, R.; Marino, Y.; Franco, G.A.; Cuzzocrea, S.; Spano, N.; Gugliandolo, E.; Peritore, A.F. Environmental Toxicity Assessment of Sodium Fluoride and Platinum-Derived Drugs Co-Exposure on Aquatic Organisms. *Toxics* **2022**, *10*, 272. [CrossRef]

79. Di Paola, D.; Natale, S.; Iaria, C.; Cordaro, M.; Crupi, R.; Siracusa, R.; D'Amico, R.; Fusco, R.; Impellizzeri, D.; Cuzzocrea, S.; et al. Intestinal Disorder in Zebrafish Larvae (*Danio rerio*): The Protective Action of N-Palmitoylethanolamide-oxazoline. *Life* **2022**, *12*, 125. [CrossRef]

80. Genovese, T.; Impellizzeri, D.; D'Amico, R.; Fusco, R.; Peritore, A.F.; Di Paola, D.; Interdonato, L.; Gugliandolo, E.; Crupi, R.; Di Paola, R.; et al. Role of Bevacizumab on Vascular Endothelial Growth Factor in Apolipoprotein E Deficient Mice after Traumatic Brain Injury. *Int. J. Mol. Sci.* **2022**, *23*, 4162. [CrossRef]

81. Cordaro, M.; Siracusa, R.; D'Amico, R.; Genovese, T.; Franco, G.; Marino, Y.; Di Paola, D.; Cuzzocrea, S.; Impellizzeri, D.; Di Paola, R.; et al. Role of Etanercept and Infliximab on Nociceptive Changes Induced by the Experimental Model of Fibromyalgia. *Int. J. Mol. Sci.* **2022**, *23*, 6139. [CrossRef]

82. D'Amico, R.; Gugliandolo, E.; Cordaro, M.; Fusco, R.; Genovese, T.; Peritore, A.F.; Crupi, R.; Interdonato, L.; Di Paola, D.; Cuzzocrea, S.; et al. Toxic Effects of Endocrine Disruptor Exposure on Collagen-Induced Arthritis. *Biomolecules* **2022**, *12*, 564. [CrossRef]

83. D'Amico, R.; Gugliandolo, E.; Siracusa, R.; Cordaro, M.; Genovese, T.; Peritore, A.F.; Crupi, R.; Interdonato, L.; Di Paola, D.; Cuzzocrea, S.; et al. Toxic Exposure to Endocrine Disruptors Worsens Parkinson's Disease Progression through NRF2/HO-1 Alteration. *Biomedicines* **2022**, *10*, 1073. [CrossRef]

84. Fusco, R.; D'Amico, R.; Cordaro, M.; Gugliandolo, E.; Siracusa, R.; Peritore, A.F.; Crupi, R.; Impellizzeri, D.; Cuzzocrea, S.; Di Paola, R. Absence of formyl peptide receptor 1 causes endometriotic lesion regression in a mouse model of surgically-induced endometriosis. *Oncotarget* **2018**, *9*, 31355–31366. [CrossRef]

85. Di Paola, R.; Impellizzeri, D.; Fusco, R.; Cordaro, M.; Siracusa, R.; Crupi, R.; Esposito, E.; Cuzzocrea, S. Ultramicronized palmitoylethanolamide (PEA-um((R))) in the treatment of idiopathic pulmonary fibrosis. *Pharmacol. Res.* **2016**, *111*, 405–412. [CrossRef]

86. Fusco, R.; Siracusa, R.; D'Amico, R.; Peritore, A.F.; Cordaro, M.; Gugliandolo, E.; Crupi, R.; Impellizzeri, D.; Cuzzocrea, S.; Di Paola, R. Melatonin Plus Folic Acid Treatment Ameliorates Reserpine-Induced Fibromyalgia: An Evaluation of Pain, Oxidative Stress, and Inflammation. *Antioxidants* **2019**, *8*, 628. [CrossRef]

87. Kang, H.B.; Kim, S.H.; Uhm, S.H.; Kim, D.K.; Lee, N.S.; Jeong, Y.G.; Sung, N.Y.; Kim, D.S.; Han, I.J.; Yoo, Y.C.; et al. Perilla frutescens Leaf Extract Attenuates Vascular Dementia-Associated Memory Deficits, Neuronal Damages, and Microglial Activation. *Curr. Issues Mol. Biol.* **2022**, *44*, 257–272. [CrossRef]

88. Young, M.M.; Kester, M.; Wang, H.G. Sphingolipids: Regulators of crosstalk between apoptosis and autophagy. *J. Lipid Res.* **2013**, *54*, 5–19. [CrossRef]

89. Iadecola, C. The pathobiology of vascular dementia. *Neuron* **2013**, *80*, 844–866. [CrossRef]

90. Skoog, I. Status of risk factors for vascular dementia. *Neuroepidemiology* **1998**, *17*, 2–9. [CrossRef]

91. Wang, X.X.; Zhang, B.; Xia, R.; Jia, Q.Y. Inflammation, apoptosis and autophagy as critical players in vascular dementia. *Eur. Rev. Med. Pharmacol. Sci.* **2020**, *24*, 9601–9614. [CrossRef] [PubMed]

92. Barichella, M.; Cereda, E.; Pezzoli, G. Major nutritional issues in the management of Parkinson's disease. *Mov. Disord.* **2009**, *24*, 1881–1892. [CrossRef] [PubMed]

93. Seidl, S.E.; Santiago, J.A.; Bilyk, H.; Potashkin, J.A. The emerging role of nutrition in Parkinson's disease. *Front. Aging Neurosci.* **2014**, *6*, 36. [CrossRef] [PubMed]

94. Chen, W.W.; Zhang, X.; Huang, W.J. Role of neuroinflammation in neurodegenerative diseases (Review). *Mol. Med. Rep.* **2016**, *13*, 3391–3396. [CrossRef]

95. De Almeida Magalhaes, T.S.S.; De Oliveira Macedo, P.C.; Converti, A.; Neves de Lima, A.A. The Use of *Euterpe oleracea* Mart. As a New Perspective for Disease Treatment and Prevention. *Biomolecules* **2020**, *10*, 813. [CrossRef]

96. Denzer, I.; Munch, G.; Friedland, K. Modulation of mitochondrial dysfunction in neurodegenerative diseases via activation of nuclear factor erythroid-2-related factor 2 by food-derived compounds. *Pharmacol. Res.* **2016**, *103*, 80–94. [CrossRef]

97. Kovacs, G.G. Molecular Pathological Classification of Neurodegenerative Diseases: Turning towards Precision Medicine. *Int. J. Mol. Sci.* **2016**, *17*, 189. [CrossRef]

98. Lewerenz, J.; Maher, P. Chronic Glutamate Toxicity in Neurodegenerative Diseases-What is the Evidence? *Front. Neurosci.* **2015**, *9*, 469. [CrossRef]

99. Machado, A.K.; Andreazza, A.C.; Da Silva, T.M.; Boligon, A.A.; Do Nascimento, V.; Scola, G.; Duong, A.; Cadona, F.C.; Ribeiro, E.E.; Da Cruz, I.B. Neuroprotective Effects of Acai (*Euterpe oleracea* Mart.) against Rotenone In Vitro Exposure. *Oxid. Med. Cell Longev.* **2016**, *2016*, 8940850. [CrossRef]

100. Torma, P.D.; Brasil, A.V.; Carvalho, A.V.; Jablonski, A.; Rabelo, T.K.; Moreira, J.C.; Gelain, D.P.; Flores, S.H.; Augusti, P.R.; Rios, A.O. Hydroethanolic extracts from different genotypes of acai (*Euterpe oleracea*) presented antioxidant potential and protected human neuron-like cells (SH-SY5Y). *Food Chem.* **2017**, *222*, 94–104. [CrossRef]

101. Xie, C.; Kang, J.; Li, Z.; Schauss, A.G.; Badger, T.M.; Nagarajan, S.; Wu, T.; Wu, X. The acai flavonoid velutin is a potent anti-inflammatory agent: Blockade of LPS-mediated TNF-alpha and IL-6 production through inhibiting NF-kappaB activation and MAPK pathway. *J. Nutr. Biochem.* **2012**, *23*, 1184–1191. [CrossRef]

102. Gauthier, S.; Ferris, S. Outcome measures for probable vascular dementia and Alzheimer's disease with cerebrovascular disease. *Int. J. Clin. Pract. Suppl.* **2001**, 29–39.

103. Bathgate, D.; Snowden, J.S.; Varma, A.; Blackshaw, A.; Neary, D. Behaviour in frontotemporal dementia, Alzheimer's disease and vascular dementia. *Acta Neurol. Scand.* **2001**, *103*, 367–378. [CrossRef]

104. Yang, Y.; Jiang, S.; Yan, J.; Li, Y.; Xin, Z.; Lin, Y.; Qu, Y. An overview of the molecular mechanisms and novel roles of Nrf2 in neurodegenerative disorders. *Cytokine Growth Factor Rev.* **2015**, *26*, 47–57. [CrossRef]

105. Liu, D.D.; Yuan, X.; Chu, S.F.; Chen, C.; Ren, Q.; Luo, P.; Lin, M.Y.; Wang, S.S.; Zhu, T.B.; Ai, Q.D.; et al. CZ-7, a new derivative of Claulansine F, ameliorates 2VO-induced vascular dementia in rats through a Nrf2-mediated antioxidant responses. *Acta Pharmacol. Sin.* **2019**, *40*, 425–440. [CrossRef]

106. Zhang, D.; Xiao, Y.; Lv, P.; Teng, Z.; Dong, Y.; Qi, Q.; Liu, Z. Edaravone attenuates oxidative stress induced by chronic cerebral hypoperfusion injury: Role of ERK/Nrf2/HO-1 signaling pathway. *Neurol. Res.* **2018**, *40*, 1–10. [CrossRef]

107. O'Brien, C.E.; Bonanno, L.; Zhang, H.; Wyss-Coray, T. Beclin 1 regulates neuronal transforming growth factor-beta signaling by mediating recycling of the type I receptor ALK5. *Mol. Neurodegener.* **2015**, *10*, 69. [CrossRef]

108. Diniz, L.P.; Matias, I.; Siqueira, M.; Stipursky, J.; Gomes, F.C.A. Astrocytes and the TGF-beta1 Pathway in the Healthy and Diseased Brain: A Double-Edged Sword. *Mol. Neurobiol.* **2019**, *56*, 4653–4679. [CrossRef]

109. Caraci, F.; Gulisano, W.; Guida, C.A.; Impellizzeri, A.A.; Drago, F.; Puzzo, D.; Palmeri, A. A key role for TGF-beta1 in hippocampal synaptic plasticity and memory. *Sci. Rep.* **2015**, *5*, 11252. [CrossRef]

110. Kandasamy, M.; Anusuyadevi, M.; Aigner, K.M.; Unger, M.S.; Kniewallner, K.M.; De Sousa, D.M.B.; Altendorfer, B.; Mrowetz, H.; Bogdahn, U.; Aigner, L. TGF-beta Signaling: A Therapeutic Target to Reinstate Regenerative Plasticity in Vascular Dementia? *Aging Dis.* **2020**, *11*, 828–850. [CrossRef]

111. Liu, H.X.; Zhang, J.J.; Zheng, P.; Zhang, Y. Altered expression of MAP-2, GAP-43, and synaptophysin in the hippocampus of rats with chronic cerebral hypoperfusion correlates with cognitive impairment. *Brain Res. Mol. Brain Res.* **2005**, *139*, 169–177. [CrossRef]

112. Gomazkov, O.; Lagunin, A. Vascular dementia: Molecular targets of neuroprotective therapy. *Biol. Bull. Rev.* **2017**, *7*, 528–536.

113. Bertelli, S.; Remigante, A.; Zuccolini, P.; Barbieri, R.; Ferrera, L.; Picco, C.; Gavazzo, P.; Pusch, M. Mechanisms of Activation of LRRC8 Volume Regulated Anion Channels. *Cell Physiol. Biochem.* **2021**, *55*, 41–56. [CrossRef]

114. Remigante, A.; Morabito, R.; Spinelli, S.; Trichilo, V.; Loddo, S.; Sarikas, A.; Dossena, S.; Marino, A. d-Galactose Decreases Anion Exchange Capability through Band 3 Protein in Human Erythrocytes. *Antioxidants* **2020**, *9*, 689. [CrossRef]

Article

Açaì (*Euterpe oleracea*) Extract Protects Human Erythrocytes from Age-Related Oxidative Stress

Alessia Remigante [1,†], Sara Spinelli [1,2,†], Elisabetta Straface [3], Lucrezia Gambardella [3], Daniele Caruso [4], Giuseppe Falliti [4], Silvia Dossena [2], Angela Marino [1,‡] and Rossana Morabito [1,*,‡]

[1] Department of Chemical, Biological, Pharmaceutical and Environmental Sciences, University of Messina, 98166 Messina, Italy; aremigante@unime.it (A.R.); sspinelli@unime.it (S.S.); marinoa@unime.it (A.M.)
[2] Institute of Pharmacology and Toxicology, Paracelsus Medical University, 5020 Salzburg, Austria; silvia.dossena@pmu.ac.at
[3] Biomarkers Unit, Center for Gender-Specific Medicine, Istituto Superiore di Sanità, 00161 Rome, Italy; elisabetta.straface@iss.it (E.S.); lucrezia.gambardella@iss.it (L.G.)
[4] Complex Operational Unit of Clinical Pathology of Papardo Hospital, 98166 Messina, Italy; daniele.caruso1985@libero.it (D.C.); peppefal@tin.it (G.F.)
* Correspondence: rmorabito@unime.it
† These authors contributed equally to this work.
‡ These authors shared senior authorship.

Citation: Remigante, A.; Spinelli, S.; Straface, E.; Gambardella, L.; Caruso, D.; Falliti, G.; Dossena, S.; Marino, A.; Morabito, R. Açaì (*Euterpe oleracea*) Extract Protects Human Erythrocytes from Age-Related Oxidative Stress. *Cells* **2022**, *11*, 2391. https://doi.org/10.3390/cells11152391

Academic Editor: Alexander E. Kalyuzhny

Received: 8 July 2022
Accepted: 2 August 2022
Published: 3 August 2022

Abstract: Aging is a process characterised by a general decline in physiological functions. The high bioavailability of reactive oxygen species (ROS) plays an important role in the aging rate. Due to the close relationship between aging and oxidative stress (OS), functional foods rich in flavonoids are excellent candidates to counteract age-related changes. This study aimed to verify the protective role of Açaì extract in a D-Galactose (D-Gal)-induced model of aging in human erythrocytes. Markers of OS, including ROS production, thiobarbituric acid reactive substances (TBARS) levels, oxidation of protein sulfhydryl groups, as well as the anion exchange capability through Band 3 protein (B3p) and glycated haemoglobin (A1c) have been analysed in erythrocytes treated with D-Gal for 24 h, with or without pre-incubation for 1 h with 0.5–10 μg/mL Açaì extract. Our results show that the extract avoided the formation of acanthocytes and leptocytes observed after exposure to 50 and 100 mM D-Gal, respectively, prevented D-Gal-induced OS damage, and restored alterations in the distribution of B3p and CD47 proteins. Interestingly, D-Gal exposure was associated with an acceleration of the rate constant of SO_4^{2-} uptake through B3p, as well as A1c formation. Both alterations have been attenuated by pre-treatment with the Açaì extract. These findings contribute to clarify the aging mechanisms in human erythrocytes and propose functional foods rich in flavonoids as natural antioxidants for the treatment and prevention of OS-related disease conditions.

Keywords: Açaí berry; D-Galactose; aging; oxidative stress; glycation; plasma membrane; band 3 protein function; erythrocytes

1. Introduction

Aging is a dynamic chronological process characterized by the gradual accumulation of damage to cells, progressive functional decline, and increased susceptibility to disease [1]. A causal hypothesis that gained considerable interest in recent years postulates that pathophysiological changes during aging are due to progressive oxidative damage to cellular macromolecules [2–6]. In physiological conditions, the production of reactive oxygen and nitrogen species generated during cellular metabolism in biological systems is balanced by the ability of the latter to defend through their sophisticated antioxidant machinery. Nevertheless, when oxidants are produced in excess, or when the antioxidant defenses that regulate them are ineffective, this balance can be perturbed, thus resulting in oxidative stress (OS) [4,7–10]. In these conditions, biomolecules can be altered through oxidation to an extent that exceeds repair capacity. Specifically, OS induces lipid peroxidation and glycoxidation reactions, which lead to the formation of highly reactive species that attack free amino groups in proteins, causing their covalent modifications, thus resulting in the generation of advanced glycation and lipo-oxidation end products (AGEs and ALEs) [11,12]. AGEs impair the protein structure due to covalent cross-linking, resulting in protein oligomerization and aggregation. These modifications lead to alterations in the cellular structure and/or function, and ultimately cell death through apoptosis or necrosis. Abnormal levels of reactive species may be the common denominator underlying aging in several acute and chronic pathologies, including systemic sclerosis, cardiovascular diseases, or chronic obstructive pulmonary disease, although the specific mechanisms contributing to OS-induced damage are poorly investigated [13].

Although many cellular models have been used to study the biochemical alterations during aging, erythrocytes get superiority amongst them [3]. In fact, red blood cells are unique, highly specialized, and the most abundant cells in different organisms [14]. Although their primary function is the transportation of the respiratory gases O_2 and CO_2 between the lungs and tissues, these circulatory cells are equipped with effective antioxidative systems that make them mobile free radical scavengers providing antioxidant protection not only to themselves, but also to other tissues and organs in the body [15]. In fact, erythrocytes can contribute to detoxify reactive species, thus rescuing or partially protecting cells in regions of increased OS. However, erythrocytes face particular metabolic challenges regarding the generation and mitigation of oxidative damage. First, erythrocytes have a unique source of OS because of the high load of iron associated with haemoglobin, which drives radical-generating Fenton reactions [16]. Second, unlike most other tissues, the repair of OS in erythrocytes cannot involve *de novo* protein synthesis owing to the lack of nuclei and ribosomes. As such, elucidating the pathways by which red blood cells counteract OS can provide unique clues on cellular responses to oxidant injury [17,18]. Alterations in morphological properties of erythrocytes, such as adhesivity, aggregability, and deformability, have been detected in aging, as well as a number of human pathologic conditions displaying systemic OS as a hallmark [19].

During aging, the erythrocyte volume decreases with time, with an increase in density and a decrease in the haemoglobin content, especially during the first and the second part of the lifespan, respectively. These changes are associated with a loss of cholesterol and phospholipids and a decrease in the mean surface area of 20%, thus indicating a loss of the membrane constituents during aging. Membrane loss occurs through the formation of vesicles containing haemoglobin, haem–Fe, and iron, which act as oxidants able to modulate blood functions [20]. Consequently, aging erythrocytes lose the membrane domain, their normal deformability, and consequently, their classic biconcave disc shape. Moreover, phosphatidylserine, which is normally exposed to the intracellular side of the plasma membrane, becomes externalised [21]. The removal of erythrocytes from circulation is known to be accelerated by age-related changes in cell membrane composition, which, in turn, alters the rheological and immunological properties of these cells [22].

Functional foods and natural products with antioxidant properties can potentially improve the life functions of erythrocytes and, consequently, the homeostasis of the whole body. Euterpe oleracea, an Amazonian Brazilian fruit popularly known as Açaì [23], presents several bioactive molecules with different properties, including antioxidant, anti-inflammatory, and analgesic activities, and also modulates calcium homeostasis. The biological effects of Açaì are related to its chemical matrix, which includes numerous phytochemicals components, such as flavonoids [24]. These molecules can directly neutralize reactive oxygen species (ROS) and/or inactivate molecules with a pro-oxidant capacity [25]. Since aging of the circulating erythrocytes is tightly related to OS, we hypothesized that an Açaì extract might have beneficial effects against aging-related processes in these cells. Among the experimental models of aging, long-term D-Galactose (D-Gal) exposure is the most similar to natural aging [26,27]. Here, we explore the possible protective effect of an Açaì extract on age-related events such as oxidative damage and glycation in a D-Gal-induced model of aging in human erythrocytes.

2. Materials and Methods

2.1. Solutions and Chemicals

All chemicals were purchased from Sigma (Milan, Italy). 4,4′-diisothiocyanatostilbene-2,2′-disulfonate (DIDS) stock solution (10 mM) was prepared in dimethyl sulfoxide (DMSO). D-Galactose stock solution (1 M) was prepared in distilled water. Freeze-dried Açaì extract was dissolved in distilled water. This substance (Cas Number: 879496-95, 906351-38-0) was originally purchased from Farmalabor Srl (Canosa di Puglia, Barletta, Italy), and then kindly provided to us by Professor M. Cordaro from University of Messina. N-ethylmaleimide (NEM) (310 mM) stock solution was prepared in ethanol. H_2O_2 was diluted in distilled water from a 30% w/w stock solution. Ethanol never exceeded 0.001% v/v in the experimental solutions and was previously tested on erythrocytes to exclude haemolysis.

2.2. Erythrocyte Preparation

This study was prospectively reviewed and approved by a duly constitute Ethics Committee (prot.52-22, 20-04-2022). Upon informed consent, whole human blood from healthy volunteers was collected in test tubes containing ethylenediaminetetraacetic acid (EDTA). Plasma concentration of glycated haemoglobin (A1c) was less than 5%. Erythrocytes were washed in isotonic solution (composition in mM: NaCl 150, 4-(2-hydroxyethyl)-1-piperazineethanesulfonic acid (HEPES) 5, Glucose 5, pH 7.4, osmotic pressure 300 mOsm/kgH$_2$O) and centrifuged thrice (Neya 16R, 1200× g, 5 min) to remove plasma and buffy coat. Erythrocytes were then suspended at specific haematocrits in isotonic solution and addressed to downstream analysis.

2.3. Haemolysis Measurement

To verify the % haemolysis, erythrocytes (35% haematocrit) were incubated with or without Açaì extract in isotonic solution, suspended at 0.5% haematocrit in isotonic solution, centrifuged (Neya 16R, 1200× g, 5 min), and resuspended at 0.05% haematocrit in a 0.9% v/v NaCl solution [28,29]. Haemoglobin absorbance was measured at 405 nm wavelength and subtracted for the absorbance of blank (0.9% v/v NaCl solution).

2.4. Thiobarbituric-Acid-Reactive Substances (TBARS) Level Measurement

TBARS levels were measured as described by Mendanha and collaborators [30], with minor modifications. TBARS are derived from the reaction between thiobarbituric acid (TBA) and malondialdehyde (MDA), which is the end-product of lipid peroxidation [30]. Erythrocytes were suspended at 20% haematocrit and pre-incubated in the presence or absence of different concentrations of Açaì extract for 1 h at 37 °C. Successively, samples were incubated with 50 or 100 mM D-Gal for 24 h at 25 °C. Then, samples were centrifuged (Neya 16R, 1200× g, 5 min) and suspended in isotonic solution. Erythrocytes (1.5 mL) were treated with 10% (w/v) trichloroacetic acid (TCA) and centrifuged (Neya 16R, 3000× g, 10 min). TBA (1% in hot distilled water, 1 mL) was added to the supernatant and the mixture was incubated at 95 °C for 30 min. At last, TBARS levels were obtained by subtracting 20% of the absorbance at 453 nm from the absorbance at 532 nm (Onda Spectrophotometer, UV-21). Results are indicated as μM TBARS levels (1.56×10^5 M^{-1} cm^{-1} molar extinction coefficient).

2.5. Total Sulfhydryl Group Content

Measurement of total -SH groups was carried out according to the method of Aksenov and Markesbery [31], with minor modifications. In short, erythrocytes (35% haematocrit), left untreated or exposed to D-Gal-containing solutions with or without pre-incubation with Açaì extract, were centrifuged (Neya 16R, 1200× g, 5 min) and 100 μL haemolysed in 1 mL of distilled water. A 50 μL aliquot was added to 1 mL of phosphate-buffered saline (PBS, pH 7.4) containing EDTA (1 mM). 5,5′-Dithiobis (2-nitrobenzoic acid) (DTNB, 10 mM,

30 µL) was added to initiate the reaction and the samples were incubated for 30 min at 25 °C, protected from light. Control samples, without cell lysate or DTNB, were processed concurrently. After incubation, sample absorbance was measured at 412 nm (Onda spectrophotometer, UV-21) and 3-thio-2-nitro-benzoic acid (TNB) levels were detected after subtraction of blank absorbance (samples containing only DTNB). To achieve full oxidation of -SH groups, an aliquot of erythrocytes (positive control) was incubated with 2 mM NEM for 1 h at 25 °C [32,33]. Data were normalised to protein content and results reported as µM TNB/mg protein.

2.6. Analysis of Cell Shape by Scanning Electron Microscopy (SEM)

Samples, left untreated or exposed to D-Gal-containing solutions with or without pre-incubation with Açaì extract, were collected, plated on poly-l-lysine-coated slides and fixed with 2.5% glutaraldehyde in 0.1 M cacodylate buffer (pH 7.4) at room temperature for 20 min. Then, samples were post-fixed with 1% OsO_4 in 0.1 M sodium cacodylate buffer and dehydrated through a graded series of ethanol solutions (from 30% to 100%). Then, absolute ethanol was gradually substituted by a 1:1 solution of hexamethyldisilazane (HMDS)/absolute ethanol and successively by pure HMDS. Successively, HMDS was completely removed and samples were dried in a desiccator. Dried samples were mounted on stubs, coated with gold (10 nm), and analyzed by a Cambridge 360 scanning electron microscope (Leica Microsystem, Wetzlar, Germany) [34]. Altered erythrocyte shape was evaluated by counting $\geq$500 cells (50 erythrocytes for each different SEM field at a magnification of 3000×) from samples in triplicate.

2.7. Detection of Reactive Oxygen Species (ROS)

To evaluate intracellular reactive oxygen intermediates, erythrocytes, left untreated or exposed to D-Gal-containing solutions with or without pre-incubation with Açaì extract, were incubated in Hanks' balanced salt solution, pH 7.4, containing dihydrorhodamine 123 (DHR 123; Molecular Probes, Milan, Italy), and then analyzed with a FACScan flow cytometer (Becton-Dickinson, Mountain View, CA, USA). At least 20,000 events were acquired. The median values of fluorescence intensity histograms were used to provide a semi-quantitative analysis of ROS production [35].

2.8. Detection of Apoptotic Erythrocytes

Erythrocytes, left untreated or exposed to D-Gal-containing solutions with or without pre-incubation with Açaì extract, were processed to detect apoptosis by using the FITC-conjugated Annexin V apoptosis detection kit (Biovision, CA, USA.) and Trypan blue staining (0.05% Trypan blue for 10 min at room temperature) [36]. Then, erythrocytes were analyzed with a FACScan flow cytometer (Becton-Dickinson, Mountain View, CA, USA) equipped with a 488 nm argon laser.

2.9. Analytical Cytology

Erythrocytes, left untreated or exposed to D-Gal-containing solutions with or without pre-incubation with Açaì extract, were fixed with 3.7% formaldehyde in PBS (pH 7.4) for 10 min at room temperature and then washed in the same buffer. Cells were then permeabilised with 0.5% Triton X-100 in PBS for 5 min at room temperature. After washing with PBS, samples were incubated with monoclonal anti-Band 3 protein (Sigma, Milan, Italy) or monoclonal anti-CD47 (Santa Cruz Biotechnology, Dallas, TX, United States) antibodies for 30 min at 37 °C, washed, and then incubated with a fluorescein isothiocyanate (FITC)-labeled anti-mouse antibody (Sigma, Milan, Italy) for 30 min at 37 °C [20]. Cells incubated with the secondary antibody given alone were used as the negative control. Samples were analyzed by an Olympus BX51 Microphot fluorescence microscope or by a FACScan flow cytometer (Becton Dickinson, Mountain View, CA, USA) equipped with a 488 nm argon laser. At least 20,000 events have been acquired. The median values of fluorescence intensity histograms were used to provide a semiquantitative analysis. Fluorescence intensity values were normalised for those of untreated erythrocytes and expressed in %.

2.10. Measurement of Glycated Haemoglobin (%A1c) Levels

The glycated haemoglobin content (%A1c) was determined with the A1c liquidirect reagent as previously described by Sompong and collaborators [37], with minor modifications. Briefly, after incubation with D-Gal with or without pre-incubation with frozen Açaì extract, erythrocytes were lysed in hypotonic buffer and then incubated with latex reagent at 37 °C for 5 min. The absorbance of samples was measured at 610 nm (BioPhotometer Plus, Eppendorf, Manchester, United Kingdom) and the A1c content was calculated from a standard curve constructed by using known A1c concentrations and expressed in %.

2.11. $SO_4{}^{2-}$ Uptake Measurement

2.11.1. Control Condition

$SO_4{}^{2-}$ uptake measurement was used to evaluate the anion exchange through B3p, as described elsewhere [38–41]. Briefly, after washing, erythrocytes were suspended to 3% haematocrit in 35 mL $SO_4{}^{2-}$ medium (composition in mM: Na_2SO_4 118, HEPES 10, glucose 5, pH 7.4, osmotic pressure 300 mOsm/kgH_2O) and incubated at 25 °C. After 5, 10, 15, 30, 45, 60, 90, and 120 min, DIDS (10 µM), which is an inhibitor of B3p activity [42], was added to 5 mL sample aliquots, which were kept on ice. Subsequently, samples were washed three times in cold isotonic solution and centrifuged (Neya 16R, 4 °C, 1200× g, 5 min) to eliminate $SO_4{}^{2-}$ from the external medium. Distilled water (1 mL) was added to induce osmotic lysis of erythrocytes and perchloric acid (4% v/v) was used to precipitate proteins. After centrifugation (Neya 16R, 4 °C, 2500× g, 10 min), the supernatant containing $SO_4{}^{2-}$ trapped by erythrocytes was directed to the turbidimetric analysis. Supernatant (500 µL from each sample) was sequentially mixed to 500 µL glycerol diluted (1:1) in distilled water, 1 mL 4 M NaCl, and 500 µL 1.24 M $BaCl_2 \bullet 2H_2O$. Finally, the absorbance of each sample was measured at 425 nm (Spectrophotometer, UV-21, Onda Spectrophotometer, Carpi, Modena, Italy). By means of a calibrated standard curve previously obtained by precipitating known $SO_4{}^{2-}$ concentrations, the absorbance was converted to [$SO_4{}^{2-}$] L cells × 10^{-2}. The rate constant of $SO_4{}^{2-}$ uptake (min^{-1}) was derived from the following equation: $C_t = C_\infty (1 - e^{-rt}) + C_0$, where C_t, C_∞, and C_0 indicate the intracellular $SO_4{}^{2-}$ concentrations measured at time t, ∞, and 0, respectively, e represents the Neper number (2.7182818), r indicates the rate constant accounting for the process velocity, and t is the specific time at which the $SO_4{}^{2-}$ concentration was measured. The rate constant is the inverse of the time needed to reach ~63% of total $SO_4{}^{2-}$ intracellular concentration [38] and [$SO_4{}^{2-}$] L cells × 10^{-2} reported in figures represents $SO_4{}^{2-}$ micromolar concentration internalized by 5 mL erythrocytes suspended at 3% haematocrit.

2.11.2. Experimental Conditions

After 1 h pre-incubation with or without freeze-dried Açaì extract at 37 °C, erythrocytes (3% haematocrit) were exposed to D-Gal (50 or 100 mM) for 24 h at 25 °C. Successively, samples were centrifuged (Neya 16R, 4 °C, 1200× g, 5 min) to replace the supernatant with $SO_4{}^{2-}$ medium. The rate constant of $SO_4{}^{2-}$ uptake was then determined as described for the control condition.

2.12. Experimental Data and Statistics

All data are expressed as arithmetic means ± standard error of the mean. For statistical analysis and graphics, GraphPad Prism (version 8.0, GraphPad Software, San Diego, CA, USA) and Excel (Version 2019, Microsoft, Redmond, WA, USA) software were used. Data normality was verified with the D'Agostino and Pearson Omnibus normality test. Significant differences between mean values were determined by one-way analysis of variance (ANOVA), followed by Bonferroni's multiple comparison post-test or ANOVA with Dunnet's post-test, as appropriate. Statistically significant differences were assumed at $p < 0.05$; (n) corresponds to the number of independent measurements.

3. Results

3.1. Effect of Increasing Concentrations of Freeze-Dried Açaì Extract on Haemolysis, Thiobarbituric-Acid-Reactive Substances (TBARS) Levels, and Sulfhydryl Groups Total Content

To select the concentrations of the freeze-dried Açaì extract to apply for downstream analysis, we measured the haemolysis percentage, thiobarbituric acid reactive substances (TBARS) levels, and the total content of sulfhydryl groups following the incubation of erythrocytes with increasing concentrations (0.005–1000 µg/mL) of freeze-dried Açaì extract for 1 h at 37 °C. As shown in Figure 1, 1000 µg/mL of freeze-dried Açaì extract significantly increased the haemolysis percentage compared to cells left untreated (control). As expected, in erythrocytes exposed to 20 mM H_2O_2, TBARS levels were significantly higher with respect to those of control erythrocytes (Figure 2A). The Açaì extract in the concentration range between 50 and 1000 µg/mL induced a significant increase of TBARS levels compared to control cells (Figure 2A). Conversely, treatment with 0.005–10 µg/mL did not significantly affect TBARS levels compared to the control (Figure 2A). Figure 2B shows the total content of sulfhydryl groups (µM TNB/µg protein) of erythrocytes left untreated or treated with either the oxidising compound NEM (2 mM for 1 h, as the positive control) or different concentrations of freeze-dried Açaì extract for 1 h at 37 °C. As expected, exposure to NEM led to a significant reduction in sulfhydryl groups content. The Açaì extract in the range from 50 to 1000 µg/mL also significantly reduced sulfhydryl group abundance with respect to control, while lower concentrations were ineffective. Based on these results, only concentrations of freeze-dried Açaì extract <50 µg/mL, which do not induce haemolysis or OS, have been selected for the experiments shown in the following.

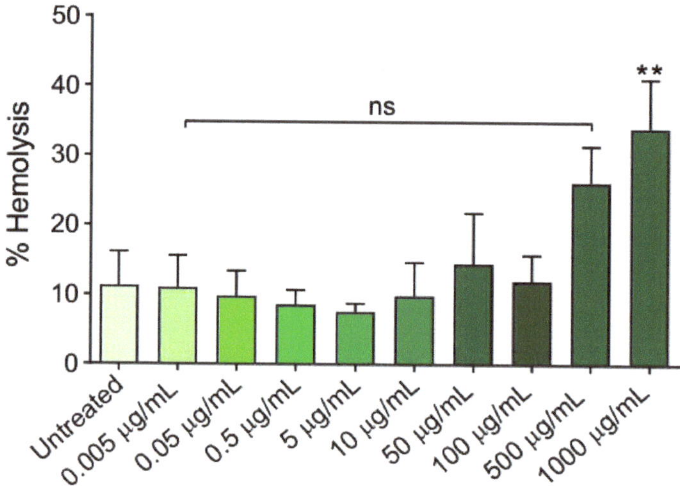

Figure 1. Haemolysis measurement. Human erythrocytes have been exposed to increasing concentrations of freeze-dried Açaì extract for 1 h at 37 °C. ns, not statistically significant versus control; **, $p < 0.01$ versus control, ANOVA with Dunnet's post-test ($n = 10$).

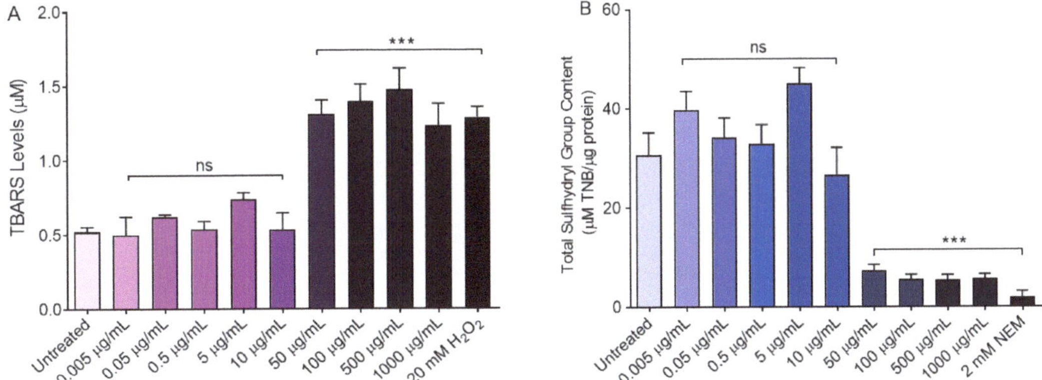

Figure 2. Oxidative stress assessment. Human erythrocytes have been exposed to increasing concentrations of freeze-dried Açaì extract for 1 h at 37 °C. (**A**) Estimation of TBARS levels—a marker of lipid peroxidation—as well as (**B**) total sulfhydryl group content (μM TNB/μg protein). ns, not statistically significant versus control; ***, $p < 0.001$ versus control, ANOVA with Dunnet's post-test ($n = 10$).

3.2. Freeze-Dried Açaì Extract Prevents Cell Shape Changes in D-Gal-Treated Erythrocytes

As depicted in Figure 3, treatment with D-Gal (50 or 100 mM) for 24 h induced morphological alterations of erythrocytes. In fact, by scanning electron microscopy analysis we detected 15.7% of acanthocytes (erythrocytes with surface blebs) in samples treated with 50 mM D-Gal and 28.2% of leptocytes (erythrocytes with a flattened shape) in samples treated with 100 mM D-Gal (Table 1). Pre-treatment with 0.5 or 10 μg/mL freeze-dried Açaì extract reduced the percentage of morphologically altered cells. Specifically, in samples treated with 50 mM D-Gal, the percentage of acanthocytes was reduced to 6.9% and 9.8% in samples pre-treated with 0.5 and 10 μg/mL Açaì, respectively. Instead, in samples treated with 100 mM D-Gal, the percentage of leptocytes was reduced to 22% and 12% after pre-treatment with 0.5 and 10 μg/mL Açaì, respectively.

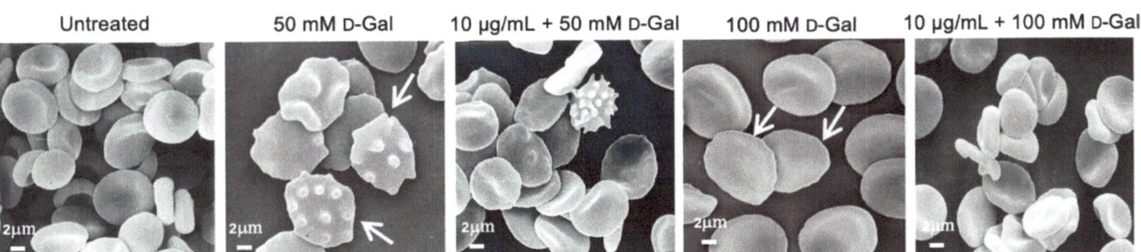

Figure 3. Erythrocyte morphology evaluation. Representative scanning electron microscopy images showing erythrocytes with a typical biconcave form (left untreated); with surface blebs (acanthocytes, arrows) (50 mM D-Gal); with a flattened shape (leptocytes, arrows) (100 mM D-Gal). Pre-treatment with freeze-dried Açaì extract (10 μg/mL) attenuated the morphological changes compared to D-Gal treatment. Magnification 3000×.

Table 1. Percentage of morphological alterations in erythrocytes left untreated (control) or treated as indicated. Data are presented as means ± S.E.M. from separate three independent experiments, where [ns], not statistically significant versus untreated (biconcave shape, acanthocytes, and leptocytes); **, $p < 0.01$ versus control (biconcave shape); [°°] $p < 0.01$ versus 50 or 100 mM D-Gal (biconcave shape). [˜˜˜], $p < 0.001$ versus control (acanthocytes); [$$], [$$$], $p < 0.01$ and $p < 0.001$ versus 50 or 100 mM D-Gal (acanthocytes). [£], [£££], $p < 0.05$, and $p < 0.001$ versus control (leptocytes); [ςς], [ςςς], $p < 0.01$ versus 100 mM D-Gal (leptocytes); one-way ANOVA followed by Bonferroni's multiple comparison post-hoc test.

	Biconcave Shape	Acanthocytes	Leptocytes
Untreated (control)	84% ± 0.014	6% ± 0.012	5% ± 0.010
50 mM D-Gal	80% ± 0.011 [ns]	15.7% ± 0.011 [˜˜˜]	8.9% ± 0.011 [£]
0.5 µg/mL + 50 mM D-Gal	88.8% ± 0.010 [ns]	6.9% ± 0.008 [ns,$$$]	4.3% ± 0.009 [ns]
10 µg/mL + 50 mM D-Gal	86.3% ± 0.009 [ns]	9.8% ± 0.009 [ns,$$$]	3.9% ± 0.009 [ns]
100 mM D-Gal	60.4% ± 0.008 **	11.4% ± 0.007 [˜˜˜]	28.2% ± 0.010 [£££]
0.5 µg/mL + 100 mM D-Gal	69.3% ± 0.012 **	8.7% ± 0.0012 [ns,$$]	22% ± 0.011 [ςς]
10 µg/mL + 100 mM D-Gal	80% ± 0.0011 [ns,°°]	8% ± 0.0012 [ns,$$]	12% ± 0.014 [ςςς]

3.3. Evaluation of Intracellular ROS Levels

The evaluation of ROS species was carried out by flow cytometry in erythrocytes left untreated or, alternatively, exposed to D-Gal with or without pre-exposure to 0.5 or 10 µg/mL freeze-dried Açaì extract for 1 h. Figure 4 shows the intracellular ROS levels at different time points (0, 3, 5, and 24 h after exposure to D-Gal). Samples exposed to 50 or 100 mM D-Gal showed a significant increase of ROS levels compared to the control samples. After 3 h, levels of ROS increased by 50% in D-Gal treated samples and remained unchanged in time. In Figure 4, the effect of freeze-dried Açaì extract is also reported. In samples pre-exposed to 0.5 or 10 µg/mL Açaì extract, 50 or 100 mM D-Gal failed to significantly increase ROS levels, which remained unchanged compared to control values (Figure 4A,B).

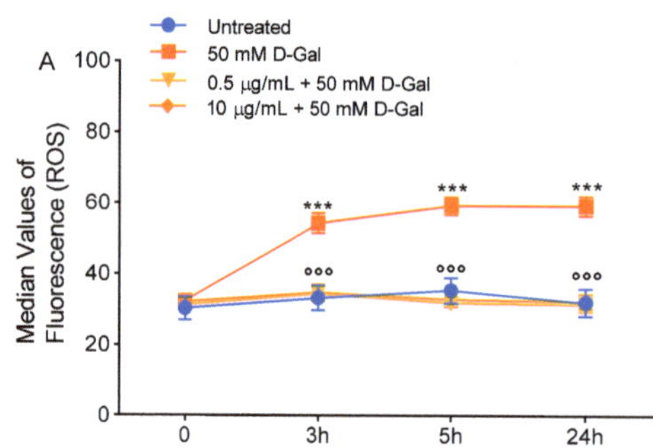

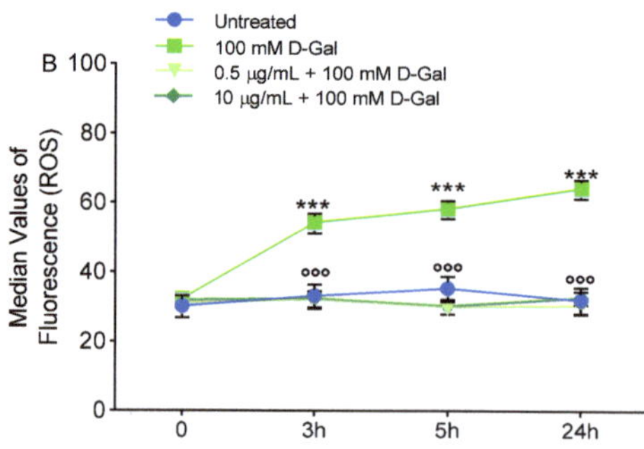

Figure 4. Detection of reactive oxygen species (ROS) levels by flow cytometry. Time course of ROS production in erythrocytes left untreated (control) or treated for 24 h with 50 (**A**) or 100 mM (**B**) D-Gal, with or without pre-exposure to 0.5 or 10 µg/mL freeze-dried Açaì extract for 1 h. ns, not statistically significant versus control; ***, $p < 0.001$ versus control; [°°°], $p < 0.001$ versus D-Gal, one-way ANOVA followed by Bonferroni's post hoc test ($n = 8$).

3.4. Measurement of Thiobarbituric-Acid-Reactive Substances (TBARS) Levels

Thiobarbituric-acid-reactive substances (TBARS) measurements in erythrocytes are reported in Figure 5. As expected, TBARS levels of erythrocytes treated with 20 mM H_2O_2 for 1 h were significantly higher with respect to those of erythrocytes left untreated (control). Similarly, after 24 h of incubation with 50 and 100 mM D-Gal, TBARS levels were significantly increased with respect to those of control erythrocytes. Importantly, in erythrocytes pre-treated with increasing concentrations of freeze-dried Açaì extract, and then exposed to 50 or 100 mM D-Gal, TBARS levels were significantly reduced compared to those measured in 50 or 100 mM D-Gal-treated erythrocytes. Of note, freeze-dried Açaì extracts alone did not significantly affect TBARS levels (Figure 2A).

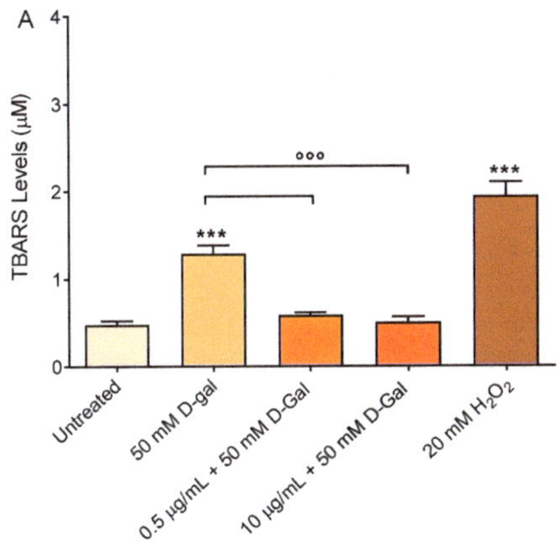

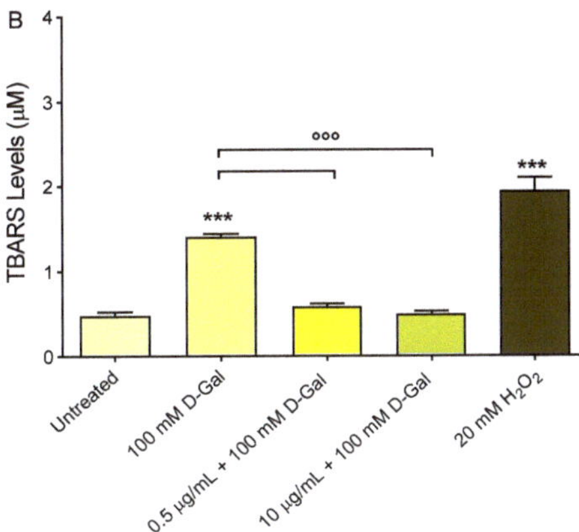

Figure 5. Detection of TBARS levels. TBARS levels (µM) in erythrocytes left untreated (control) or treated for 24 h with 50 (**A**) or 100 mM (**B**) D-Gal, with or without pre-incubation for 1 h with 0.5 or 10 µg/mL freeze-dried Açaì. H_2O_2 (20 mM, 1 h at 37 °C) was used as the positive control. ***, $p < 0.001$ versus control; ooo, $p < 0.001$ versus 50 and 100 mM D-Gal, one-way ANOVA followed by Bonferroni's post hoc test ($n = 10$).

3.5. Measurement of Total Sulfhydryl Group Content

Figure 6 shows the total content of sulfhydryl groups (µM TNB/µg protein) in erythrocytes left untreated or treated with either the oxidising compound NEM (2 mM for 1 h, as the positive control), or 50 and 100 mM D-Gal for 24 h with or without pre-treatment with freeze-dried Açaì extract. As expected, exposure to NEM led to a significant reduction in the sulfhydryl groups' content. Sulfhydryl groups in 50 and 100 mM D-Gal-treated erythrocytes were also significantly reduced with respect to the control. Importantly, pre-treatment with freeze-dried Açaì extract (0.5 or 10 µg/mL) significantly restored the total content of the sulfhydryl groups in 50 or 100 mM D-Gal-treated erythrocytes (Figure 6). Freeze-dried Açaì extracts alone did not significantly affect the total sulfhydryl groups' content (Figure 2B).

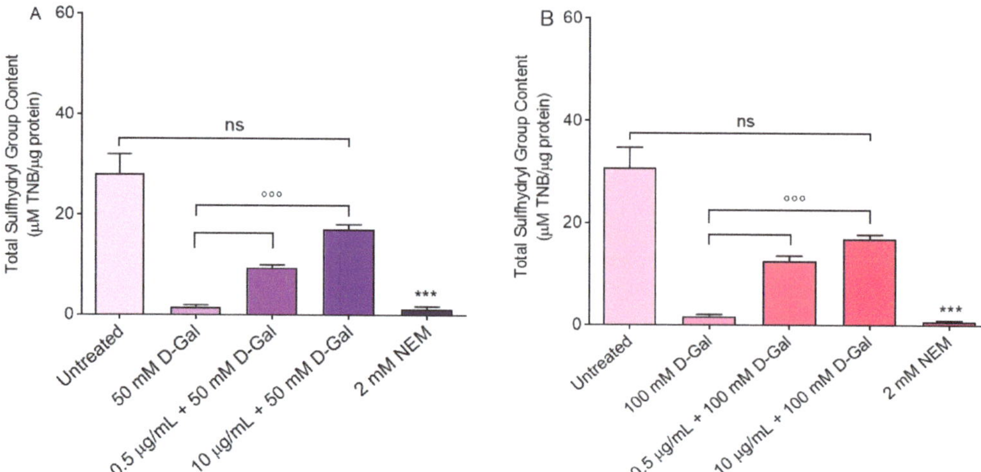

Figure 6. Measurement of total SH group content. Sulfhydryl group content (µM TNB/µg protein) in erythrocytes left untreated (control) or treated with 50 (**A**) or 100 (**B**) mM D-Gal for 24 h with or without pre-incubation for 1 h with 0.5 or 10 µg/mL freeze-dried Açaì extract, or alternatively, with 2 mM NEM (positive control). ns, not statistically significant versus control; °°°, $p < 0.001$ versus 50 and 100 mM D-Gal; ***, $p < 0.001$ versus control, one-way ANOVA followed by Bonferroni's multiple comparison post-hoc test ($n = 10$).

3.6. Determination of Aging Markers

Phosphatidylserine (PS) externalisation, CD47 protein, and B3p have been selected as aging markers. Regarding the percentage of erythrocytes with PS externalisation (apoptosis) no significant difference was detected after treatment with 50 and 100 mM D-Gal (Figure 7).

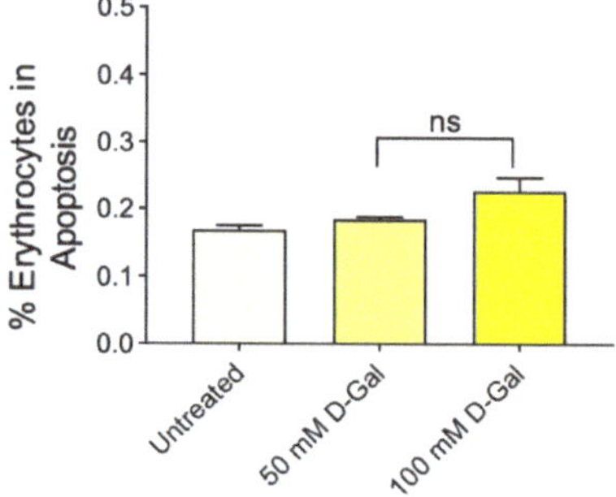

Figure 7. Detection of apoptosis by flow cytometry. Percentage of apoptotic erythrocytes positive to Annexin 5 and Trypan blue detected in samples treated with 50 and 100 mM D-Gal. ns, not statistically significant versus control, ANOVA with Dunnet's post-test ($n = 8$).

Regarding CD47, flow cytometry analysis has shown a significantly decreased expression in samples treated with 50 or 100 mM D-Gal for 24 h compared to untreated (control) samples (Figure 8A,B). Pre-treatment (1 h) with 0.5 µg/mL freeze-dried Açaì extract did not restore CD47 expression in 50 mM D-Gal-treated erythrocytes, while a total restoration of CD47 expression was evident in 100 D-Gal-treated erythrocytes. Conversely, the expression of this protein was significantly restored by pre-treatment with 10 µg/mL freeze-dried Açaì in both 50 and 100 mM D-Gal-treated erythrocytes. Freeze-dried Açaì extract alone did not significantly affect CD47 expression (data not shown). Data obtained by flow cytometry were confirmed by immunofluorescence analyses, which showed a dramatic rearrangement and redistribution of this protein (Figure 8C,D).

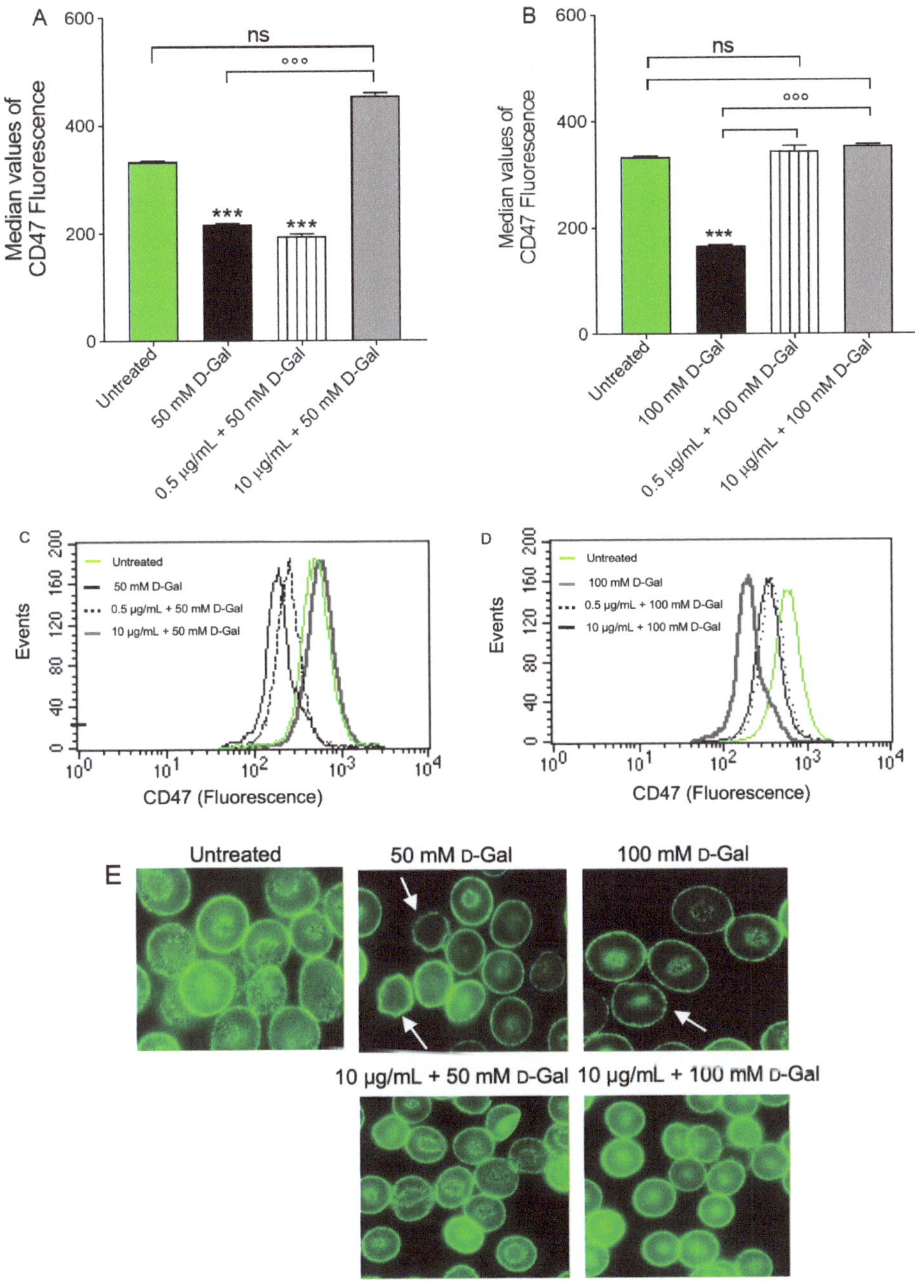

Figure 8. Flow cytometry immunofluorescence of CD47 protein expression. Erythrocytes were treated for 24 h with 50 (**A**) or 100 mM (**B**) D-Gal, with or without pre-incubation for 1 h with 0.5 or 10 µg/mL freeze-dried Açaì extract. Histograms represent median values of fluorescence intensity. In (**C,D**), typical flow cytometry measurements of CD47 expression of a representative experiment are shown. In (**E**), representative images of CD47 expression obtained by flow cytometry immunofluorescence are shown. Samples were observed with a 100× objective. Note the significant morphological changes in both 50 and 100 mM d-Gal (arrows). ns, not statistically significant versus control; ***, $p < 0.001$ versus control; °°°, $p < 0.001$ versus 50 or 100 mM d-Gal, one-way ANOVA followed by Bonferroni's post hoc test ($n = 8$).

B3p protein expression was found significantly decreased in human erythrocytes treated with 50 or 100 mM D-Gal for 24 h with respect to those left untreated (control) (Figure 9). Freeze-dried Açaì extract (0.5 or 10 µg/mL) pre-treatment did not restore B3p expression in erythrocytes treated with 50 mM D-Gal (Figure 9A). Conversely, B3p expression was significantly restored in erythrocytes pre-treated with 10 µg/mL freeze-dried Açaì extract (Figure 9). Freeze-dried Açaì extracts alone did not significantly affect B3p expression (data not shown). In addition, a redistribution of B3p was detected by immunofluorescence (Figure 9E). In particular, B3p was mainly localised in blebs (arrows) of acanthocytes after treatment with 50 mM D-Gal, or alternatively, clustered (arrows) in leptocytes after treatment with 100 mM D-Gal, with respect to untreated erythrocytes. In the latter, these changes were attenuated by freeze-dried Açaì extract (10 µg/mL) pre-treatment.

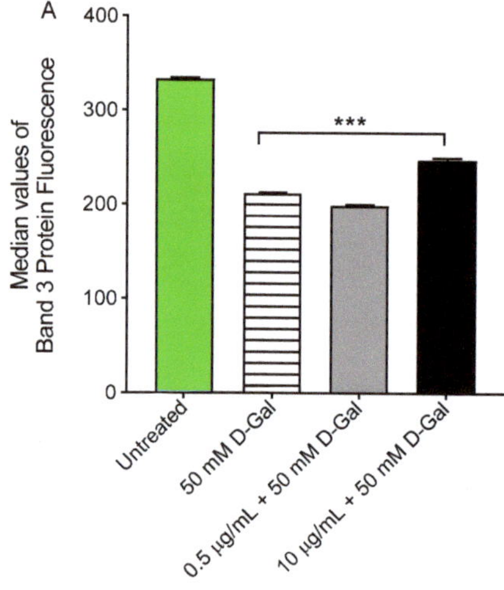

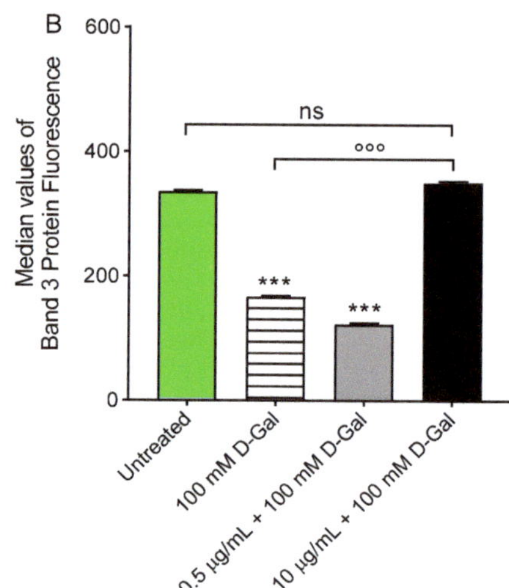

Figure 9. *Cont.*

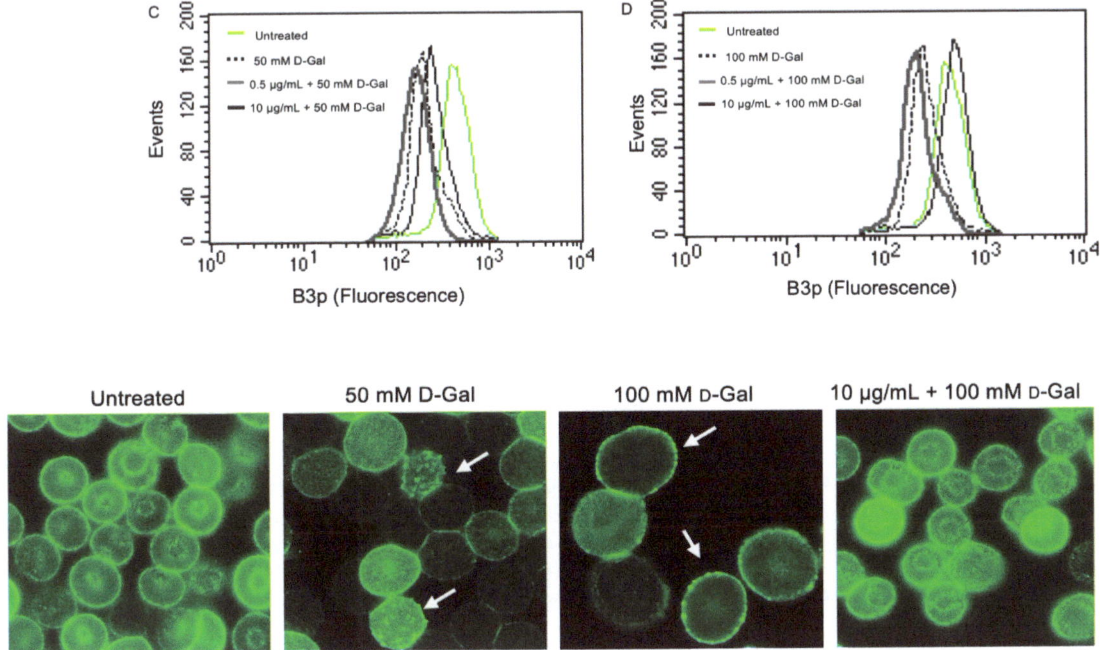

Figure 9. Flow cytometry immunofluorescence of B3p protein expression. Erythrocytes were treated for 24 h with 50 (**A**) or 100 mM (**B**) D-Gal, with or without pre-incubation for 1 h with 0.5 or 10 µg/mL freeze-dried Açaì extract. Histograms represent median values of fluorescence intensity. In (**C,D**), typical flow cytometry measurements of B3p expression of a representative experiment are shown. In (**E**), representative micrographs obtained by flow cytometry immunofluorescence showing B3p distribution in erythrocytes left untreated, treated with 50 mM or 100 mM D-Gal, or alternatively, pre-treated with 10 µg/mL freeze-dried Açaì extract, and then exposed to 100 mM D-Gal are shown. Samples were observed with a 100x objective. Note the significant morphological changes in both 50 and 100 mM D-Gal (arrows). ns, not statistically significant versus untreated; ***, $p < 0.001$ versus untreated (control); °°°, $p < 0.001$ versus 100 mM D-Gal, one-way ANOVA with Bonferroni's multiple comparison post-hoc test ($n = 8$).

3.7. Measurement of Glycated Haemoglobin (%A1c) Levels

Figure 10 shows the glycated haemoglobin levels (%A1c) measured in erythrocytes left untreated or treated with 50 or 100 mM D-Gal for 24 h with or without pre-treatment with 0.5 or 10 µg/mL freeze-dried Açaì extract for 1 h at 37 °C. The %A1c levels measured following exposure to 50 or 100 mM D-Gal were significantly increased with respect to those of erythrocytes left untreated (control). Pre-incubation with freeze-dried Açaì extract (0.5 or 10 µg/mL) for 1 h significantly reduced the %A1c levels in both 50 or 100 mM D-Gal-treated erythrocytes towards values that did not differ from control values. Freeze-dried Açaì extracts alone did not significantly affect the %A1c content (data not shown).

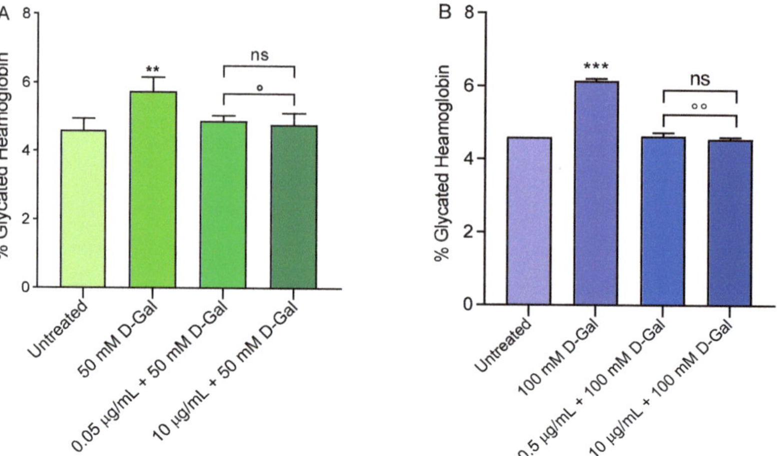

Figure 10. Glycated haemoglobin content (%A1c). Erythrocytes were left untreated or incubated for 24 h with 50 (**A**) or 100 mM (**B**) D-Gal, with or without pre-exposure to 0.5 or 10 μg/mL freeze-dried Açaì extract (pre-incubation for 1 h). ns, not statistically significant versus untreated; **, ***, $p < 0.01$, and $p < 0.001$ versus untreated (control); °, $p < 0.05$ versus 50 mM D-Gal; °°, $p < 0.01$ versus 100 mM D-Gal, one-way ANOVA with Bonferroni's multiple comparison post-hoc test ($n = 10$).

3.8. SO_4^{2-} Uptake Measurement

Figure 11 describes the SO_4^{2-} uptake as a function of time in erythrocytes left untreated (control) and in erythrocytes treated with 50 or 100 mM D-Gal for 24 h with or without pre-incubation with 0.5 or 10 μg/mL freeze-dried Açaì extract for 1 h at 37 °C. In control conditions, SO_4^{2-} uptake progressively increased and reached equilibrium within 45 min (rate constant of SO_4^{2-} uptake = 0.059 ± 0.001 min^{-1}). Erythrocytes treated with 0.5 or 10 μg/mL freeze-dried Açaì extract showed a rate constant of SO_4^{2-} uptake not significantly different with respect to the control (Supplementary Materials). On the contrary, the rate constant value in erythrocytes treated with 50 or 100 mM D-Gal ($0.111/0.113 \pm 0.001$ min^{-1}) was significantly increased with respect to the control (*** $p < 0.001$).

In erythrocytes pre-incubated with 0.5 and 10 μg/mL of freeze-dried Açaì extract and then exposed to 50 mM D-Gal, the rate constant (0.058 and 0.055 ± 0.001 min^{-1}) was significantly lower than that of erythrocytes treated with 50 mM D-Gal (0.111 ± 0.001 min^{-1}), but was not significantly different with respect to the control (Table 2). In erythrocytes pre-incubated with 0.5 or 10 μg/mL of freeze-dried Açaì extract and then exposed to 100 mM D-Gal, the rate constant (0.058 and 0.080 ± 0.001 min^{-1}) was significantly lower than that of erythrocytes treated with 100 mM D-Gal (0.113 ± 0.001 min^{-1}), but was not significantly different with respect to the control (Table 2). SO_4^{2-} uptake was almost completely blocked by 10 μM DIDS applied at the beginning of incubation in the SO_4^{2-} medium (0.017 ± 0.001 min^{-1}, *** $p < 0.001$, Table 2). Additionally, the SO_4^{2-} amount internalized by 50 or 100 mM D-Gal-treated erythrocytes after 45 min of incubation in the SO_4^{2-} medium was not significantly different compared to the control (Table 2). In DIDS-treated cells, the SO_4^{2-} amount internalized (5.39 ± 2.50) was significantly lower than that determined in both the control or treated erythrocytes (*** $p < 0.001$, Table 2).

Table 2. Rate constant of SO_4^{2-} uptake and amount of SO_4^{2-} trapped in erythrocytes left untreated (control) and erythrocytes treated as indicated. Data are presented as means $\pm$ S.E.M. from separate (n) experiments, where [ns], not statistically significant versus untreated; ***, $p < 0.001$ versus control; [°°°], $p < 0.01$ versus 50 or 100 mM D-Gal, one-way ANOVA followed by Bonferroni's multiple comparison post-hoc test.

Experimental Conditions	Rate Constant (min^{-1})	Time (min)	n	SO_4^{2-} Amount Trapped after 45 min of Incubation in SO_4^{2-} Medium [SO_4^{2-}] l Cells $\times$ 10^{-2}
Untreated (control)	0.059 ± 0.001	16.67	10	318.60 ± 18.63
50 mM D-Gal	0.111 ± 0.002 ***	8.95	10	306.35 ± 19.90 [ns]
0.5 µg/mL Açaì Extract + 50 mM D-Gal	0.058 ± 0.001 [ns,°°°]	17.09	10	301 ± 15.90 [ns]
10 µg/mL Açaì Extract + 50 mM D-Gal	0.055 ± 0.001 [ns,°°°]	17.83	10	314 ± 15.80 [ns]
100 mM D-Gal	0.113 ± 0.001 ***	8.75	9	306.35 ± 16.50 [ns]
0.5 µg/mL Açaì Extract + 100 mM D-Gal	0.056 ± 0.001 [ns,°°°]	17.70	9	301 ± 19.80 [ns]
10 µg/mL Açaì Extract + 100 mM D-Gal	0.062 ± 0.001 [ns,°°°]	15.79	8	349.97 ± 11.15 [ns]
10 µM DIDS	0.017 ± 0.001 ***	62.50	16	5.49 ± 3.50 ***

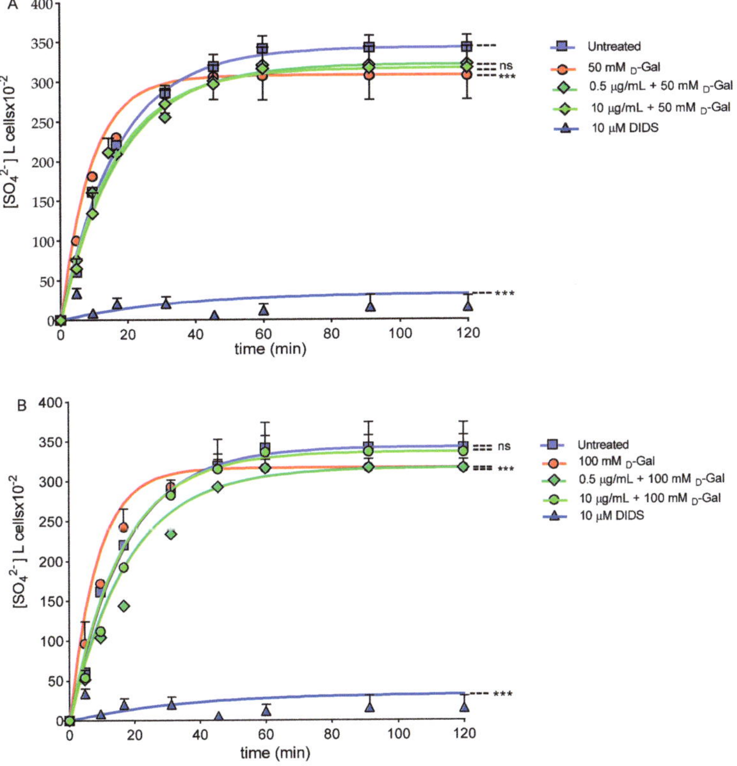

Figure 11. Time course of SO_4^{2-} uptake. Erythrocytes were left untreated (control) or treated with (**A**) 50 or (**B**) 100 mM D-Gal with or without pre-exposure to 0.5 or 10 µg/mL freeze-dried Açaì extract (pre-incubation for 1 h), or 10 µM DIDS. ns, not statistically significant versus control; ***, $p < 0.001$ versus control, one-way ANOVA followed by Bonferroni's post hoc test.

4. Discussion

In recent years, an increasing body of research has focused on natural antioxidants and their ability to counteract OS-induced pathological conditions. In fact, the supplementation of phytochemicals in one's diet has been found to provide numerous health benefits [25,43,44]. The Açai berry represents a good supplementation candidate since it shows several beneficial activities, including antioxidant properties in various experimental models [45–47]. In this regard, the present study describes the important protective effects of freeze-dried Açaì extract in a D-Gal-induced aging model in human erythrocytes. This cell-based model has been validated in a previous study, which demonstrates that D-Gal is an efficient tool to induce chronic OS and accelerated aging in erythrocytes [26]. The results shown here indicate that the chemical matrix of Açaì fruit could improve both the structure and function of human erythrocytes as well as pathological events attributable to aging.

There are many studies describing the multiple activities of flavonoids, however, the effects of Açaì fruits on human erythrocytes have not yet been evaluated. Thus, the first step of this research was to test different concentrations of Açaí extract (from 0.005 µg/mL to 1000 µg/mL) to exclude a possible haemolytic power, as well as the ability to induce lipoperoxidation and/or oxidation of proteins. As shown in the Figure 1, 1000 µg/mL freeze-dried Açaì extract significantly increased the percentage of haemolysis after 1 h of incubation compared to cells left untreated. Vice versa, concentrations ranging from 0.005 µg/mL to 500 µg/mL did not cause haemolytic events (Figure 1). Similarly, in erythrocytes treated with increasing concentrations of Açaì extract (from 50 µg/mL to 1000 µg/mL), the estimation of TBARS levels as well as the total sulfhydryl group content were statistically increased compared to those of erythrocytes left untreated. These findings denote that antioxidant compounds in excessive amounts can manifest pro-oxidant activity, in agreement with other studies [48]. On the contrary, lower concentrations (from 0.005 µg/mL to 10 µg/mL) of Açaì extract induced neither an increase in TBARS levels nor in the oxidation of the total sulfhydryl groups, respectively (Figure 2A,B). Based on these findings, 0.5 and 10 µg/mL freeze-dried Açaí extract were selected to be tested for their possible beneficial activity in our OS-related model of aging. These concentrations were in accordance with former studies performed on different cellular types [49,50].

The susceptibility of erythrocytes to D-Gal exposure was investigated in terms of morphological alterations by scanning electron microscopy (SEM). The images (Figure 3) revealed dramatic changes in the erythrocyte shape. The typical biconcave shape was lost in a significant number of cells, which showed surface blebs (acanthocytes, 50 mM D-Gal), or a flattened shape (leptocytes, 100 mM D-Gal). Based on these results, it can be assumed that erythrocytes may undergo dose-dependent damage through characteristic changes in shape. In fact, the percentage of acanthocytes was higher (15.7%) with respect to leptocytes (8.9%) in 50 mM D-Gal-treated erythrocytes (Table 1), while the percentage of leptocytes was higher (28.2%) with respect to acanthocytes (11.4%) in 100 mM D-Gal-treated erythrocytes (Table 1). However, pre-treatment with freeze-dried Açaì extract attenuated the morphological changes (Figure 3), with a reduction of the percentage of both acanthocytes and leptocytes. The study of erythrocyte morphology is of great importance in the field of hemorheology. The morphology of the circulating cells has a fundamental influence on the rheological properties of the blood, and changes in morphology can lead to decreased deformability and increased aggregation [51,52]. These cells may respond to any form of insult by changing their morphology following changes in their membrane or biochemical composition. Certain phenomena such as the oxidation of sulfhydryl groups of membrane proteins, oxidation of membrane fatty acid residues, or oxidation of haemoglobin could alter membrane properties and the cell shape. Since the oxidation of biological macromolecules, such as lipids and proteins, emanates from the deleterious effects of ROS generated during cellular metabolism, intracellular ROS levels have firstly been evaluated. Our findings show that pre-treatment with freeze-dried Açaí (0.5 or 10 µg/mL) extract induced a reduction of ROS production caused by 50 or 100 mM D-Gal incubation (Figure 4A,B). This evidence is supported by many authors and suggests that dietary supplementation with polyphenols and phytochemicals has antioxidant activity, which is generally attributed to their ability to directly neutralise ROS [53,54].

To better explore the oxidation of biological macromolecules, the estimation of TBARS levels—a marker of lipid peroxidation—as well as the sulfhydryl group content of the total proteins have been considered, respectively. Our results show that 1 h pre-treatment with 0.5 and 10 µg/mL freeze-dried Açaì extract avoided the lipid peroxidation of membranes induced by treatment with 50 or 100 D-Gal (Figure 5A,B). Similarly, the measurement of the sulfhydryl group content was also evaluated. Açaì extract (0.5 and 10 µg/mL) also protected erythrocyte proteins from oxidative damage (Figure 6A,B). In a previous investigation [55], we evaluated the potential protective role of quercetin (Q), a polyphenolic flavonoid compound, on an aging model represented by human erythrocytes treated with 100 mM D-Gal for 24 h. In this regard, TBARS levels, which were augmented following 100 mM D-Gal exposure (24 h), were completely restored by 10 µM Q pre-treatment (1 h). Similarly, the freeze-dried Açaì extract pre-treatment (0.5–10 µg/mL for 1 h) completely restored TBARS levels in this experimental model. On the contrary, the oxidation of protein sulfhydryl groups was partially restored after pre-treatment (1 h) with Q or Açaì extract in erythrocytes treated with 100 mM D-Gal (24 h). Although the activity of these two antioxidants can depend on their concentration as well as phytochemical components, we can conclude that Q and Açaì extract show a similar antioxidant power in our experimental model and could, therefore, play a crucial role in counteracting oxidative stress increases in erythrocytes. These data are in line with what has been previously demonstrated by other authors. For example, the effects of Açaì have been studied on anaerobic exercise-induced changes in the blood antioxidant defense system in junior hurdlers. Specifically, six weeks of daily consumption of a diet supplemented with Açaì pulp resulted in significant increases in plasma total polyphenols and blood GSH content compared to baseline levels, as well as beneficial changes in the lipid profile (TBARS levels) [56]. In addition, experiments in rats revealed that a diet supplemented with 2% Açai pulp for six weeks caused a reduction in erythrocyte protein oxidation compared to control animals [57]. In fact, compared with other antioxidant-rich fruits, Açaì extract has 4.8, 6.1, and 7.5-fold the antioxidant capacity of blackberries, blueberries, and strawberries, respectively [23]. The antioxidant effect of Açaì extract has been attributed to its phytochemical composition comprising hydroxybenzoic acids and flavanols, along with cyanidin 3-O-rutinoside and cyanidin 3-O-glucoside as the predominant anthocyanins [58].

Traditionally, erythrocyte phagocytosis has been proposed to be the result of the accumulation of "eat me" signals on the membrane of aged erythrocytes. Several "eat me" signals have been identified to be important for the clearance of aged erythrocytes by macrophages residing in the spleen, or alternatively, in the liver [59]. Aging has been associated with eryptosis, a controversial process that closely mimics the programmed cell death of nucleated cells (apoptosis). This phenomenon is characterized by a progressive increase in membrane phospholipid asymmetry, owing to the consumption of ATP reservoirs, which results in the apoptosis-like externalization of PS to the outer leaflet of the plasma membrane [60,61]. The externalization of PS on the surface of eryptotic cells can have two major pathophysiological implications: on one hand, it initiates the phagocytosis of erythrocytes; on the other hand, it mediates the adherence of erythrocytes to vascular endothelium cells, which also express PS receptors. Excessive eryptosis initiating the phagocytosis of many red blood cells may, therefore, result in the acute loss of erythrocytes. Exposure to 50 and 100 mM D-Gal did not induce the translocation of PS at the outer plasma membrane leaflet (Figure 7), thus demonstrating that, in this aging model, erythrocytes remain in an early phase of the aging process. A similar condition could be detected in OS-related pathologies, including anaemia, cardiovascular complications, chronic kidney disease, and diabetes [62,63]. Hence, the beneficial effects of Açaì extract shown here apply to an early stage of aging.

In human erythrocytes, CD47 protein is part of the mechanisms regulating membrane architecture, but can also function as an important marker of "self" [64]. In fact, it has been demonstrated that the expression of CD47 on the red blood cell surface could be one of the mechanisms that regulate the removal of senescent cells from the blood flow by phagocytosis. Treatment with 50 mM D-Gal for 24 h did not induce a dramatic loss of this protein, but rather its rearrangement and redistribution (Figure 8E), which could be linked to the progressive plasma membrane blebbing and vesiculation shown by SEM images (Figure 3). Instead, 100 mM D-Gal treatment for 24 h induced both a loss and a redistribution of CD47 expression on the plasma membrane (Figure 8E). This latter finding was paralleled by the appearance of cells with an atypical shape, identified as leptocytes, in SEM images (Figure 3).

During their lifetime, erythrocytes release both exosomes and plasma membrane-derived ectosomes. During the development of erythrocytes, exosomes are generated through a process that involves the double invagination of the endosomal membrane to form multivesicular bodies containing intraluminal vesicles. This process is followed by the fusion of the multivesicular bodies to the plasma membrane to produce exosomes. By contrast, ectosomes (micro-vesicles) and large vesicles are generated by the outward budding of the plasma membrane, followed by vesicle shedding [65,66]. Vesiculation is a way to get rid of dangerous molecules, such as oxidized proteins or denatured haemoglobin. Multilevel changes in lipids and proteins of erythrocyte membranes, caused by aging, cause a decrease in erythrocyte deformability and unfavourable changes in blood flow, which promote additional OS, proneness to atherosclerotic lesions, and an increase in blood viscosity. As a result of ectosome formation, the protein composition of erythrocytes varies among circulating erythrocytes, with new erythrocytes being larger with a full complement of membrane proteins and old erythrocytes smaller and denser with a significantly lower content in membrane proteins, including CD47 and B3p [67].

In contrast to the "eat me" signals, CD47 could inhibit phagocytosis of erythrocytes by macrophages during early aging. It is probable that CD47 exerts its inhibitory effect through binding to the signal regulatory protein (SIRP) alpha on macrophages, which induces inhibitory signaling by the immunoreceptor tyrosine-based inhibition motifs (ITIMs) residing in the cytoplasmic tail of the SIRP alpha. Upon the binding of CD47 to the SIRP alpha, the tyrosine phosphatases SHP-1 and SHP-2 are recruited to the ITIMs and activated, which in turn regulates downstream signaling pathways and effector functions, generally in a negative fashion [59]. However, a total loss of CD47 expression associated to PS externalisation, as well as the clustering of the extracellular regions of B3p, might be proposed as a removal mechanism of aging erythrocytes. In addition, the potential conversion of CD47 from a "do not eat me" to an "eat me" signal is due to a conformational change in CD47. Intriguingly, erythrocyte phagocytosis after this switch seems to be mediated by the same receptor that normally signals the inhibition of phagocytosis, that is, the SIRP alpha [68]. In our model of aging, CD47 expression and distribution were completely re-established in erythrocytes pre-incubated with 0.5 and 10 μg/mL of freeze-dried Açaí extract. This evidence is also confirmed by a recovery of erythrocyte morphological properties in SEM images (Figure 3, Table 1).

As mentioned above, CD47 can be modified by an OS increase, or alternatively, changes in the organisation of the B3p complex; both are potential extracellular indicators of intracellular damage and aging in human erythrocytes [68]. To better clarify the role of B3p, the expression of this protein was also investigated. Specifically, data show that B3p re-arranges in surface blebs, favouring its reorganisation under 50 mM D-Gal treatment (Figure 9E). Vice versa, the treatment with 100 mM D-Gal induced both a loss and a redistribution of the protein, most probably caused by vesicle shedding [67] (Figure 9E).

Band 3 protein is an integral membrane protein that accounts for approximately 25% of the erythrocyte membrane surface. It has several crucial functions and it is composed by two rather different domains of a similar size [69]. The C-terminal domain mediates the chloride–bicarbonate Cl^-/HCO_3^- exchange across the plasma membrane [70,71]. Instead, the N-terminal cytoplasmic domain contains binding sites for cytoskeletal and cytoplasmic proteins, including haemoglobin [72,73]. Haemoglobin glycation is the first example of the non-enzymatic glycation of proteins, which can contribute to the development of complications in various diseases associated with aging [74]. To better study the molecular interaction between B3p and haemoglobin, A1c levels were evaluated. Our results indicated that exposure to 50 or 100 mM D-Gal for 24 h increased the content of A1c levels (Figure 10A,B). These modifications can initiate a cascade of biochemical and structural transformations, including the clustering of B3p regions. When aging processes are advanced, these clusters could provide a recognition site for antibodies directed against aging cells, thus triggering the premature removal of senescent erythrocytes from the circulation at the end of their 120-days life span. Pre-treatment with 0.5 or 10 µg/mL of Açaí extract prevented A1c formation (Figure 10A,B).

At last, our focus has been addressed to assay the anion exchange capability through B3p [15,75–77]. Hence, the SO_4^{2-} uptake was measured after treatment with D-Gal (50 or 100 mM) with or without pre-exposure to 0.5 or 10 µg/mL of freeze-dried Açaí extract [26,29]. In erythrocytes incubated with D-Gal (50 and 100 mM), the rate constant for SO_4^{2-} uptake was accelerated compared to the control (Figure 11A,B, Table 2). However, one-hour pre-treatment with 0.5 or 10 µg/mL of Açaí extract completely restored the rate constant of SO_4^{2-} uptake, which demonstrated a protective effect of the Açaí extract on B3p function. Previously, we reported that the exposure of erythrocytes to 10 mM D-Gal for 24 h induced a reduction of the rate constant of SO_4^{2-} uptake, which was not, however, associated with an OS increase [78]. Therefore, the different effect on the SO_4^{2-} transport kinetics observed in the present study is not surprising. Indeed, a reduction of the transport rate in a former study could be most likely linked to mechanisms other than OS, putatively, to the formation of glycated haemoglobin. The evidence that B3p exhibits modifications in the rate constant for SO_4^{2-} uptake following the exposure of human erythrocytes to OS has also been previously demonstrated. Specifically, H_2O_2-induced OS provoked a reduction in the rate constant for SO_4^{2-} uptake [33,79], whereas treatment with high glucose concentrations induced an acceleration of the anion exchange [29]. Therefore, it is tempting to speculate that such a two-sided effect on anion exchange velocity depends on the specific structure targeted by the stressors and the underlying pathways.

5. Conclusions

We conclude that exposure to 50 or 100 mM D-Gal induced aging in human erythrocytes, and manifested as increased OS, dose-dependent distinct morphological changes, as well as an alteration of B3p and CD47 protein surface expression. Pre-treatment with 0.5 or 10 µg/mL Açaí extract avoided the formation of acanthocytes or leptocytes observed after exposure to 50 and 100 mM D-Gal, respectively, prevented D-Gal-induced OS damage, including ROS production, lipid peroxidation, as well as sulfhydryl group oxidation of total proteins, and restored the distribution of B3p and CD47 on the plasma membrane. Moreover, D-Gal exposure was associated with an acceleration of the rate constant of SO_4^{2-} uptake through B3p, as well as A1c formation. Both alterations have been attenuated by pre-treatment with the Açaì extract. The present study provides mechanistic insights into the benefits of polyphenol-rich extracts on cells exposed to OS. In this light, future investigations are needed to clarify the signaling underlying restoration of a normal anion exchange ability by the Açaí extract, including possible effects on the interaction between B3p and the cytoskeletal proteins ankirin and spectrin, their potential post-translation modifications, as well as a possible influence on the endogenous antioxidant machinery.

Supplementary Materials: The following supporting information can be downloaded at: https://www.mdpi.com/article/10.3390/cells11152391/s1, Figure S1: Time course of SO_4^{2-} uptake in erythrocytes left untreated (control) or treated with increasing concentrations (0.5 and 10 µg/mL) of freeze-dried Açaì extract (pre-incubation for 1 h), or 10 µM DIDS. ns, not statistically significant versus control; ***, $p < 0.001$ versus control, one-way ANOVA followed by Bonferroni's post hoc test.

Author Contributions: A.R. and R.M. conceived and designed the research; A.R., S.S., E.S., L.G., D.C. and G.F. performed the experiments and analyzed the data; A.R., S.S., S.D., A.M. and R.M. interpreted the results of the experiments; A.R. and R.M. prepared the figures; A.R. and R.M. drafted the manuscript; A.R., S.D., A.M. and R.M. edited and revised the manuscript. All authors have read and agreed to the published version of the manuscript.

Funding: This research received no external funding.

Institutional Review Board Statement: The study was conducted in accordance with the Declaration of Helsinki and approved by the Institutional Review Board (or Ethics Committee) of University of Messina (prot.52-22, date of approval 20 April 2022).

Informed Consent Statement: Informed consent was obtained from all subjects involved in the study.

Data Availability Statement: The data that support the findings of this study are available from the corresponding author upon reasonable request.

References

1. Luo, J.; Mills, K.; le Cessie, S.; Noordam, R.; van Heemst, D. Ageing, age-related diseases and oxidative stress: What to do next? *Ageing Res. Rev.* **2020**, *57*, 100982. [CrossRef] [PubMed]
2. Finkel, T.; Holbrook, N.J. Oxidants, oxidative stress and the biology of ageing. *Nature* **2000**, *408*, 239–247. [CrossRef] [PubMed]
3. Pandey, K.B.; Rizvi, S.I. Markers of oxidative stress in erythrocytes and plasma during aging in humans. *Oxidative Med. Cell. Longev.* **2010**, *3*, 565920. [CrossRef]
4. Akki, R.; Siracusa, R.; Cordaro, M.; Remigante, A.; Morabito, R.; Errami, M.; Marino, A. Adaptation to oxidative stress at cellular and tissue level. *Arch. Physiol. Biochem.* **2019**, *128*, 521–531. [CrossRef] [PubMed]
5. Akki, R.; Siracusa, R.; Morabito, R.; Remigante, A.; Campolo, M.; Errami, M.; La Spada, G.; Cuzzocrea, S.; Marino, A. Neuronal-like differentiated SH-SY5Y cells adaptation to a mild and transient H_2O_2-induced oxidative stress. *Cell Biochem. Funct.* **2018**, *36*, 56–64. [CrossRef] [PubMed]
6. Dossena, S.; Marino, A. Cellular Oxidative Stress. *Antioxidants* **2021**, *10*, 399. [CrossRef]
7. Pizzino, G.; Irrera, N.; Cucinotta, M.; Pallio, G.; Mannino, F.; Arcoraci, V.; Squadrito, F.; Altavilla, D.; Bitto, A. Oxidative Stress: Harms and Benefits for Human Health. *Oxidative Med. Cell. Longev.* **2017**, *2017*, 8416763. [CrossRef]
8. Guo, Q.; Li, F.; Duan, Y.; Wen, C.; Wang, W.; Zhang, L.; Huang, R.; Yin, Y. Oxidative stress, nutritional antioxidants and beyond. *Sci. China Life Sci.* **2020**, *63*, 866–874. [CrossRef] [PubMed]
9. Bertelli, S.; Remigante, A.; Zuccolini, P.; Barbieri, R.; Ferrera, L.; Picco, C.; Gavazzo, P.; Pusch, M. Mechanisms of Activation of LRRC8 Volume Regulated Anion Channels. *Cell. Physiol. Biochem.* **2021**, *55*, 41–56. [CrossRef]
10. Ferrera, L.; Barbieri, R.; Picco, C.; Zuccolini, P.; Remigante, A.; Bertelli, S.; Fumagalli, M.R.; Zifarelli, G.; La Porta, C.A.M.; Gavazzo, P.; et al. TRPM2 Oxidation Activates Two Distinct Potassium Channels in Melanoma Cells through Intracellular Calcium Increase. *Int. J. Mol. Sci.* **2021**, *22*, 8359. [CrossRef]
11. Moldogazieva, N.T.; Mokhosoev, I.M.; Mel'nikova, T.I.; Porozov, Y.B.; Terentiev, A.A. Oxidative Stress and Advanced Lipoxidation and Glycation End Products (ALEs and AGEs) in Aging and Age-Related Diseases. *Oxidative Med. Cell. Longev.* **2019**, *2019*, 3085756. [CrossRef] [PubMed]
12. Remigante, A.; Morabito, R.; Marino, A. Natural Antioxidants Beneficial Effects on Anion Exchange through Band 3 Protein in Human Erythrocytes. *Antioxidants* **2019**, *9*, 25. [CrossRef] [PubMed]
13. Liguori, I.; Russo, G.; Curcio, F.; Bulli, G.; Aran, L.; Della-Morte, D.; Gargiulo, G.; Testa, G.; Cacciatore, F.; Bonaduce, D.; et al. Oxidative stress, aging, and diseases. *Clin. Interv. Aging* **2018**, *13*, 757–772. [CrossRef] [PubMed]
14. Crupi, R.; Morabito, R.; Remigante, A.; Gugliandolo, E.; Britti, D.; Cuzzocrea, S.; Marino, A. Susceptibility of erythrocytes from different sources to xenobiotics-induced lysis. *Comp. Biochem. Physiol. C Toxicol. Pharmacol.* **2019**, *221*, 68–72. [CrossRef] [PubMed]
15. Remigante, A.; Morabito, R.; Marino, A. Band 3 protein function and oxidative stress in erythrocytes. *J. Cell Physiol.* **2021**, *236*, 6225–6234. [CrossRef] [PubMed]
16. Fujii, J.; Homma, T.; Kobayashi, S.; Warang, P.; Madkaikar, M.; Mukherjee, M.B. Erythrocytes as a preferential target of oxidative stress in blood. *Free Radic. Res.* **2021**, *55*, 781–799. [CrossRef]
17. Minetti, M.; Malorni, W. Redox control of red blood cell biology: The red blood cell as a target and source of prooxidant species. *Antioxid. Redox Signal.* **2006**, *8*, 1165–1169. [CrossRef]
18. D'Alessandro, A.; Hay, A.; Dzieciatkowska, M.; Brown, B.C.; Morrison, E.J.; Hansen, K.C.; Zimring, J.C. Protein-L-isoaspartate O-methyltransferase is required for in vivo control of oxidative damage in red blood cells. *Haematologica* **2021**, *106*, 2726–2739. [CrossRef]
19. Dinarelli, S.; Longo, G.; Dietler, G.; Francioso, A.; Mosca, L.; Pannitteri, G.; Boumis, G.; Bellelli, A.; Girasole, M. Erythrocyte's aging in microgravity highlights how environmental stimuli shape metabolism and morphology. *Sci. Rep.* **2018**, *8*, 5277. [CrossRef]
20. Giovannetti, A.; Gambardella, L.; Pietraforte, D.; Rosato, E.; Giammarioli, A.M.; Salsano, F.; Malorni, W.; Straface, E. Red blood cell alterations in systemic sclerosis: A pilot study. *Cell. Physiol. Biochem.* **2012**, *30*, 418–427. [CrossRef] [PubMed]
21. Hoehn, R.S.; Jernigan, P.L.; Chang, A.L.; Edwards, M.J.; Pritts, T.A. Molecular mechanisms of erythrocyte aging. *Biol. Chem.* **2015**, *396*, 621–631. [CrossRef] [PubMed]

22. Caprari, P.; Scuteri, A.; Salvati, A.M.; Bauco, C.; Cantafora, A.; Masella, R.; Modesti, D.; Tarzia, A.; Marigliano, V. Aging and red blood cell membrane: A study of centenarians. *Exp. Gerontol.* **1999**, *34*, 47–57. [CrossRef]

23. Baptista, S.L.; Copetti, C.L.K.; Cardoso, A.L.; Di Pietro, P.F. Biological activities of acai (*Euterpe oleracea* Mart.) and jucara (*Euterpe edulis* Mart.) intake in humans: An integrative review of clinical trials. *Nutr. Rev.* **2021**, *79*, 1375–1391. [CrossRef] [PubMed]

24. de Oliveira, N.K.S.; Almeida, M.R.S.; Pontes, F.M.M.; Barcelos, M.P.; Silva, G.M.; de Paula da Silva, C.H.T.; Cruz, R.A.S.; da Silva Hage-Melim, L.I. Molecular Docking, Physicochemical Properties, Pharmacokinetics and Toxicity of Flavonoids Present in Euterpe oleracea Martius. *Curr. Comput.-Aided Drug Des.* **2021**, *17*, 589–617. [CrossRef]

25. Lopez, J.G. Flavonoids in Health and Disease. *Curr. Med. Chem.* **2019**, *26*, 6972–6975. [CrossRef]

26. Remigante, A.; Spinelli, S.; Trichilo, V.; Loddo, S.; Sarikas, A.; Pusch, M.; Dossena, S.; Marino, A.; Morabito, R. d-Galactose induced early aging in human erythrocytes: Role of band 3 protein. *J. Cell. Physiol.* **2021**, *237*, 1586–1596. [CrossRef] [PubMed]

27. Azman, K.F.; Zakaria, R. D-Galactose-induced accelerated aging model: An overview. *Biogerontology* **2019**, *20*, 763–782. [CrossRef]

28. Goodhead, L.K.; MacMillan, F.M. Measuring osmosis and hemolysis of red blood cells. *Adv. Physiol. Educ.* **2017**, *41*, 298–305. [CrossRef] [PubMed]

29. Morabito, R.; Remigante, A.; Spinelli, S.; Vitale, G.; Trichilo, V.; Loddo, S.; Marino, A. High Glucose Concentrations Affect Band 3 Protein in Human Erythrocytes. *Antioxidants* **2020**, *9*, 365. [CrossRef]

30. Mendanha, S.A.; Anjos, J.L.; Silva, A.H.; Alonso, A. Electron paramagnetic resonance study of lipid and protein membrane components of erythrocytes oxidized with hydrogen peroxide. *Braz. J. Med. Biol. Res.* **2012**, *45*, 473–481. [CrossRef]

31. Aksenov, M.Y.; Markesbery, W.R. Changes in thiol content and expression of glutathione redox system genes in the hippocampus and cerebellum in Alzheimer's disease. *Neurosci. Lett.* **2001**, *302*, 141–145. [CrossRef]

32. Morabito, R.; Falliti, G.; Geraci, A.; Spada, G.L.; Marino, A. Curcumin Protects -SH Groups and Sulphate Transport after Oxidative Damage in Human Erythrocytes. *Cell. Physiol. Biochem.* **2015**, *36*, 345–357. [CrossRef] [PubMed]

33. Morabito, R.; Romano, O.; La Spada, G.; Marino, A. H_2O_2-Induced Oxidative Stress Affects $SO_4^=$ Transport in Human Erythrocytes. *PLoS ONE* **2016**, *11*, e0146485. [CrossRef] [PubMed]

34. Straface, E.; Rivabene, R.; Masella, R.; Santulli, M.; Paganelli, R.; Malorni, W. Structural changes of the erythrocyte as a marker of non-insulin-dependent diabetes: Protective effects of N-acetylcysteine. *Biochem. Biophys. Res. Commun.* **2002**, *290*, 1393–1398. [CrossRef] [PubMed]

35. Lucantoni, G.; Pietraforte, D.; Matarrese, P.; Gambardella, L.; Metere, A.; Paone, G.; Bianchi, E.L.; Straface, E. The red blood cell as a biosensor for monitoring oxidative imbalance in chronic obstructive pulmonary disease: An ex vivo and in vitro study. *Antioxid. Redox Signal.* **2006**, *8*, 1171–1182. [CrossRef]

36. Kupcho, K.; Shultz, J.; Hurst, R.; Hartnett, J.; Zhou, W.; Machleidt, T.; Grailer, J.; Worzella, T.; Riss, T.; Lazar, D.; et al. A real-time, bioluminescent annexin V assay for the assessment of apoptosis. *Apoptosis* **2019**, *24*, 184–197. [CrossRef]

37. Sompong, W.; Cheng, H.; Adisakwattana, S. Protective Effects of Ferulic Acid on High Glucose-Induced Protein Glycation, Lipid Peroxidation, and Membrane Ion Pump Activity in Human Erythrocytes. *PLoS ONE* **2015**, *10*, e0129495. [CrossRef]

38. Romano, L.; Peritore, D.; Simone, E.; Sidoti, A.; Trischitta, F.; Romano, P. Chloride-sulphate exchange chemically measured in human erythrocyte ghosts. *Cell. Mol. Biol.* **1998**, *44*, 351–355.

39. Romano, L.; Passow, H. Characterization of anion transport system in trout red blood cell. *Am. J. Physiol.* **1984**, *246*, C330–C338. [CrossRef]

40. Morabito, R.; Remigante, A.; Marino, A. Protective Role of Magnesium against Oxidative Stress on $SO_4^=$ Uptake through Band 3 Protein in Human Erythrocytes. *Cell. Physiol. Biochem.* **2019**, *52*, 1292–1308. [CrossRef] [PubMed]

41. Morabito, R.; Remigante, A.; Arcuri, B.; Giammanco, M.; La Spada, G.; Marino, A. Effect of cadmium on anion exchange capability through Band 3 protein in human erythrocytes. *J. Biol. Res.* **2018**, *91*, 1–7. [CrossRef]

42. Jessen, F.; Sjoholm, C.; Hoffmann, E.K. Identification of the anion exchange protein of Ehrlich cells: A kinetic analysis of the inhibitory effects of 4,4'-diisothiocyano-2,2'-stilbene-disulfonic acid (DIDS) and labeling of membrane proteins with 3H-DIDS. *J. Membr. Biol.* **1986**, *92*, 195–205. [CrossRef]

43. Li, G.; Ding, K.; Qiao, Y.; Zhang, L.; Zheng, L.; Pan, T.; Zhang, L. Flavonoids Regulate Inflammation and Oxidative Stress in Cancer. *Molecules* **2020**, *25*, 5628. [CrossRef] [PubMed]

44. Hritcu, L.; Ionita, R.; Postu, P.A.; Gupta, G.K.; Turkez, H.; Lima, T.C.; Carvalho, C.U.S.; de Sousa, D.P. Antidepressant Flavonoids and Their Relationship with Oxidative Stress. *Oxidative Med. Cell. Longev.* **2017**, *2017*, 5762172. [CrossRef]

45. Petruk, G.; Illiano, A.; Del Giudice, R.; Raiola, A.; Amoresano, A.; Rigano, M.M.; Piccoli, R.; Monti, D.M. Malvidin and cyanidin derivatives from acai fruit (*Euterpe oleracea* Mart.) counteract UV-A-induced oxidative stress in immortalized fibroblasts. *J. Photochem. Photobiol. B* **2017**, *172*, 42–51. [CrossRef] [PubMed]

46. Shibuya, S.; Toda, T.; Ozawa, Y.; Yata, M.J.V.; Shimizu, T. Acai Extract Transiently Upregulates Erythropoietin by Inducing a Renal Hypoxic Condition in Mice. *Nutrients* **2020**, *12*, 533. [CrossRef] [PubMed]

47. Carey, A.N.; Miller, M.G.; Fisher, D.R.; Bielinski, D.F.; Gilman, C.K.; Poulose, S.M.; Shukitt-Hale, B. Dietary supplementation with the polyphenol-rich acai pulps (*Euterpe oleracea* Mart. and *Euterpe precatoria* Mart.) improves cognition in aged rats and attenuates inflammatory signaling in BV-2 microglial cells. *Nutr. Neurosci.* **2017**, *20*, 238–245. [CrossRef]

48. Giordano, M.E.; Caricato, R.; Lionetto, M.G. Concentration Dependence of the Antioxidant and Prooxidant Activity of Trolox in HeLa Cells: Involvement in the Induction of Apoptotic Volume Decrease. *Antioxidants* **2020**, *9*, 1058. [CrossRef]

49. Wong, D.Y.; Musgrave, I.F.; Harvey, B.S.; Smid, S.D. Acai (*Euterpe oleraceae* Mart.) berry extract exerts neuroprotective effects against beta-amyloid exposure in vitro. *Neurosci. Lett.* **2013**, *556*, 221–226. [CrossRef] [PubMed]

50. Poulose, S.M.; Fisher, D.R.; Bielinski, D.F.; Gomes, S.M.; Rimando, A.M.; Schauss, A.G.; Shukitt-Hale, B. Restoration of stressor-induced calcium dysregulation and autophagy inhibition by polyphenol-rich acai (*Euterpe* spp.) fruit pulp extracts in rodent brain cells in vitro. *Nutrition* **2014**, *30*, 853–862. [CrossRef]

51. Gyawali, P.; Richards, R.S.; Bwititi, P.T.; Nwose, E.U. Association of abnormal erythrocyte morphology with oxidative stress and inflammation in metabolic syndrome. *Blood Cells Mol. Dis.* **2015**, *54*, 360–363. [CrossRef]

52. Gyawali, P.; Richards, R.S.; Uba Nwose, E. Erythrocyte morphology in metabolic syndrome. *Expert. Rev. Hematol.* **2012**, *5*, 523–531. [CrossRef]

53. Bernatoniene, J.; Kopustinskiene, D.M. The Role of Catechins in Cellular Responses to Oxidative Stress. *Molecules* **2018**, *23*, 965. [CrossRef] [PubMed]

54. Ola, M.S.; Al-Dosari, D.; Alhomida, A.S. Role of Oxidative Stress in Diabetic Retinopathy and the Beneficial Effects of Flavonoids. *Curr. Pharm. Des.* **2018**, *24*, 2180–2187. [CrossRef] [PubMed]

55. Remigante, A.; Spinelli, S.; Basile, N.; Caruso, D.; Falliti, G.; Dossena, S.; Marino, A.; Morabito, R. Oxidation Stress as a Mechanism of Aging in Human Erythrocytes: Protective Effect of Quercetin. *Int. J. Mol. Sci.* **2022**, *23*, 7781. [CrossRef]

56. Sadowska-Krepa, E.; Klapcinska, B.; Podgorski, T.; Szade, B.; Tyl, K.; Hadzik, A. Effects of supplementation with acai (*Euterpe oleracea* Mart.) berry-based juice blend on the blood antioxidant defence capacity and lipid profile in junior hurdlers. A pilot study. *Biol. Sport* **2015**, *32*, 161–168. [CrossRef]

57. de Souza, M.O.; Silva, M.; Silva, M.E.; Rde, P.O.; Pedrosa, M.L. Diet supplementation with acai (*Euterpe oleracea* Mart.) pulp improves biomarkers of oxidative stress and the serum lipid profile in rats. *Nutrition* **2010**, *26*, 804–810. [CrossRef] [PubMed]

58. Schauss, A.G.; Wu, X.; Prior, R.L.; Ou, B.; Patel, D.; Huang, D.; Kababick, J.P. Phytochemical and nutrient composition of the freeze-dried amazonian palm berry, *Euterpe oleraceae* mart. (acai). *J. Agric. Food Chem.* **2006**, *54*, 8598–8603. [CrossRef]

59. D'Alessandro, A.; Kriebardis, A.G.; Rinalducci, S.; Antonelou, M.H.; Hansen, K.C.; Papassideri, I.S.; Zolla, L. An update on red blood cell storage lesions, as gleaned through biochemistry and omics technologies. *Transfusion* **2015**, *55*, 205–219. [CrossRef]

60. Bosman, G.J.; Cluitmans, J.C.; Groenen, Y.A.; Werre, J.M.; Willekens, F.L.; Novotny, V.M. Susceptibility to hyperosmotic stress-induced phosphatidylserine exposure increases during red blood cell storage. *Transfusion* **2011**, *51*, 1072–1078. [CrossRef]

61. Yasin, Z.; Witting, S.; Palascak, M.B.; Joiner, C.H.; Rucknagel, D.L.; Franco, R.S. Phosphatidylserine externalization in sickle red blood cells: Associations with cell age, density, and hemoglobin F. *Blood* **2003**, *102*, 365–370. [CrossRef] [PubMed]

62. Massaccesi, L.; Galliera, E.; Corsi Romanelli, M.M. Erythrocytes as markers of oxidative stress related pathologies. *Mech. Ageing Dev.* **2020**, *191*, 111333. [CrossRef] [PubMed]

63. Foller, M.; Lang, F. Ion Transport in Eryptosis, the Suicidal Death of Erythrocytes. *Front. Cell Dev. Biol.* **2020**, *8*, 597. [CrossRef] [PubMed]

64. Oldenborg, P.A. Role of CD47 in erythroid cells and in autoimmunity. *Leuk. Lymphoma* **2004**, *45*, 1319–1327. [CrossRef] [PubMed]

65. Buttari, B.; Profumo, E.; Rigano, R. Crosstalk between red blood cells and the immune system and its impact on atherosclerosis. *Biomed. Res. Int.* **2015**, *2015*, 616834. [CrossRef] [PubMed]

66. Thangaraju, K.; Neerukonda, S.N.; Katneni, U.; Buehler, P.W. Extracellular Vesicles from Red Blood Cells and Their Evolving Roles in Health, Coagulopathy and Therapy. *Int. J. Mol. Sci.* **2020**, *22*, 153. [CrossRef] [PubMed]

67. Kuo, W.P.; Tigges, J.C.; Toxavidis, V.; Ghiran, I. Red Blood Cells: A Source of Extracellular Vesicles. *Methods Mol. Biol.* **2017**, *1660*, 15–22. [CrossRef] [PubMed]

68. Burger, P.; de Korte, D.; van den Berg, T.K.; van Bruggen, R. CD47 in Erythrocyte Ageing and Clearance—The Dutch Point of View. *Transfus Med. Hemother.* **2012**, *39*, 348–352. [CrossRef]

69. Arakawa, T.; Kobayashi-Yurugi, T.; Alguel, Y.; Iwanari, H.; Hatae, H.; Iwata, M.; Abe, Y.; Hino, T.; Ikeda-Suno, C.; Kuma, H.; et al. Crystal structure of the anion exchanger domain of human erythrocyte band 3. *Science* **2015**, *350*, 680–684. [CrossRef]

70. Reithmeier, R.A.; Casey, J.R.; Kalli, A.C.; Sansom, M.S.; Alguel, Y.; Iwata, S. Band 3, the human red cell chloride/bicarbonate anion exchanger (AE1, SLC4A1), in a structural context. *Biochim. Biophys. Acta* **2016**, *1858*, 1507–1532. [CrossRef]

71. Remigante, A.; Spinelli, S.; Pusch, M.; Sarikas, A.; Morabito, R.; Marino, A.; Dossena, S. Role of SLC4 and SLC26 solute carriers during oxidative stress. *Acta Physiol.* **2022**, *235*, e13796. [CrossRef] [PubMed]

72. Anong, W.A.; Franco, T.; Chu, H.; Weis, T.L.; Devlin, E.E.; Bodine, D.M.; An, X.; Mohandas, N.; Low, P.S. Adducin forms a bridge between the erythrocyte membrane and its cytoskeleton and regulates membrane cohesion. *Blood* **2009**, *114*, 1904–1912. [CrossRef] [PubMed]

73. Wu, F.; Satchwell, T.J.; Toye, A.M. Anion exchanger 1 in red blood cells and kidney: Band 3's in a pod. *Biochem. Cell Biol.* **2011**, *89*, 106–114. [CrossRef] [PubMed]

74. Luevano-Contreras, C.; Chapman-Novakofski, K. Dietary advanced glycation end products and aging. *Nutrients* **2010**, *2*, 1247–1265. [CrossRef]

75. Morabito, R.; Remigante, A.; Cavallaro, M.; Taormina, A.; La Spada, G.; Marino, A. Anion exchange through band 3 protein in canine leishmaniasis at different stages of disease. *Pflüg. Arch.* **2017**, *469*, 713–724. [CrossRef] [PubMed]

76. Morabito, R.; Remigante, A.; Cordaro, M.; Trichilo, V.; Loddo, S.; Dossena, S.; Marino, A. Impact of acute inflammation on Band 3 protein anion exchange capability in human erythrocytes. *Arch. Physiol. Biochem.* **2020**, 1–7. [CrossRef] [PubMed]

77. Morabito, R.; Remigante, A.; Marino, A. Melatonin Protects Band 3 Protein in Human Erythrocytes against H_2O_2-Induced Oxidative Stress. *Molecules* **2019**, *24*, 2741. [CrossRef]
78. Remigante, A.; Morabito, R.; Spinelli, S.; Trichilo, V.; Loddo, S.; Sarikas, A.; Dossena, S.; Marino, A. d-Galactose Decreases Anion Exchange Capability through Band 3 Protein in Human Erythrocytes. *Antioxidants* **2020**, *9*, 689. [CrossRef]
79. Morabito, R.; Remigante, A.; Di Pietro, M.L.; Giannetto, A.; La Spada, G.; Marino, A. $SO_4^{=}$ uptake and catalase role in preconditioning after H_2O_2-induced oxidative stress in human erythrocytes. *Pflüg. Arch.* **2017**, *469*, 235–250. [CrossRef]

MDPI AG
Grosspeteranlage 5
4052 Basel
Switzerland
Tel.: +41 61 683 77 34

Cells Editorial Office
E-mail: cells@mdpi.com
www.mdpi.com/journal/cells